CURATIVE AND PREVENTIVE PROPERTIES OF MEDICINAL PLANTS

Research on Disease Management and Animal Model Studies

Innovations in Plant Science for Better Health: From Soil to Fork

CURATIVE AND PREVENTIVE PROPERTIES OF MEDICINAL PLANTS

Research on Disease Management and Animal Model Studies

Edited by

Megh R. Goyal, PhD

Junaid Ahmad Malik, PhD

Ademola Olabode Ayeleso, PhD

AAP | APPLE ACADEMIC PRESS

First edition published 2023

Apple Academic Press Inc.
1265 Goldenrod Circle, NE,
Palm Bay, FL 32905 USA

760 Laurentian Drive, Unit 19,
Burlington, ON L7N 0A4, CANADA

CRC Press
6000 Broken Sound Parkway NW,
Suite 300, Boca Raton, FL 33487-2742 USA

4 Park Square, Milton Park,
Abingdon, Oxon, OX14 4RN UK

Library and Archives Canada Cataloguing in Publication

Title: Curative and preventive properties of medicinal plants : research on disease management and animal model studies / edited by Megh R. Goyal, PhD, Junaid Ahmad Malik, PhD, Ademola Olabode Ayeleso, PhD.
Names: Goyal, Megh R., editor. | Malik, Junaid Ahmad, 1987- editor. | Ayeleso, Ademola Olabode, editor.
Series: Innovations in plant science for better health.
Description: First edition. | Series statement: Innovations in plant science for better health : from soil to fork | Includes bibliographical references and index.
Identifiers: Canadiana (print) 20230142028 | Canadiana (ebook) 20230142060 | ISBN 9781774911952 (hardcover) | ISBN 9781774911969 (softcover) | ISBN 9781003328544 (ebook)
Subjects: LCSH: Diseases—Animal models. | LCSH: Materia medica, Vegetable.
Classification: LCC RB125 C87 2023 | DDC 616.02/7—dc23

Library of Congress Cataloging-in-Publication Data

CIP data on file with US Library of Congress

ISBN: 978-1-77491-195-2 (hbk)
ISBN: 978-1-77491-196-9 (pbk)
ISBN: 978-1-00332-854-4 (ebk)

OTHER BOOKS ON PLANT SCIENCE FOR BETTER HEALTH BY APPLE ACADEMIC PRESS, INC.

Book Series: *Innovations in Plant Science for Better Health: From Soil to Fork*
Editor-in-Chief: Hafiz Ansar Rasul Suleria, PhD

- **Assessment of Medicinal Plants for Human Health: Phytochemistry, Disease Management, and Novel Applications**
  Editors: Megh R. Goyal, PhD, and Durgesh Nandini Chauhan, MPharm
- **Bioactive Compounds of Medicinal Plants: Properties and Potential for Human Health**
  Editors: Megh R. Goyal, PhD, and Ademola O. Ayeleso
- **Bioactive Compounds from Plant Origin: Extraction, Applications, and Potential Health Claims**
  Editors: Hafiz Ansar Rasul Suleria, PhD, and Colin Barrow, PhD
- **Bioactive Compounds from Multifarious Natural Foods for Human Health**
  Editors: Hafiz Ansar Rasul Suleria, PhD, Megh R. Goyal, PhD, and Huma Bader Ul Ain
- **Cereals and Cereal-Based Foods**
  Editors: Megh R. Goyal, PhD, Kamaljit Kaur, PhD, Jaspreet Kaur, PhD
- **Curative and Preventive Properties of Medicinal Plants: Research on Disease Management and Animal Model Studies**
  Editors: Megh R. Goyal, PhD, Junaid Ahmad Malik, PhD, and Ademola Olabode Ayeleso, PhD
- **Health Benefits of Secondary Phytocompounds from Plant and Marine Sources**
  Editors: Hafiz Ansar Rasul Suleria, PhD, and Megh Goyal, PhD
- **Herbs, Spices, and Medicinal Plants for Human Gastrointestinal Disorders**
  Editors: Megh R. Goyal, PhD, Preeti Birwal, PhD, and Durgesh Nandini Chauhan, MPharm
- **Human Health Benefits of Plant Bioactive Compounds: Potentials and Prospects**
  Editors: Megh R. Goyal, PhD, and Hafiz Ansar Rasul Suleria, PhD
- **Phytochemicals and Medicinal Plants in Food Design: Strategies and Technologies for Improved Healthcare**
  Editors: Megh R. Goyal, PhD, Preeti Birwal, PhD, and Santosh K. Mishra, Ph

- **Phytochemicals from Medicinal Plants: Scope, Applications, and Potential Health Claims**
  Editors: Hafiz Ansar Rasul Suleria, PhD, Megh R. Goyal, PhD, and
  Masood Sadiq Butt, PhD
- **Plant- and Marine-Based Phytochemicals for Human Health: Attributes, Potential, and Use**
  Editors: Megh R. Goyal, PhD, and Durgesh Nandini Chauhan, MPharm
- **Plant-Based Functional Foods and Phytochemicals: From Traditional Knowledge to Present Innovation**
  Editors: Megh R. Goyal, PhD, Arijit Nath, PhD, and Hafiz Ansar Rasul Suleria, PhD
- **Plant Secondary Metabolites for Human Health: Extraction of Bioactive Compounds**
  Editors: Megh R. Goyal, PhD, P. P. Joy, PhD, and Hafiz Ansar Rasul Suleria, PhD
- **The Role of Phytoconstitutents in Healthcare: Biocompounds in Medicinal Plants**
  Editors: Megh R. Goyal, PhD, Hafiz Ansar Rasul Suleria, PhD,
  and Ramasamy Harikrishnan, PhD
- **The Therapeutic Properties of Medicinal Plants: Health-Rejuvenating Bioactive Compounds of Native Flora**
  Editors: Megh R. Goyal, PhD, PE, Hafiz Ansar Rasul Suleria, PhD,
  Ademola Olabode Ayeleso, PhD, T. Jesse Joel, and Sujogya Kumar Panda

About Senior-Editor-in-Chief

Megh R. Goyal, PhD, PE

Retired Professor in Agricultural and Biomedical Engineering, University of Puerto Rico, Mayaguez Campus; Senior Acquisitions Editor, Biomedical Engineering and Agricultural Science, Apple Academic Press, Inc.

Megh R. Goyal, PhD, PE, is a Retired Professor in Agricultural and Biomedical Engineering from the General Engineering Department in the College of Engineering at the University of Puerto Rico–Mayaguez Campus. During his professional career, he has worked as a Soil Conservation Inspector; Research Assistant at Haryana Agricultural University and Ohio State University; Research Agricultural Engineer/Professor at the Department of Agricultural Engineering of UPRM; and Professor of Agricultural and Biomedical Engineering in the General Engineering Department of UPRM. He spent a one-year sabbatical leave in 2002–2003 at the Biomedical Engineering Department of Florida International University, Miami, USA.

Dr. Goyal was the first agricultural engineer to receive the professional license in agricultural engineering from the College of Engineers and Surveyors of Puerto Rico. In 2005, he was proclaimed the "Father of Irrigation Engineering in Puerto Rico for the Twentieth Century" by the American Society of Agricultural and Biological Engineers, Puerto Rico Section, for his pioneering work on micro irrigation, evapotranspiration, agroclimatology, and soil and water engineering.

During his professional career of 52 years, he has received many awards, including Scientist of the Year, Membership Grand Prize for the American Society of Agricultural Engineers Campaign, Felix Castro Rodriguez Academic Excellence Award, Man of Drip Irrigation by the Mayor of Municipalities of Mayaguez/Caguas/Ponce and Senate/Secretary of Agriculture of ELA, Puerto Rico, and many others. He has been recognized as one of the experts "who rendered meritorious service for the development of [the] irrigation sector in India" by the Water Technology Centre of Tamil Nadu Agricultural University in Coimbatore, India, and ASABE who bestowed on him the 2018 Netafim Microirrigation Award.

Dr. Goyal has authored more than 200 journal articles and edited more than 100 books. AAP has published many of his books, including *Management of Drip/Trickle or Micro Irrigation; Evapotranspiration: Principles and Applications for Water Management;* ten-volume set on *Research Advances in Sustainable Micro Irrigation.* He has also authored the textbooks *Elements of Agroclimatology* (Spanish) by UNISARC, Colombia, and two *Bibliographies on Drip Irrigation.* Dr. Goyal has also developed several book series with AAP, including Innovations in Agricultural & Biological Engineering, Innovations and Challenges in Micro Irrigation; and Innovations in Plant Science for Better Health: From Soil to Fork.

Dr. Goyal received his BSc degree in Engineering from Punjab Agricultural University, Ludhiana, India, and his MSc and PhD degrees from the Ohio State University, Columbus, Ohio, USA. He also earned a Master of Divinity degree from the Puerto Rico Evangelical Seminary, Hato Rey, Puerto Rico, USA.

Readers may contact him at goyalmegh@gmail.com.

About the Editors

Junaid Ahmad Malik, PhD
Lecturer, Department of Zoology, Government Degree College, Bijbehara, Kashmir, Jammu and Kashmir, India

Junaid Ahmad Malik, PhD, is a Lecturer with the Department of Zoology at Government Degree College, Bijbehara, Kashmir (J&K), India, and is actively involved with teaching and research activities. He has more than eight years of research experience. His areas of interest are ecology, soil macrofauna, wildlife biology, conservation biology, etc. Dr. Malik has published over 20 research articles and technical papers in international peer-reviewed journals and has authored and edited books, book chapters, and more than 10 popular editorial articles. He also serves as an editor and reviewer of several journals. He has participated in several state, national, and international conferences, seminars, workshops, and symposia and has more than 20 conference papers to his credit. He is a life member of the Society for Bioinformatics and Biological Sciences.

Dr. Malik received a BSc (2008) in Science from the University of Kashmir, Srinagar, J & K; MSc (2010) in Zoology from Barkatullah University, Bhopal, Madhya Pradesh, India; and PhD (2015) in Zoology from the same university. He completed his BEd program in 2017 at the University of Kashmir, Srinagar, J&K.

Readers may contact him at: malik.junaidahmad@gmail.com.

Ademola Olabode Ayeleso, PhD

Professor, Department of Biochemistry, Bowen University, Iwo, Nigeria; Honorary Professor, Department of Life and Consumer Sciences, University of South Africa, South Africa

Ademola Olabode Ayeleso, PhD, is a Professor of Biochemistry at Bowen University, Iwo, Nigeria, and Honorary Professor in the Department of Life and Consumer Sciences, University of South Africa, South Africa. He was appointed a member of the Adeleke University 2020–2025 strategic plan drafting committee to appraise all research and academic activities of the university in 2019. At Adeleke University, he established a research group named DrosophilaAU Research Group, in which fruit flies (*Drosophila melanogaster*) are being used as a model to study diabetes and obesity. He has worked as a Postdoctoral Research Fellow at three different universities in South Africa. Dr. Ayeleso is actively involved in life science research, especially in the niche of diabetes/obesity and phytomedicine. He has taught several courses in biochemistry at both the undergraduate and postgraduate levels. Prof. Ayeleso is a prolific writer with about forty publications to his credit and has participated in several national and international scientific conferences. He is also a member of some notable professional bodies and an alumnus of a Nobel laureates meeting in medicine/ physiology held in Germany in 2011.

Prof. Ayeleso earned his doctoral degree in Biomedical Technology at the Department of Biomedical Sciences, Cape Peninsula University of Technology (CPUT), Bellville, in 2013, South Africa. He was awarded a Master of Science degree in Biochemistry from the University of Port Harcourt, Port Harcourt, Nigeria, in 2007 and Bachelor of Technology degree in Biochemistry from the Federal University of Technology, Akure, Nigeria in 2002.

Contents

Contributors

Babawale Emmanuel Abiola
Department of Pure and Applied Chemistry, Ladoke Akintola University of Technology, Ogbomoso, Oyo State, Nigeria; E-mail: babawale2008@yahoo.com

Falana Benedict Abiola
Department of Anatomy, Faculty of Basic Medical Sciences, University of Western Cape, Robert Sobukwe Road, Bellville 7535, Republic of South Africa; E-mail: benabiola@yahoo.com

Temitope Abiola
Faculty of Basic Medical Sciences, Department of Biochemistry, Redeemer's University, Ede, Osun State, Nigeria; E-mail:debotem100@yahoo.com

Yapo Guillaume Aboua
Department of Health Sciences, Faculty of Health and Applied Sciences, Namibia University of Science and Technology, Private Bag 13388, Windhoek 9000, Namibia; E-mail: yaboua@nust.na

Olugbenga Obajimi Adebawo
Department of Biochemistry, School of Basic Medical Sciences, Babcock University, Ogun State, Nigeria; E-mail: adebawoo@babcock.edu.ng

Ifeoluwa Dorcas Adegbite
Department of Biochemistry, Faculty of Science, Adeleke University, P.M.B. 250, Ede 232101, Osun State, Nigeria; E-mail: dorcasifeoluwaadegbite@gmail.com

Mosebolatan Victoria Adegbola
Department of Biochemistry, Faculty of Science, Adeleke University, P.M.B. 250, Ede 232101, Osun State, Nigeria; E-mail: mosebolatanadegbola@gmail.com

Ayodeji Emmanuel Adepoju
Department of Biochemistry, Faculty of Science, Adeleke University, P.M.B. 250, Ede 232, Nigeria; E-mail: ayodejiadepoju@adelekeuniversity.edu.ng

Adekunle Moses Adetayo
Department of Surgery, School of Clinical Sciences, Babcock University, Ogun State, Nigeria; E-mail: adetayoa@babcock.edu.ng

Modupe Olusola Adetayo
Department of Biochemistry, School of Basic Medical Sciences, Babcock University, Ilishan-Remo, Ogun State, Nigeria; E-mail: adetayom@babcock.edu.ng

Oluwaseyi Adegoke Adetunji
Department of Biochemistry, School of Basic Medical Sciences, Babcock University, Ilishan-Remo, Ogun State, Nigeria; E-mail: adetunjiol@babcock.edu.ng

Samuel Adegoke Adewole
Department of Agriculture and Industrial Technology, Babcock University, Ogun State, Nigeria; E-mail: adewoles@babcock.edu.ng

Babajide Oluwaseun Ajayi
Biochemistry Division, Department of Chemical Sciences, Faculty of Natural Sciences, Ajayi Crowther University, PMB 1066 Oyo, Nigeria; E-mail: bo.ajayi@acu.edu.ng; ajayibabajide11@yahoo.com

Lydia Oluwatoyin Ajayi
Department of Biochemistry, Faculty of Science, Adeleke University, P.M.B. 250, Ede 232101, Osun State, Nigeria; E-mail: ajayity82@gmail.com

Toyosi S. Ajiboye
Department of Biochemistry, School of Basic Medical Sciences, Babcock University, Ilishan-Remo, Ogun State, Nigeria; E-mail: adetayom@babcock.edu.ng

Larai Aku-Akai
Department of Health Sciences, Faculty of Health and Applied Sciences, Namibia University of Science and Technology, Private Bag 13388, Windhoek 9000, Namibia; E-mail: lakuakai@nust.na

Oluseyi Adeboye Akinloye
Department of Biochemistry, College of Biosciences, Federal University of Agriculture, PMB 2240 Abeokuta, Nigeria; E-mail: akinloyeoa@funaab.edu.ng; oaakin@yahoo.com

Omowunmi O. Akinnawo
Department of Biochemistry, School of Basic Medical Sciences, Babcock University, Ilishan-Remo, Ogun State, Nigeria; E-mail: akinnawoo@babcock.edu.ng

Joseph Fadeyemi Akinrinmade
Department of Veterinary Surgery and Radiology, Faculty of Veterinary Medicine, University of Ibadan, Ibadan, Nigeria; E-mail: fadeyemiakinrinmade@yahoo.com

Akinleye Stephen Akinrinde
Department of Veterinary Physiology and Biochemistry, Faculty of Veterinary Medicine, University of Ibadan, Ibadan, Nigeria; E-mail: as.akinrinde@ui.edu.ng

Olakunle Akinyode
Department of Chemical Sciences, Biochemistry Unit, College of Natural and Applied Sciences, Oduduwa University, Ipetumodu, Ile-Ife, Osun State, Nigeria; E-mail:olakunle2308@gmail.com

Gogonte Hezekiah Amah
Department of Biochemistry, School of Basic Medical Sciences, Babcock University, Ilishan-Remo, Ogun State, Nigeria; E-mail: amahg@babcock.edu.ng

Adefolaju Gbenga Anthony
Department of Pre-Clinical Sciences, University of Limpopo, Private Bag x1106, Sovenga 0727, South Africa; E-mail: Gbenga.Adefolaju@ul.ac.za

Mwakikunga Anthony
School of Life Sciences and Allied Health Professions, Biomedical Sciences Department, Anatomy Division, Kamuzu University of Health Sciences, Blantyre, Malawi;
E-mail: amwakikunga@medcol.mw

Folake Olubukola Asejeje
Department of Chemical Sciences, Faculty of Natural Sciences, Ajayi Crowther University Oyo, Nigeria; E-mail: olu4la@gmail.com

GbolahanIyiola Asejeje
Department of Chemistry, University of Ibadan, Ibadan, Nigeria; E-mail: asejejeg@yahoo.co.uk

Ademola Olabode Ayeleso
Professor, Department of Biochemistry, Bowen University, Iwo, Nigeria
Honorary Professor, Department of Life and Consumer Sciences, University of South Africa, South Africa; Email: ademola.ayeleso@gmail.com; ademola.ayeleso@bowen.edu.ng

Adebola Oluwayemisi Ayeni
Department of Biochemistry, Federal University of Technology, P.M.B 704, Ondo State, Nigeria;
E-mail: ayeniadebola2014@gmail.com

Odukoya Samson Ayodeji
Department of Anatomy and Cell Biology, Obafemi Awolowo University, Ile-Ife, Nigeria;
E-mail: ayoodukoya75@gmail.com

Kawkab Mujahidah Babalola
Department of Anatomy, University of Lagos, Lagos, Nigeria; E-mail: kawkabbabalolam@gmail.com

Samuel Olufolarin Babarinde
Department of Biochemistry, Faculty of Science, Adeleke University, P.M.B. 250, Ede 232101,
Osun State, Nigeria; E-mail: samuel.babarinde@adelekeuniversity.edu.ng

Maria Onomaghuan Ebesunun
Department of Chemical Pathology, College of Health Sciences, Olabisi Onabanjo University,
Ogun State, Nigeria; E-mail: ebesununm@oou.edu.ng

Olusola Olalekan Elekofehinti
Bioinformatics and Molecular Biology Unit, Department of Biochemistry, Federal University of
Technology, P.M.B 704, Ondo State, Nigeria; E-mail: solaeleko90@gmail.com

Joseph Enya
Department of Anatomy, PAMO University of Medical Sciences, Port Harcourt, Nigeria;
E-mail: joetechnice@gmail.com

Oghale O. Godwin
Department of Biological Sciences, Faculty Science, University of Medical Sciences, PMB 536,
Ondo City, Ondo State, Nigeria; E-mail: ooghale@unimed.edu.ng

Megh R. Goyal
Agricultural and Biomedical Engineering, University of Puerto Rico-Mayaguez, Mayaguez—PR-USA,
USA; E-mail: goyalmegh@gmail.com

Hertha Kasiku Haikera
School of Nursing & Public Health, Faculty of Health Sciences and Veterinary Medicine, University of
Namibia, Private bag 88, Rundu Campus, Kavango Region, Namibia; E-mail: hhaikera@unam.na

Aminat Omolola Imam-Fulani
Department of Physiology, University of Ilorin, Ilorin 1515, Nigeria; E-mail: faminat2@gmail.com

MacDonald Idu
Department of Plant Biology and Biotechnology, University of Benin, PMB 1154 Benin City, Edo State,
Nigeria; E-mail: mcdonald@uniben.edu

Festus Igbe
Department of Biochemistry, Federal University of Technology, P.M.B 704, Ondo State, Nigeria;
E-mail: ask4festus2001@yahoo.co.nz

Olagunju Doris Imuetiyanosa
Department of Biochemistry, Faculty of Science, Adeleke University, P.M.B. 250, Ede, 232101,
Osun State, Nigeria; E-mail: deprecious2014@gmail.com

Bisola Jimoh
Department of Chemical Sciences, Biochemistry Unit, College of Natural and Applied Sciences,
Oduduwa University, Ipetumodu, Ile-Ife, Osun State, Nigeria; E-mail: bisolajimoh47@gmail.com

Gregory Joseph
2Department of Medical Laboratory Sciences, Faculty of Basic Medical Sciences, Adeleke University, P.M.B. 250, Ede 232, Nigeria

Junaid Ahmad Malik
Department of Zoology, Government Degree College, Bijbehara 192124, Kashmir, J&K, India; E-mail: malik.junaidahmad@gmail.com

Adetoun Elizabeth Morakinyo
Department of Biochemistry, Faculty of Science, Adeleke University, P.M.B. 250, Ede 232101, Osun State, Nigeria; E-mail: adetounadejumobios@adelekeuniversity.edu.ng

Esther Emem Nwanna
Department of Biochemistry, Functional food and Nutraceutical Unit, Federal University of Technology, P.M.B 704, Ondo State, Nigeria; E-mail: esthernwanna@gmail.com; eenwaana@futa.edu.ng

Samuel Chibueze Nzekwe
Department of Biochemistry, Faculty of Science, Adeleke University, P.M.B. 250, Ede 232101, Osun State, Nigeria; E-mail: samuel.nzekwe@adelekeuniversity.edu.ng

Favour Oghenefejiro
Department of Biochemistry, Faculty of Science, Adeleke University, P.M.B. 250, Ede 232101, Osun State, Nigeria; E-mail: favourgift2099@gmail.com

Babafemi Tosin Ogunbiyi
Department of Biochemistry, School of Basic Medical Sciences, Babcock University, Ilishan-Remo, Ogun State, Nigeria; E-mail: ogunbiyib@babcock.edu.ng

Laide O. Olajide
Department of Biochemistry, School of Basic Medical Sciences, Babcock University, Ilishan-Remo, Ogun State, Nigeria; E-mail: olajidel@babcock.edu.ng

Ebenezer Tunde Olayinka
Biochemistry Division, Department of Chemical Sciences, Faculty of Natural Sciences, Ajayi Crowther University, PMB 1066 Oyo, Nigeria;
E-mail: et.olayinka@acu.edu.ng; ebenezertundeolayinka@gmail.com

Oluwatosin Elijah Olopade
Department of Biochemistry, Faculty of Science, Adeleke University, P.M.B. 250, Ede 232101, Osun State, Nigeria; E-mail: oluwatosin.olopade@adelekeuniversity.edu.ng

Adedamola Aminat Bayo-Olugbami
Neuroscience Unit, Department of Physiology, Osun State University, Osogbo 232101, Nigeria; E-mail: adedamola.bayo@gmail.com
Neuroscience Unit, Department of Physiology, Adeleke University, 232101, Ede, Nigeria; E-mail: adedamola.bayo@gmail.com

Adedayo Olukanni
Faculty of Basic Medical Sciences, Department of Biochemistry, Redeemer's University, Ede, Osun State, Nigeria; E-mail: olukannia@run.edu.ng

Olumide Olukanni
Faculty of Basic Medical Sciences, Department of Biochemistry, Redeemer's University, Ede, Osun State, Nigeria; E-mail: olukannio@run.edu.ng

OlufemiIdowu Oluranti
Bowen University, Iwo, Osun state, Nigeria;
E-mail: olufemi.oluranti@bowen.edu.ng, olufemi2008@gmail.com

Ayokanmi Ore
Biochemistry Division, Department of Chemical Sciences, Faculty of Natural Sciences, Ajayi Crowther University, PMB 1066 Oyo, Nigeria;
E-mail: a.ore@acu.edu.ng; oreayokanmi@gmail.com

Ifeoluwa Opeyemi Alao-Owunna
Department of Economics, Faculty of Business and Social Sciences, Adeleke University, Osun State, Nigeria; E-mail: ifeoluwa.alao-owunna@adelekeuniversity.edu.ng

Oluokun Oluboade Oyedapo
Department of Biochemistry and Molecular Biology, Faculty of Science, Obafemi Awolowo University, P.M.B. 13, Ile-Ife, 220282, Osun State, Nigeria; E-mail: ooyedapo@yahoo.co.uk

Temitope Adenike Oyedepo
Department of Biochemistry, Faculty of Science, Adeleke University, P.M.B. 250, Ede 232101, Osun State, Nigeria; E-mail: topeoyedepo@adelekeuniversity.edu.ng

Ibukun Temitope Sossou
Faculty of Basic Medical Sciences, Department of Nursing Sciences, Redeemer's University, Ede, Osun State, Nigeria; E-mail: sossoui@run.edu.ng

Olorunfemi Samuel Tokunbo
Department of Anatomy, Osun State University, Osogbo, Nigeria; E-mail: tokunbo@uniosun.edu.ng

Abraham Yakubu
M.Sc. Student, Department of Biochemistry, Faculty of Science, Adeleke University, P.M.B. 250, Ede 232101, Osun State, Nigeria; E-mail: yakubuabrahamfp@gmail.com

Abbreviations

ABTS	2,2-azino-bis-(3-ethylbenzothiazoline-6-sulfonic acid)
AChE	acetylcholinesterase
Actb	actin beta
ACTH	adrenocorticotropic hormone
AD	Alzheimer's disease
ADHD	attention deficit hyperactivity disorder
AF-EESYT	active fraction of ethanol extract of *Scirpus yagara* tuber
AFR	aqueous extract of fully ripe *C. papaya*
AGE	aged garlic extract
AGEL	agarose gel electrophoresis
AGTE	aqueous green tea extract
ALF	acute liver failure
ALI	acute liver injury
ALP	alkaline phosphatase
ALS	amyotrophic lateral sclerosis
ALT	alanine aminotransferase
AMG	amygdalin
ANOVA	one-way analysis of variance
AP-1	activator protein 1
Apaf1	apoptotic protease activating factor 1
APR	aqueous extract of partly ripe C. papaya
ARE	aqueous rooibos extract
ARSA	ABTS radical scavenging activity
ASC	adaptor apoptosis-associated speck-like protein
AST	aspartate transaminase
ATP	adenosine triphosphate
AUR	aqueous extract of unripe C. papaya
AWR	abdominal withdrawal reflex
Aβ	amyloid-β
B2m	beta-2-microglobulin
Bad	BCL2-associated agonist of cell death
BAT	brown adipose tissue
Bax	B-cell lymphoma 2 (Bcl-2)-associated x-protein
BBB	blood-brain barrier

BCh	butylcholine
Bcl	B-cell lymphoma
Bcl-2	B-cell lymphoma 2
Bcl-xL	B-cell lymphoma extra large
BDNF	brain-derived neurotrophic factors
BIL	bilirubin
BMI	body mass index
BrDU	bromodeoxyuridine
BVEE	*B. vulgaris* ethanolic extract
bw	body weight
bw/d	body weight per day
C. papaya	Carica papaya
CAN	carnosic acid
CAS	chronic acute combined stress
Caspases	cysteine aspartases (cysteine-aspartic proteases)
CAT	catalase
CD14	cluster of differentiation 14
c-DNA	complementary deoxyribonucleic acid
CE	catechin equivalent
CGA	chlorogenic acid
Ch	choline
ChEIs	cholinesterase inhibitors
Chlof	chloroform
CK-MB	creatinine kinase myocardial band
CMA	cranial mesenteric artery
CMS	chronic mild stress
CNS	central nervous SYSTEM
CORT	corticosterone
COX-2	cyclooxygenase-2
CpG	5'-cytosine-phosphate-guanine-3'
CPT1α	carnitine palmitoyltransferase 1-alpha
CREB	camp responsive element binding
CRP	C-reactive protein
CSD	cortical spreading depression
CUMS	chronic unpredictable mild stress
CUR	curcumin
CurNPs	curcumin nanoparticles
CVD	cardiovascular disease
CVEE	*Calluna vulgaris* ethanolic extracts

Cyt-c	cytochrome c
DAD	damaurone D
DC	diabetic control
DCT	distal convoluted tube
dH2O	distilled water
DHI	Danhong injection
DI	diabetes insipidus
DIC	disseminated intravascular coagulation
DIO	dioscin
DM	diabetes mellitus
DMRT	Duncan Multiple Range Test
DNA	deoxyribonucleic acid
DOB	*Dombeya buettneri*
DPPH	1-1,diphenyl, 2-picryl hydrazine
DPPH	2, 2- diphenyl-1-picrylhydrazyl
DRT	Dopamine Replacement Therapy
EC_{50}	effective concentration on 50% of test experiment
EDHF	endothelial-derived hyperpolarizing factor
EDTA	ethylenediaminetetraacetic acid
EE	energy expenditure
EEIH	ethanol extract of *Illiciumhenryi*
EESBR	ethanolic extracts of S. baicalensis root
EFR	ethyl acetate fraction of fully ripe C. papaya
EGCG	epigallocatechin-3-gallate
EMD	emodin
EMT	epithelial-mesenchymal transition pathway
Enl	ethanol
EPR	ethyl acetate fraction of partly ripe C. papaya
ET	electron transfer
EtOH	ethanol
EUR	ethyl acetate fraction of unripe C. papaya
FasL	Fas ligand
FBG	fasting blood glucose
FBS	fasting blood sugar
$Fecl_3$	ferric chloride
FFA	free fatty acid
FODMAP	fermentable oligosaccharides, disaccharides, monosaccharide, and polyols
FOS	fructo oligosaccharides

FRAP	ferric-reducing antioxidant power
FSE	Forsythia suspensa extract
FST	forced swimming test
G6PDD	glucose-6-phosphate dehydrogenase
GABA	gamma amino butyric acid
GAE	gallic acid equivalent
GJWE	Gardenia jasminoides water extract
glo 1	glyoxalase 1
glo 2	glyoxalase 2
GLP-1	glucagon-like peptide-1
GPx	glutathione peroxidase
GSH	glutathione
GSH-Px	glutathione peroxidase
GSH-Rd	glutathione reductase
GSPE	grape seed procyanidin extract
GSR	glutathione S-reductase
GSSG	oxidized glutathione
GTPs	green tea polyphenols
H_2O_2	hydrogen peroxide
HA	hyaluronic acid
HAT	hydrogen atom transfer
HB/Hb	hemoglobin
HbS	hemoglobin S
HbSS	sickle cell anemia
HCL	hydrochloric acid
HCT	hematocrit
HDL	high-density lipoprotein
HDL-C	high-density lipoprotein cholesterol
HDN	hesperidin
HEBCS	hydroethanolic extract of (defatted) *Buchholzia coriacea* seed
HEEPN	hydroethanolic extract of *Parquetina nigrescens*
HELLP syndrome	hemolysis-elevated liver enzymes, low-platelets syndrome
HGB	hemoglobin
HIF-1α	hypoxia-inducible factor-1α
Ho1	heme oxygenase 1
HPLC	high-performance liquid chromatography
Hyp	hydroxyproline

i.g.	intragastric
i.p.	intraperitoneal
i.v.	intravenous
IARC	International Agency for Research on Cancer
IBS	irritable bowel syndrome
IC_{50}	inhibition concentration
ICAM-1	interstitial cell adhesion molecule-1
ICR	imprinting control region
ICU	intensive care unit
IDDM	insulin-dependent diabetes mellitus
IDF	International Diabetes Federation
IE	inhibitory effect
IIRI	intestinal ischemia-reperfusion injury
IkB	inhibitor of nuclear factor of kappa light polypeptide gene enhancer in B-cells
IKK	IκB kinase
IL	interleukin
IL-1β	interleukin 1 beta
IL-10	interleukin 10
IL-6	interleukin 6
IL-8	interleukin 8
iNOS	inducible nitric oxide synthase
iNOS	inducible nitric oxide synthase
IR	ischemia-reperfusion
IRAK4	interleukin-1 receptor-associated kinase 4
ISO	isoproterenol
IUFD	intrauterine fetal death
JAK2	Janus kinase 2
KV	Kolaviron
LBP	LPS-binding protein
LCBE	*Lonicera caerulea* berry extract
LDH	lactate dehydrogenase
LDL	low-density lipoprotein
LDL-C	low-density lipoprotein cholesterol
L-NAME	N^{ω}-nitro-l-arginine methyl ester
LPS	lipopolysaccharide
LSE	lotus seedpod extract
LYM	lymphocytes
MAL	MyD88 adapter like

MAO-A	monoamine oxidase A
MAPK	mitogen-activated protein kinase
MCH	mean corpuscular hemoglobin
MCHC	mean corpuscular hemoglobin concentration
Mcp1	monocyte chemotactic protein 1
MCV	mean corpuscular volume
MD	mental disorder
MDA	malondialdehyde
METo	methanol extraction of Telfairia occidentalis
MG	methylglyoxal
$MgCO_3^+$	magnesium carbonate
MHC	major histocompatibility complex
MiRNA	micro ribonucleic acid
MMP	matrix metalloproteinases
MMP-3	matrix metalloproteinase-3
MO	Moringa oleifera
MODS	multiple organ dysfunction syndrome
MPO	myeloperoxidase
MPTP	1-methyl-4-phenyl-1, 2, 3, 6-tetrahydropyridine
MPV	mean platelets volume
mRNA	messenger RNA
MS	multiple sclerosis
MyD88	myeloid differentiation protein 88
Naq	nonaqueous extract (methanol)
NDDs	neurodegenerative diseases
NDHS	Namibia Demographic Health Survey
NF-Kb	nuclear factor kappa B
NF-kB	nuclear factor kappa-light-chain-enhancer of activated B cells
NFT	neurofibrillary tangles
NF-κB	nuclear factor kappa B
NH_3	ammonia
NIDDM	noninsulin-dependent diabetes mellitus
NIH	National Institute of Health
NLRP3	Nod-like receptor family, pyrin domain containing 3
NMDA	N-methyl-D-aspartate
nNOS	neuronal nitric oxide synthase
NO	nitric oxide
NPY	neuropeptide Y

Nrf2	nuclear factor E2-related factor 2
NSFT	novelty-suppressed feeding test
OFT	open field test
OME	*Origanum majorana* Extract
p.o.	per os (orally)
P_LCR	platelets larger cell ratio
p53	transformation-related protein 3
p-ACC	phospho-acetyl-CoA carboxylase
PAF	platelet-activating factor
p-AMPK	phosphorylated adenosine monophosphate activated protein kinase
PBS	phosphate-buffered saline
PCR	polymerase chain reaction
PCT	proximal convoluted tubule
PCV	packed cell volume
PD	Parkinson's disease
PDW	platelet distribution width
PFC	prefrontal cortex
PGC1α/β	proliferator-activated receptor gamma coactivator 1-alpha/beta
PGE2	prostaglandin E2
PI3K	phosphoinositide 3-kinase
PKC	protein kinase C
P-LCR	platelets larger cell ratio
PLT	platelets
POMC	Proopiomelanocortin
PPAR	peroxisome proliferator-activated receptor
PPME	*Pulicaria petiolaris* methanol extract
PYY	peptide YY
QCT	quercetin
QE	quercetin equivalent
qPCR	quantitative polymerase chain reaction
RAGE	receptors for advanced glycation end products
RBC	red blood cell
RDW-CV	red cell distribution width coefficient variation
RDW-SD	red cell distribution width standard deviation
RGAE	red ginseng alcohol extract
RGWE	red ginseng water extract
RNA	ribonucleic acid

RNS	reactive nitrogen species
ROS	reactive oxygen specie
RP-LC	reverse phase liquid chromatography
RVS	*Rhus verniciflua* stokes
RW-SD	red cell width standard deviation
SAC	S-allyl cysteine
SCD	sickle cell disease
SCFAs	short-chain fatty acids
SD	standard deviation/Sprague-Dawley
Sdha	succinate dehydrogenase complex flavoprotein subunit A
SDS	sodium dodecyl sulphate
SEM	standard error of mean
sFas	soluble FAS antigen
SGME	sun ginseng methanol extract
SHR	spontaneous hypertensive rat
SIRS	systemic inflammatory response syndrome
SMA	superior mesenteric artery
SNP	sodium nitroprusside
SNpc	substantia nigra pars compacta
SOD	superoxide dismutase
SPIONs	superparamagnetic Iron Oxide Nanoparticles
SPSS	statistical package for social sciences
ssDNA	single-stranded DNA
SSRI	selective serotonin reuptake inhibitor
STAT3	signal transducer and activator of transcription 3
STE	sativoside
STE-EXT	stevia extract
STZ	streptozotocin
T2DM	type-2diabetes mellitus
TAC	total antioxidant capacity
TAG	triacylglycerol
T-AOC	total antioxidant capacity
TBA	thiobarbituric acid
TBARS	thiobarbituric acid reactive species
TBAs	traditional birth attendants
TC	total cholesterol
TG	triglyceride
TLR-4	toll-like receptor 4
TMQ	thymoquinone

TNF-α	tumor necrosis factor alpha
TO	*Taraxacum officinale*
TRAF6/2	TNF-receptor associated factor 6/2
TRAIL	tumor necrosis factor-related apoptosis-inducing ligand
TRAM	toll-like receptor adapter molecule
TRIF	TIR domain-containing adapter-including interferon-β
TST	tail suspension test
UAE	urinary albumin excretion
UCMS	unpredictable chronic mild stress
UCPs	uncoupling proteins
USD	United States dollars
VAO	virgin argan oil
VEGF	vascular endothelial growth factor
WAT	white adipose tissue
WBC	white blood cell
WHO	World Health Organization
XDH	xanthine dehydrogenase
XO	xanthine oxidase
α-Syn	α-synuclein
3-NP	3-nitropropionic acid
5HT	5-hydroxytryptamine
6-OHDA	6-hydroxydopamine

Preface

Plants, with their different parts (such as: leaves, sap, fruits, stem bark, and root bark), have been acclaimed to be alternative therapies in the management of various diseases. Medicinal plants are known for their health benefits due to their phytochemical constituents (such as flavonoids, saponins, and alkaloids). Scientific research has evolved because processed medicinal plant products have now come with dosage recommendations and specialized curative potential. These recent advances have made available plant phytochemicals to target the exact disease condition, thereby improving human health.

Animal models varying from mammals to flies and worms are widely used in studying human diseases. These models share homolog genes, as expressed in a human disease condition, Thus making it easier to elucidate mechanisms and pathways of these diseases in different organs of the animals, which might be difficult in clinical-based research. This book provides scientific reports on therapeutic potential of medicinal plants in different animal models and provides information on the beneficial roles of the medicinal plants on human diseases. Some of the disease conditions reported in the book include diabetes, cardiovascular disease, obesity, and neurodegenerative diseases. The therapeutic mechanisms of some phytoconstituents are also discussed.

The book is the output of the passion of researchers and academics around the globe. who positively responded the call and provided the soul to this compendium. The names of the individual contributors and the corresponding authors are given in the list of contributors. We would like to thank all the contributors who indeed devoted their valuable time for this brainstorming activity and to whom we also feel indebted to for sharing their vision with the highest level of dedication and enthusiasm to complete this draft.

We also extend our sincere thanks to the editorial office of Apple Academic Press, Inc.

We hereby appeal our readers their productive suggestions that may form the basis for improving future editions.

We take this opportunity to thank: (1) our families for their continuous support and motivation with the showers of blessings in counteracting every obstacle coming in the way during this assignment; (2) The entire research community for setting the highest standards for budding scientists, for their continuous encouragement and endless scientific support during the compilation of this book.

—Editors

PART I

POTENTIAL OF MEDICINAL PLANTS IN METABOLIC DISEASES

CHAPTER 1

Leaf Extract of *Dombeya buettneri* for Inhibition of Inflammatory Cytokines Aggregation in Type-2 Diabetes

Esther Emem Nwanna, Olusola Olalekan Elekofehinti, Festus Igbe, and Adebola Oluwayemisi Ayeni

ABSTRACT

Dombeya buettneri (DOB) is a wild tropical tree plant used locally in Africa to manage some ailments with little or no scientific information supporting its usage. This study investigated the bioactive content of the leafy part using the chromatography technique. The effect of the aqueous extract inhibition (10 mg/mL) on α-amylase, α-glucosidase enzymes was determined, while their antioxidative potential on nitric oxide (NO, DPPH) radicals and peroxides were assessed. Furthermore, the effect of the extract (600 mg/kg wt) on TNF-α, IL-1β, and IL-6 expression in type-2 diabetes (T2DM) rats was also assessed. The results revealed polyphenol characterized with HPLC was in this order [(quercetin (57.31 mg) > kaempferol (33.06 mg) > caffeic acid (20.59 mg) > chlorogenic acid (20.43 mg) > Protocatechoic acid (17.12 mg) > Rutin (12.51 mg) > Hydroxybenzoic acid (11.56 mg)]/100 g. The extract administered in diabetic rats lowered the blood glucose from (305.50 mg/dl) to (100.25 mg/dl) and also downregulated significantly ($p<0.05$) [(TNF-α (81.94%), IL-1β (100.34%) IL-6 (89.71%)] expressed genes from (140%) in diabetic rat organs without treatment. The enzyme activities were inhibited due to the antioxidant property of the extract which might have altered the functions of immune system viz-a-viz controlled blood glucose level.

Curative and Preventive Properties of Medicinal Plants: Research on Disease Management and Animal Model Studies. Megh R. Goyal, Junaid Ahmad Malik, & Ademola Olabode Ayeleso (Eds.)

Conclusively, DOB might have contributed to the observed immunomodulatory property in the diabetic rats.

1.1 INTRODUCTION

Diabetes mellitus whether type-1 or type-2 diabetes comes with various complications which have caused short life span and generated negative impact on the economy as well as social burdens when the basic of the incidence are not properly managed.[41] Along with infections with other diseases, T2DM have also been shown to stimulate the production of inflammatory cytokines.[16] These cytokines are known to be a group of soluble extracellular proteins with key intercellular controllers, which are mainly chemokines, interferons, and interleukins with differences in the receptors sites.[38,40] They are currently known to be of importance in several natural processes such as cell progression, repair, and differentiation, in innate and adaptive inflammatory responses. They are secreted or expressed in nucleated cell types as response to an injured tissue.[32]

It is a well-established fact that TNF-α serves as a key intermediary of pro-inflammatory cytokine[37] which is capable of inducing multiple signaling cascades in response to defense in the host, but uncontrolled signaling can cause tissue injury or inflammation.[34] IL-6 is also known to have an important role in normal physiology, hormones production from various glands.[34] These cytokines in diabetic patients were reported to be highly expressed, also in endothelial dysfunction.[17,18,21,33] Furthermore, cytokines have been found to offer a link between organ to system, providing molecular information on inflammatory processes for maintaining physiological stability.[21,33]

Presently, scientists are searching for alternatives from herbal plants which are rich in potent phytochemicals with antioxidant, free radical scavenging and holistically promoting health benefits[21,31,34,38] that is a natural-based drug which can treat or manage several health issues. Control of blood glucose, as well as that of cytokines secretion, could offer a unique strategy in the handling of T2DM using plant-derived natural products as immune-modulators.[10] Plant products have been reported to afford protection in immune system and also alter cytokines activities via dynamic regulation of informational molecules.[34]

Dombeya buettneri K. Schum is a tree plant of tropical origin found in West and central Africa (known as niama), which belongs to the division Spermatophyta (seed-bearing plants).[7] Its aqueous extract has been found to cause significant reduction in the histamine-stimulated gastric acid secretion in rats.[30] Also, it was found to lower blood pressure in albino Wistar rats.[31] Aside these few scientific report findings indicate no other detailed information on the use of the phytochemical constitutes and mechanism underlying the observed therapeutic activities of this plant.

This chapter investigates the effectiveness of the DOB aqueous extract on in vitro α-amylase, α-glucosidase, as well as TNF-α, IL-1β, and IL-6 expression in STZ-induced Wistar rats. Also, the extract effects on (NO*), (DPPH*) scavengers, and peroxides were determined, while the leafy part was characterized and quantified using chromatography (HPLC) and colorimetric techniques.

1.2 MATERIALS AND METHODS

Equipment and chemicals used in this study were: Thermocycler, Centrifuge, Electrophoresis machine, JENWAY 6305 Spectrophotometer, Oven Chloroform, Trizol, Ethanol, RNAase-free water, Agarose gel, Ethidium bromide, Primers kits for specific gene Agarose gel, Tris buffer.

1.2.1 SAMPLE PREPARATION

Dombeya buettneri leaves were collected within the FUTA environment. The plant identification was done in the Forestry and Wood Technology Department (FUTA) with a voucher number IFE 17692 and deposited in the herbarium. The leaves were plucked, rinsed under running H_2O, and then air-dried before pulverized into powder. Hundred grams (100 g) of ground powder was dissolved in 1 L of dH_2O, shaken for 4 h, and the mixture was filtered using a cheese cloth sieve and filter paper until every little particle was removed to obtain a clear supernatant. 100 mg/mL stock was prepared for in vitro study, while 600 mg/kg body weight extract was used for the bioassay after standardization.

1.2.2 QUALITATIVE SCREENING OF PHYTOCOMPOUNDS

1.2.2.1 SAPONIN

Observation of frothing was used as a test for saponin. The 0.5 g of the sample was dissolved in 100 mL dH_2O and shaken (AOAC, 1990).

1.2.2.2 TANNIN

Blue black green or blue green precipitate of DOB extract with $Fecl_3$ reagent indicated tannin presence.[4]

1.2.2.3 ANTHRAQUINONE

A pink red or violet coloration presented free anthraquinone,[4] when the DOB sample was mixed with benzene and 10% $NH_{3.}$

1.2.2.4 FLAVONOID

The extract was stirred with 20 mL of diluted NH_3, which gives a yellowish color, then the disappearance of the color happens when 1 mL of concentrated H_2SO_4 was added, which indicated the presence of flavonoid.[6]

1.2.2.5 TERPENOID

A reddish-brown color layer indicated the presence of terpenoid[4] when the chemical was shaken with the extract and then H_2SO_4 was added to the filtrate.

1.2.3 QUANTITATIVE DETERMINATION

1.2.3.1 FLAVONOID DETERMINATION

The method described by Meda et al.,[23] was used to quantify for total flavonoid.

1.2.3.2 SAPONIN DETERMINATION

The 2 g of DOB was mixed with 100 mL of Isobutyl alcohol,[9] and it was shaken within 5 h. It was filtered while 20 mL of 40% saturated $MgCO_3$ solution was added to the filtrate to obtain colorless solution; thereafter 5% $FeCl_3$ with dH_20 was added. The final mixture stayed for 30 min after which the Abs was taken at 380 nm using a spectrophotometer.

1.2.3.3 TERPENOID DETERMINATION

The method described in Sofowora[35] was used, while the Abs was read at 510 nm.

1.2.4 PHENOLIC DETERMINATION WITH HPLC

The method described by Amaral et al.[3] was used to carry out the HPLC operations according to the standard protocol.

1.2.5 DOB ANTIOXIDANT PROPERTY

1.2.5.1 NITRIC OXIDE

The method by Mercocci et al.[24] was used for the NO scavenging property with appropriate dilution of sample extract 100–400 μL with 1 mL 25 mM of SNP-Sol. The reaction was allowed for 120 min at 37°C, after which 500 μl Griess reagent was added while the absorbance was taken at 570 nm.

1.2.5.2 DPPH SCAVENGING ABILITY

The method described by Gyamfi et al.[13] was used. Equal volume of DOB extract was mixed with DPPH solution, which was allowed to stay in the dark for 180 s, and then read at 516 nm.

1.2.6 PEROXIDES ASSAY

1.2.6.1 TISSUE HOMOGENATES PREPARATION

The methods described by Belle et al.[8] and Ohkawa et al.[29] were used to determine the peroxides generated TBARS at 532 nm when SNP prooxidant was incubated at 100°C for 60 min with tissue extracts from rats and DOB.

1.2.7 *IN VITRO* ENZYMES INHIBITION ASSAY

1.2.7.1 α-AMYLASE

A detailed method by Worthington[39] was used to decide the inhibitory capacity of DOB extract with pancreatic α-amylase solution, while the Abs was taken at 540 nm.

1.2.7.2 α-GLUCOSIDASE

A detailed method by Apostolidis et al.[5] was used to decide the inhibitory capacity of DOB extract with α-glucosidase solution, while the Abs was taken at 405 nm.

1.2.8 ANIMAL CARE AND HANDLING FOR THE T2DM MODEL

Animal handling followed the guideline by the Ethical Committee of the University, with approval number FUTA/SOS/1411. Thirty (30) male Wistar rats were obtained from Veterinary Department, University of Ibadan with an average weight between 150 and 180 g; they were acclimatized for 2 weeks under a controlled condition.[27] The animal was made diabetic using the method described by Nwanna et al.[27] and then allocated with specific treatment groups [(1) control rats, (2) diabetic rats not treated, (3) diabetic rats treated (100 mg/kg of metformin (antidiabetes drug) orally/day, (4) diabetic rats received (600 mg/kg aqueous DOB orally/day)] based on established standardization in the research laboratory.[11] This experiment took 14 days; thereafter, the animals were decapitated and the liver, pancreas, and kidney tissue removed and then homogenized using[11] while fraction obtained used for the biochemical assays.

1.2.9 TNF-α, IL-1α, AND IL-6 GENES EXPRESSION

1.2.9.1 RNA ISOLATION

The tissues (pancreas, liver, and kidney) were excised from each animal and homogenized in Epp end-off tubes containing 1000 ul Trizol reagent. 300-μl chloroform was added to the homogenate to extract the RNA; this was vortex and then centrifuged for 30 min at 1500 rpm. The upper fraction was aspirated into a new labeled container. 200 μl iso-amy alcohol (precipitating medium) was added to the supernatant in the new labeled tubes and vortex for 30 min at 1500 rpm; thereafter, the RNA pellet was recovered. 2X centrifugation at 15,000 rpm was carried out on the RNA pellet when 50 µl of 70% ENL then the filtrate was subsequently dissolved in 50 µl of nuclease-free water. The absorbance and purity of total RNA (mRNA) was taken by UV spectrophotometer.[12]

1.2.9.2 REVERSE TRANSCRIPTION METHOD

Reverse transcription was then performed according to the manufacturer kit description (Inqaba Biotech, South Africa).

1.2.9.3 cDNA SYNTHESIS

ProtoScript 1st Strand cDNA Synthesis Kit was used for the conversion of RNA to cDNA, while the PCR amplification was carried out using the following primer set (Table 1.1).

TABLE 1.1 Primers Used for the Genetic Expression of the Cytokines.

Gene	Forward primers	Reverse primers
TNF-α	5'-ACCACGCTCTTCTGTCTACTG-3'	5'-CTTGGTGGTTTGCTACGAC-3'
IL-6	5'-TCTCTCCGCAAGAGACTTCCA-3'	5'-ATACTGGTCTGTTGTGGGTGG-3'
IL-1β	5'-CCGGATGGGTAGGATAAAGTT-3'	5'-ACCCACTGAGGTAGGAAAGA-3'
GAPDH	5'-CTCCCTGGAGAAGAGCTATGA-3'	5'-AGGAAGGAAGGCTGGAAGA-3'

Thereafter, the method as described by Elekofehinti et al.[12] was used for the assessment of PCR products (amplicons) using agarose gel electrophoresis, then visualization of the expression as bands was done using Blue-light-transilluminator while intensities of bands were quantified densito-metrically with Image-J software.

The method of Zar[42] was used to analyze the data of replicate readings with significance ($p<0.05$).

1.3 RESULTS

The results of the qualitative analysis of phytochemical constituents in *Dombeya* spp are shown in Table 1.2. There was presence of saponin, tannin, flavonoids, steroid, and terpenoid, while alkaloid and anthraquinone were not detected. Table 1.2 also reveals the quantitative analysis of the total content of constituents in the following order: saponin (19.45 ± 2.50 mg/g) > terpenoids (10.52 ± 2.34 mg/g) > flavonoids (9.19 ± 2.10 mg/g) > steroids (5.72 ± 0.87 mg/g.) > tannin (3.26 ± 0.99 mg/g).

TABLE 1.2 Qualitative Analysis of Some Phytochemicals in DOB.

DOB constituents	Indicator	mg/g
Saponin	++	19.45 ± 2.50[a]
Tannin	+	3.26 ± 0.99[f]
Flavonoid	++	9.19 ± 2.10[c]
Steroid	+	5.72 ± 0.87[d]
Terpenoid	+	10.52 ± 2.34[b]
Alkaloids	–	n/a
Anthraquinone	–	n/a

KEY:

Presence +

Absence –

Table 1.3 shows the polyphenol identified and quantified in mg/100 g. A total of 218 mg/100 g of phenolic and flavonoid were found in the DOB plant. It was also observed that quercetin had the highest concentration in 100 g of DOB in the following order: quercetin (57.31 mg) > kaempferol (33.06 mg) > caffeic acid (20.59 mg) > chlorogenic acid (20.43 mg) > Protocatechoic acid (17.12 mg) >Rutin (12.51 mg) > Hydroxybenzoic acid (11.56 mg).

Table 1.4 reveals IC_{50} of antioxidant property of DOB in scavenging ability (%) against DPPH (8.13 ± 0.11), NO*(4.08 ± 0.06), and SNP-induced lipid peroxidation (8.46 ± 0.20). Table 1.5 shows the inhibitory concentration of the aqueous extract toward carbohydrate hydrolyzing enzymes: α-glucosidase (8.10 ± 0.06 mg/mL) and α-amylase (6.23 ± 0.64 mg/mL). Figure 1.1 shows a DOB extract overview on the blood glucose level. It was observed that the extract intubation for the 14-day period inhibited the diabetic enzymes (amylase and glucosidase), which might be due to the plant antioxidant, antidiabetes, and anti-inflammatory properties. The inflammatory cells were reduced/not seen in the liver histopathology analysis as depicted in the overview.

TABLE 1.3 Polyphenolic Composition in DOB Plant Using HPLC.

Compounds	RT*	DOB plant mg/100 g
3-0-Caffeoylauinic	23.475	2.92×10^{-2}
Apigenin	17.097	3.06×10^{-1}
Caffeic acid	14.156	9.20
Capsaicin	20.793	5.94×10^{-4}
Catechin	9.748	1284
Chlorogenic acid	23.970	20.43
Coumarin	15.504	1.27×10^{-4}
Coumestrol	17.367	3.06×10^{-1}
Daidzein	16.676	8.80×10^{-5}
Ellagic acid	21.825	5.63
Epicatechin	12.364	7.37×10^{-3}
Epigallocatechin	21.441	2.31
Ferulic	14.923	20.58
Gallic acid	11.354	5.67
Genistein	18.847	3.28×10^{-4}
Gingerol	23.234	1.20×10^{-4}
Isoquercitrin	26.952	2.26×10^{-5}
Isorhamnetic	24.613	3.03
Isovanillin	17.671	4.95×10^{-5}
Kaempferol	19.525	33.06
Luteolin	20.122	4.58
Myricetin	25.001	1.12
Naringenin	18.059	1.55×10^{-3}
Naringenin chalcone	18.358	3.02×10^{-3}
o-Coumaric acid	13.292	$8.58\ 10^{-4}$
p-Coumaric acid	10.649	5.88×10^{-4}
p-Hydroxybenzoic acid	13.543	11.56
Protocatechoic acid	10.161	17.12
Quercetin	22.607	37.31
Quercitrin	26.065	2.22×10^{-4}
Rutin	27.290	12.51
Shoganol	19.108	1.81×10^{-3}
Sinapinic acid	16.251	9.61
Syringic acid	15.817	4.83
Vanillic acid	12.562	3.53×10^{-1}
Total		217.61 mg/100 g

RT* = retention time (min)

TABLE 1.4 IC_{50} Values of Antioxidative Potential of the Aqueous Extract of DOB on DPPH*, NO*, and SNP-Induced Lipid Peroxidation.

Parameter	Aqueous extract
DPPH*	8.13 ± 0.11
NO*	4.08 ± 0.06
SNP- induced LPO	8.46 ± 0.20

TABLE 1.5 IC_{50} Values of Aqueous Extract of DOB on Carbohydrate Hydrolysis Enzymes (mg/mL).

Parameter	Aqueous extract
α-glucosidase	8.10 ± 0.06
α-amylase	6.23 ± 0.64

Figure 1.2 reveals peak of each constituent characterized in a chromatogram. Figure 1.3 shows the effect of DOB blood glucose in T2DM rats. The diabetic rats without treatment had an elevated blood glucose level of (305 ± 10.6 mg/dL), but the treatment with metformin and DOB reduced significantly ($p<0.05$) blood glucose to (105.75 ± 6.6 mg/dL) and (100.25 ± 5.5 mg/dL) compared with group 1 (89.25 ± 4.34 mg/dL) control. Figures 1.4 and 1.5 reveal that diabetic rats had an increased expression TNF-α in the liver and in the pancreas within the range of (107.97–119.31%), but the treatments reduced ($p<0.05$) the expression to (91.58–104.72%) with metformin treatment, while it was (81.94–103.09%) with DOB when compared with the control (92.61–98.97%). The same trend was observed with the expression of IL-6 in the kidney. In Figure 1.6 (control 96.30%, diabetic 115.63%, diabetic + metformin 102.60%, and diabetic + DOB 100.34%), the results were not significantly different ($p>0.05$) with the use of metformin (synthesized drug) and DOB (plant compounds). Figure 1.7 shows the expression of IL-6 in the pancreas with (control 89.96%, diabetic 100.38%, diabetic + metformin 86.18%, and diabetic + DOB 89.71%), with no difference ($p>0.05$) for both treatments. Overall, treatments downregulated the expression of these cytokines significantly ($p<0.05$). Figure 1.8 shows IL-1 in the kidney with (control 108.5%, diabetic 125.1%, diabetic + metformin 118.9%, and diabetic + DOB 110.4%); this however followed the same tread as the previous ones.

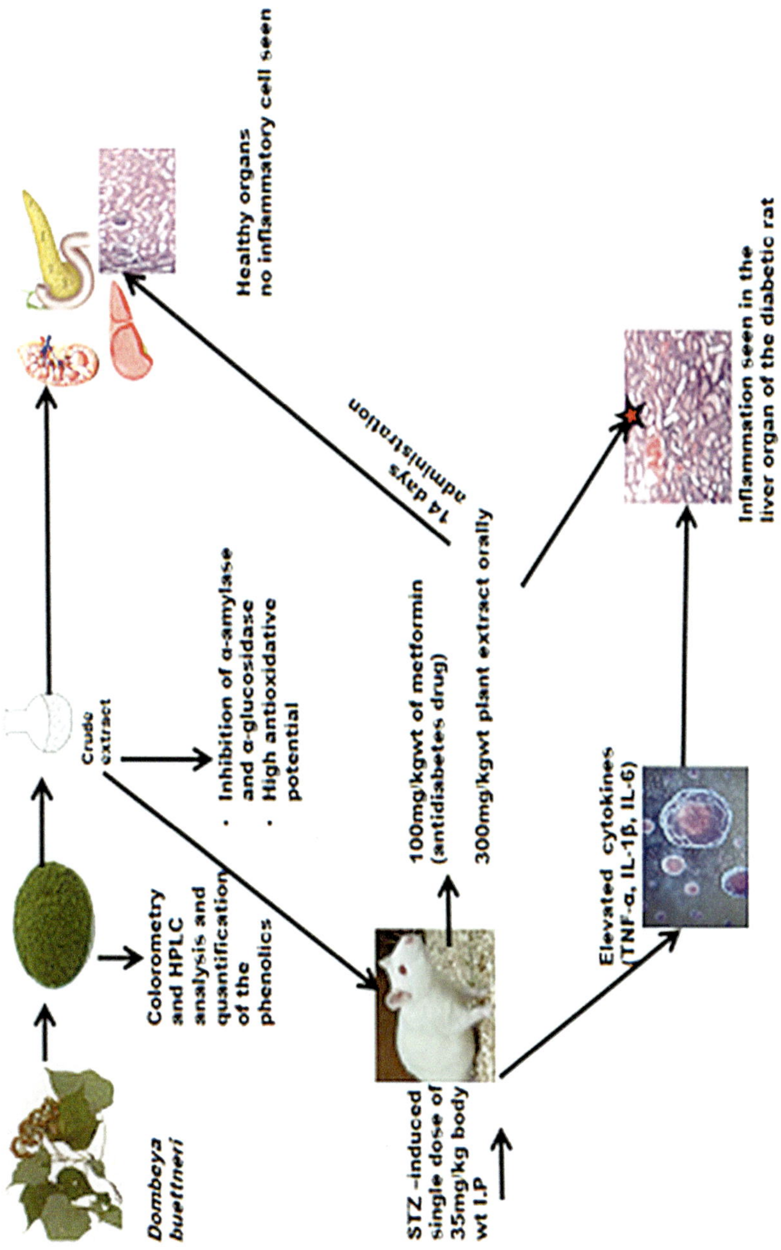

FIGURE 1.1 An overview of the effect of DOB extract in T2DM.

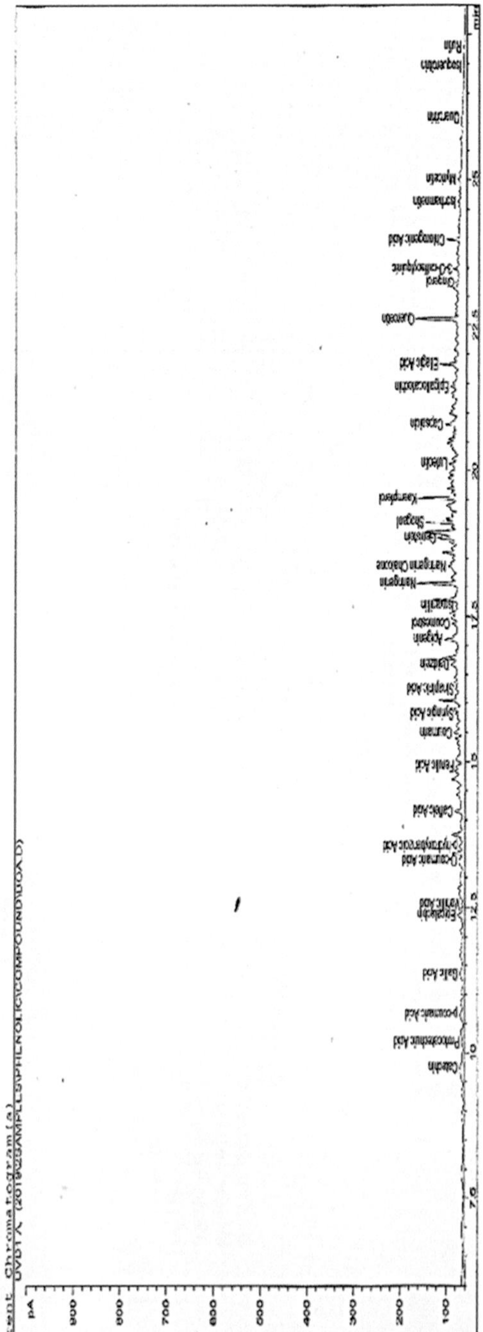

FIGURE 1.2 Chromatogram of the phenolic peaks from DOB plant.

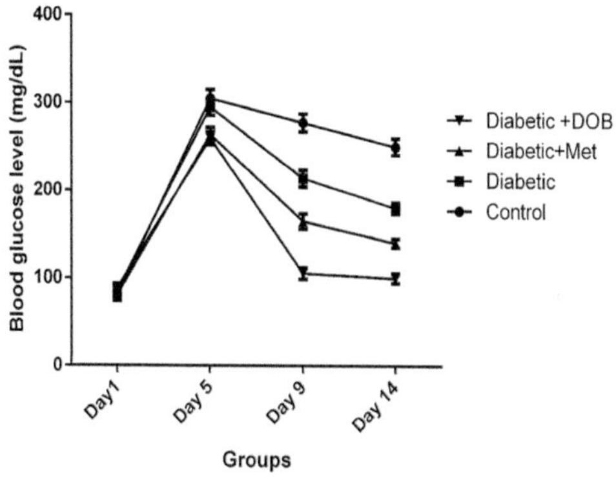

FIGURE 1.3 Blood glucose of T2DM rats treated with DOB extract. Values represent means of replicate readings.

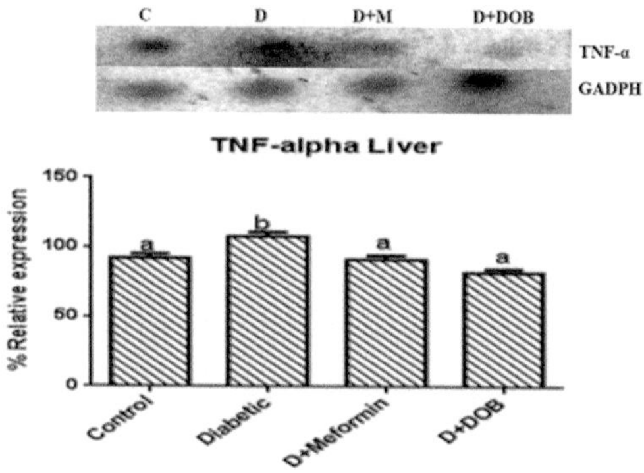

FIGURE 1.4 qPCR analysis of TNF-α expression in the liver after treatments. Different letters showed significant difference ($p<0.05$).

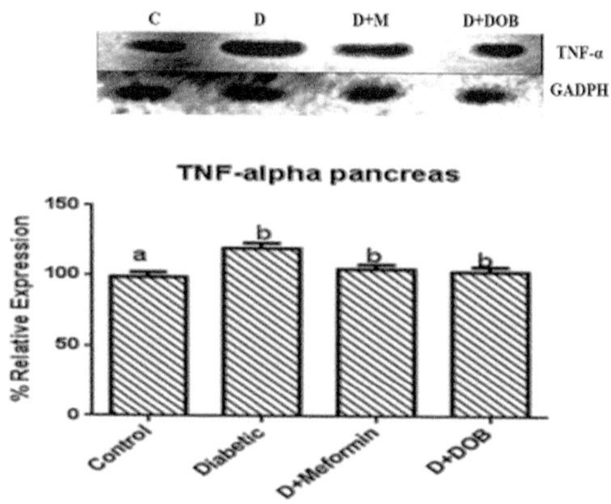

FIGURE 1.5 qPCR analysis of TNF-α expression in the pancreas after treatments. Different letters showed significant difference ($p<0.05$).

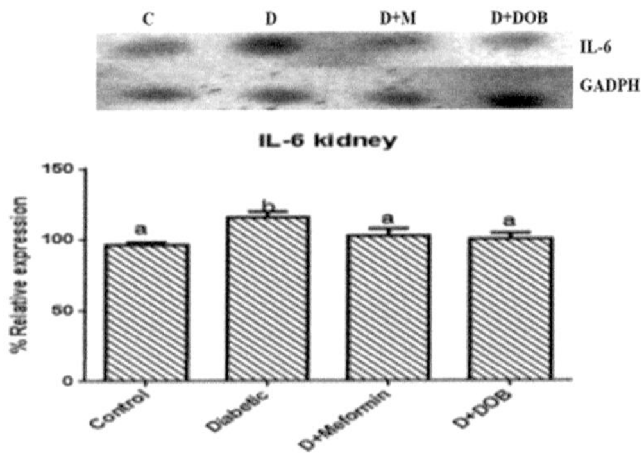

FIGURE 1.6 qPCR analysis of IL-6 expression in the kidney after treatments. Different letters showed significant difference (p<0.05).

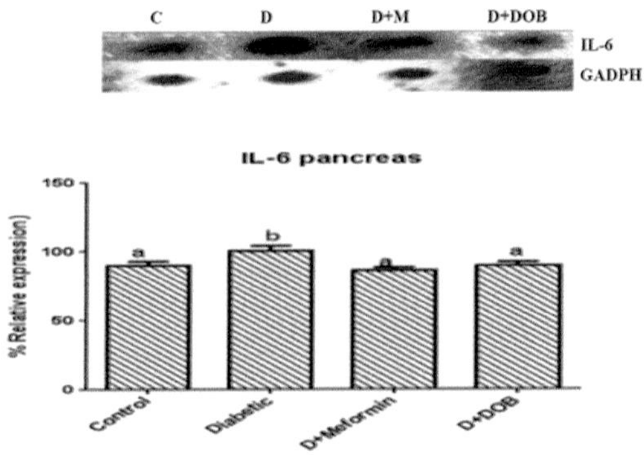

FIGURE 1.7 qPCR analysis of IL-6 expression in the pancreas after treatments. Different letters showed significant difference (p<0.05).

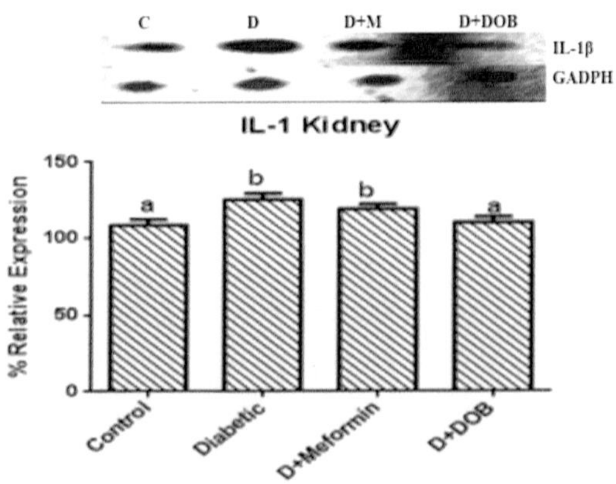

FIGURE 1.8 qPCR analysis of IL-1 expression in the kidney after treatments. Different letters showed significant difference (p<0.05).

1.4 DISCUSSION

The presence of phytoconstituent has been shown to protect and treat different diseased pathology, which includes type 2 diabetes[10,28] (Figure 1.1). The most numerous constituents in the category of phytochemicals are the phenolic and its diverse constituents[14] with various biological activities. There has been increasing evidence in the therapeutic potential of plant and its products due to minimal side effect[2] as antioxidants[25–28] because synthetic antioxidant agents have been found to possess serious health threats. Phytochemicals such as flavonoids, steroids, terpenoids, tannin, and saponin as seen in Tables 1.2 and 1.3 are potent antioxidants, chelators, and radical scavengers, with numerous health-promoting properties.[26] These compounds (Table 1.4) are reportedly known to have holistic positive attributes in the biological system.[10,34] Antioxidants can halt radicals (ROS) that kick start damage to the cell in a diseased state such as T2DM via HAT, ET, and combination of both HAT and ET.[15] HAT processes an antioxidant agent ability to extinguish free radicals as observed from this study by donating hydrogen from the phyto-compounds, while ET identifies the capacity of antioxidant to transfer one electron to lessen radicals, metals, and carbonyls.[15]

ROS tend to cause instability by changing a normal biological condition into a diseased state as observed in this study; the extract inhibited carbohydrate hydrolysis enzymes (Table 1.5) and also reduced the uncontrolled and increased blood glucose (Figure 1.3) which could have generated ROS. But, the overall result showed that the DOB extract might have served as an excellent inhibitor of ROS because of its richness in antioxidant property. In addition, the result reveals that the DOB plant (Figure 1.3) reduced the blood glucose. This study correlated with the report of Okwari et al.[31] on the DOB extract lowering the blood pressure. There are confirmed facts of connection between elevated blood glucose and blood pressure.[12,27,28] The reduced blood glucose could be associated with the concentration of active polyphenol compounds in the plant such as quercetin, kaempferol, chlorogenic acid, Protocatechoic acid, rutin to mention a few (Table 1.4 and Figure 1.2).

The hypoglycemic property of the DOB extract on amylase, glucosidase enzymes with other glucose-transporting proteins also implies that the plant constituents might have activated the signals involved with the transportation of blood glucose across the cells via various mechanisms, as observed in the in vivo study. These observations correlate with the reports of[20,22] on the glucose-lowering effect in comparison with polyphenols from plant-derived compounds. The build-up of blood glucose could have led to increased ROS generation and might have increased the oxidative stress as evidence in the elevated cytokines in the different tissues such as the liver, pancreas, and kidney (Figures 1.1 and 1.4–1.6) in T2DM. Cytokines provide information about the physiological stability of the system, while T2DM might exacerbate their production.[16,17]

Holistic control of a diseased state including cytokines exudation might serve as a positive approach in the management of T2DM. This likely offers an explanation of the immune-modulatory action of DOB extract on selected cytokine genes expression in T2DM rats which increased without treatment. But, metformin and DOB-treated groups had a reduced blood glucose level with a concomitant decreased oxidative stress, downregulated TNF-α gene expression in the liver and in the pancreas as well as IL-6 in the kidney, IL-6 in the pancreas, and IL-1 in the kidney and overall normal physiological system. The results confirm the report of Sharma and Nehru[34] on the downregulation of the expression of cytokines with the administration of curcumin extract. Decreased level of aforementioned cytokines could also have been a possible rationale of DOB extract in ameliorating gastric mucosal damage.[30] Thus, a potent antagonist from plants on TNF-α, IL-6, and IL-1 highlighted by Takashiba et al.[36] like DOB plant could also offer protection in the treatment of diseases.

1.5 SUMMARY

DOB plant, as an under-utilized tree-plant predominant in the tropics, is used locally without enough scientific information to back its usage. This study has been able to unravel some of the potential uses of DOB as antioxidant, antidiabetes, and anti-inflammatory agent. Nevertheless, DOB can be further explored as a potential natural drug compound.

KEYWORDS

- **Africa**
- **antioxidant**
- ***Dombeya buettneri***
- **downregulation**

- **flavonoids**
- **inflammatory**
- **lipid peroxidation**
- **quercetin**
- **saponins**
- **SNP-induced lipid peroxidation**
- **type-2 diabetes**
- **Wistar rats**

REFERENCES

1. Adeosun, A. M.; Oni, S. A.; Ighodaro, O. M.; Durosinlorun, O. H.; Oyedele, O. M. Phytochemical, Minerals and Free Radical Scavenging Profiles of *Phoenix dactilyfera* L. Seed Extract. *J. Taibah Univ. Medical Sci.* **2016,** *11*(1), 1–6.
2. Akinsanmi, A. O.; Nwanna, E. E. *In vitro* Comparative Studies on Antioxidant Capacities of *Gnetum africanum* ("Afang") and *Gongronema latifolium* ("Utazi") Leafy Vegetables'. *J. Pharm. Bioresour.* **2015,** *12*, 10.
3. Amaral, G. P.; Carvalho, N. R.; Barcelos, R. P.; Dobrachinski, F.; Portella, R.; da Silva, H. T.; Lugokenski, T. H.; Dias, G. R. M.; da Luz, S. C. A.; Boligon, A. A.; Athayde, M. L. Protective Action of Ethanolic Extract of *Rosmarinus officinalis* L.in Gastric Ulcer Prevention Induced by Ethanol in Rats. *Food Chem. Toxicol.* **2013,** *55*, 48–55.
4. AOAC. *Official Methods of Analysis*, 15th ed.; Association of Official Analytical Chemists (AOAC): Washington, DC, USA, 1990; pp 1250–1255.
5. Apostolidis, E.; Kwon, Y. I.; Shetty, K. Inhibitory Potential of Herb, Fruit, and Fungal Enriched Cheese Against Key Enzymes Linked to Type-2 Diabetes and Hypertension. *Innov. Food Sci. Emerg. Technol.* **2007,** *8*, 46–54.
6. Baba, S. A.; Malik, S. A.; Determination of Total Phenolic and Flavonoid Content, Antimicrobial and Antioxidant Activity of a Root Extract of *Arisaema jacquemontii* Blume. *J. Taibah Univ. Sci.* **2015,** *111*, 1–6.
7. Beentje, H. J. *Kenya Trees Shrubs and Lianas*. National Museums of Kenya: Nairobi, Kenya, 1996; pp 163–164.
8. Belle, N.; Dalmolin, G.; Fonini, G. Polyamines Reduce Lipid Peroxidation Induced by Different Pro-Oxidant Agents. *Brain Res.* **2004,** *1008*, 245–251.
9. Brunner, J. H. Direct Spectrophometer Determination of Saponin. *Anim. Chem.* **1984,** *34*, 1314–1326.
10. Chu, Y.; Sun, J.; Wu, X.; Liu, R. H. Antioxidant and Antiproliferative Activity of Common Vegetables. *J. Agric. Food Chem.* **2002,** *50*, 6910–6916.
11. Elekofehinti, O. O.; Kamdem, J. P.; Kade, I. G. Hypoglycemic, Anti- Peroxidative and Antihyperlipidemic Effects of Saponins from *Solanum anguivi* Lam. Fruits in Alloxan-Induced Diabetic Rats. *S. Afr. J. Bot.* **2017,** *88*, 56–61.

12. Elekofehinti, O. O.; Lawal, A. O.; Ejelonu, O. C.; Molehin, O. R.; Famusiwa, C. D. Involvement of Fat Mass and Obesity Action of *Annona muricta Annonaceae*: in Silico and in vivo Studies. *J. Diabetes Metab. Disord.* **2020,** *19,* 197–204.
13. Gyamfi, M. A.; Yonamine, M.; Aniya, Y. Free-Radical Scavenging Action of Medicinal Herbs from Ghana: *Thonningia sanguinea* on Experimentally-Induced Liver Injuries. *Gen. Pharmacol.* **1996,** *32*, 661–667.
14. Hahn, N I. Is Phytoestrogens Nature's Cure for What Ails Us? A Look at the Research. *J. Am. Diet. Assoc.* **1998,** *98*, 974–976.
15. Jothy, S. L.; Azlan, A. A.; Yeng, C. Y.; Sasidharan, S. Antioxidant Activity and Hepatoprotective Potential of *Polyalthia longifolia* and *Cassia spectabilis* Leaves Against Paracetamol Induced Liver Injury. *Evid. Based Complementary Altern. Med.* **2012,** *4*, 714–730.
16. Kiecolt-Glaser, J. K.; McGuire, L.; Robles, T. F. Psychoneuroimmunology and Psychosomatic Medicine: Back to the Future. *Psychosom. Med.* **2002,** *64*, 15–28.
17. Keystone, E. C. Tumor Necrosis Factor-Alpha Blockade in the Treatment of Rheumatoid Arthritis. *Rheum. Dis. Clin. North Am.* **2001,** *27*, 427–443.
18. Naitoh, Y.; Fukata, J.; Tominaga, T.; Nakai, Y.; Tamai, S.; Mori, K.; Imura, H. Interleukin-6 Stimulates the Secretion of Adrenocorticotropic Hormone in Conscious, Freely-Moving Rats. *Biochem. Biophys. Res. Communun...* **1988,** *155*, 1459–1463.
19. Makkar, H. P. S.; Blummel, M.; Bowwy, N. K.; Becker, K. Gravimetric Determination of Tannins and Their Correlations with Chemical and Protein Precipitation Methods. *J. Sci. Food Agric.* **1993,** *61*, 161–185.
20. Markowicz-Piasecka, M.; Sikora, J.; Szydłowska, A.; Skupień, A.; Mikiciuk-Olasik, E.; Huttunen, K. M. Review on Metformin: A Future Therapy for Neurodegenerative Diseases. *Pharmacol. Res.* **2007,** *17*, 2199–2205.
21. Mayadas, T. Apoptosis and Inflammation. In *Molecular Basis for Microcirculatory Disorders*; Schmid-Schonbein, G. W., Neil Granger, D., Eds.; Springer: Paris, 2013; p 259.
22. Mazibuko, S. E.; Muller, C. J.; Joubert, E.; Beer, D.; Johnson, R.; Opokub, A. R.; Louw, A. Amelioration of Palmitate-Induced Insulin Resistance in C2C12 Muscle Cells by Rooibos (*Aspalathus lineraris*). *Phytomedicine* **2013,** *20*, 813–819.
23. Meda, A.; Lamien, C. E.; Romito, M.; Millogo, J.; Nacoulma, O.G. Determination of the Total Phenolic, Flavonoid and Proline Contents in Burkina Faso Honey, as well as Their Radical Scavenging Activity. *Food Chem.* **2005,** *91*, 571–577.
24. Mercocci, L.; Maeguire, J. J.; Droy-Lefaiz, M. T.; Parker, L. The Nitric Oxide-Scavenging Properties of *Ginkgo biloba* Extract EGB761. *Biochem. Biophys. Res. Commun.* **1994,** *201*, 748–755.
25. Nwanna, E. E.; Ibukun, E. O.; Oboh, G.; Ademosun, A. O.; Boligon, A. A.; Athayde, M. HPLC-DAD Analysis and *In-vitro* Property of Polyphenols Extracts from (*Solanum aethiopium*) Fruits on α-Amylase, α-Glucosidase and Angiotensin -1- Converting Enzyme Activities'. *Int. J. Biomed. Sci.* **2014,** *10*(4), 272–281.
26. Nwanna, E. E.; Adedayo A. C. In-vitro Characterization and Biological Properties of Polyphenols Extract of Ripe Garden Egg (*Solanum gilo*)'. *J. Exp. Biol. Agric. Sci.* **2017,** *5*(1), 061–069.
27. Nwanna, E. E.; Ibukun, E. O.; Oboh, G. Effect of Some Tropical Eggplant Fruits (*Solanum* spp.) Supplemented Diet on Diabetic Neuropathy in Male Wistar Rats *in-vivo*. *Funct. Foods Health Dis.* **2016,** *6*(10), 661–676.

28. Nwanna, E. E.; Ibukun, E. O.; Oboh, G. Nutritional Content of Selected Species of Tropical Eggplant Fruit (*Solanum spp*) Diet Attenuates Hepatic Inflammation in High-Fat Fed Male Wistar Rats Induced with Streptozotocin. *Food Sci. Nutr.* **2019,** *7*(1), 109–119.
29. Ohkawa, H.; Ohishi, N.; Yagi, K. Assay for Lipid Peroxides in Animal Tissues by Thiobarbituric Acid Reaction. *Anal. Biochem.* **1979,** *95*, 351–358.
30. Okwari, O. O.; Ettarh, R. R.; Akpogomeh, B. A.; Eteng, M. U. Gastric Anti-Secretory and Anti-Ulcerogenic Effects of *Dombeya buettneri* in Rats. *J. Ethnopharmacol.* **2000,** *71*, 315–319.
31. Okwari, O.; Ofem, O.E.; Bisong, S.A.; Ettarh, R. The Aqueous Leaf Extract of *Dombeya buettneri* Lowers Blood Pressure in Albino Wistar Rats. *Int. J. Curr. Res.* **2011,** *2*(2), 92–96.
32. Oppenheim, J. J. Cytokines: Past, Present, and Future. *Int. J. Hematol.* **2001,** *74*, 3–8.
33. Saremi, A.; Anderson, R. J.; Luo, P.; Moritz, T. E.; Schwenke, D. C.; Allison, M.; Reaven, P. D. Association Between IL-6 and the Extent of Coronary Atherosclerosis in the Veterans Affairs Diabetes Trial (VADT). *Atherosclerosis* **2009,** *203*, 610–614.
34. Sharma, N.; Nehru, B. Curcumin Affords Neuroprotection and Inhibits a-Synuclein Aggregation in Lipopolysaccharide-Induced Parkinson's Disease Model. *Inflammo-pharmacology* **2018,** *26*, 349–360.
35. Sofowora, A. *Medicinal Plants and Traditional Medicine in Africa*; Spectrum Books Ltd: Ibadan, Nigeria, 1993; p 289.
36. Takashiba, S.; Naruishi, K.; Murayama, Y. Perspective of Cytokine Regulation for Periodontal Treatment: Fibroblast Biology. *J. Periodontol.* **2003,** *74*, 103–110.
37. Tan, K. C.; Chow, W. S.; Tam, S.; Bucala, R.; Betteridge, J. Association Between Acute-Phase Associated with Insulin Resistance, Acutely Enhances Colorectal Epithelial Proliferation *in vivo*. *Endocrinology* **2004,** *147*, 1830–1837.
38. Terlikowski.; S. J. Tumour Necrosis Factor and Cancer Treatment: A Historical Review and Perspectives *Rocz. Akad. Med. Bialymst.* **2001,** *46*, 5–18.
39. Worthington, V. Alpha Amylase. In *Worthington Enzyme Manual*; Worthington, K., Worthington, V., Eds.; Worthington Biochemical Corp: Freehold, NJ, 1993; pp 36–41.
40. Young, A. S.; Cummins, J. M. The History of Interferon and its Use in Animal Therapy. *East Afr. Med. J.* **1990,** *67*, 31–63.
41. Zhao, Y.; Jiang, Z.; Guo C. New Hope for Type 2 Diabetics: Targeting Insulin Resistance Through the Immune Modulation of Stem Cells. *Autoimmunol. Res.* **2011,** *11*(2), 137–142.
42. Zar J. H. *Biostatistical Analysis*; Prentice-Hall Inc.: Englewood Cliffs, USA, 1984; p 718.

CHAPTER 2

Antidiabetic Potential of African Medicinal Plants: A Brief Review Using Animal Models

Gbolahan Iyiola Asejeje and Folake Olubukola Asejeje

ABSTRACT

African medicinal plants have promising prospects in drug development for various diseases, such as diabetes. In recent times, several of these plants have been investigated in in vitro and/or in vivo settings for potential benefits to manage diabetes. This chapter reviews research on African medicinal plants in the treatment of diabetes. Electronic databases including Scopus, Google Scholar, PubMed, and ScienceDirect were searched by using the key search words, such as Africa, medicinal plants, diabetes, and antidiabetic. Then relevant references were selected/reviewed including articles published up to April 2020 for all plant species of African origin with experimental design for in vivo animal models. This chapter presents a discussion of the potential benefits of African medicinal plants in the treatment of diabetes from a review of research reports that have used different animal models to investigate the potentials of medicinal plants in the development of therapeutics for diabetes. The aim of this chapter is to discuss the potentials of medicinal plants in the treatment of diabetes with a focus on plants of African origin.

2.1 INTRODUCTION

Diabetics can be regarded as a silent killer because the onset of the disease is not often noticed until almost at the advanced stage of the disease when

Curative and Preventive Properties of Medicinal Plants: Research on Disease Management and Animal Model Studies. Megh R. Goyal, Junaid Ahmad Malik, & Ademola Olabode Ayeleso (Eds.)

symptoms and other complications begin to manifest.[45] The International Diabetes Federation (IDF), in its 9th edition Diabetes Atlas, have projected that by 2030 there will be 578 million adults with diabetes, which will rise to 700 million by 2045.[19,45] The IDF also estimates the worldwide annual incidence of diabetes in under the 20 years age group to be about 128,900 new cases. The report concludes that the number of adolescents and children with type 1 diabetes is on the increase worldwide.[42] The contribution of diabetes to global health care cost and consequent economic impact is enormous. The direct expenditure on diabetes was about 760 billion USD in 2019, and has been projected to increase to about 825 billion USD and 845 billion USD in 2030 and 2045, respectively.[51] The global health and economic burden occasioned by the incidence of diabetes have made the disease to gain global recognition and research attention. Diabetes is an endocrine metabolic disorder that presents either as the more common mellitus form or the less common insipidus form.[29]

Diabetes mellitus (DM) is associated with hyperglycemia due to deficiencies in insulin production and/or action. This form of diabetes is the most prevalent and major cause of complications, hospitalization, and fatality. Notable symptoms of DM include polyuria, polyphagia, polydipsia, pruritus, and unexplained weight loss. DM are of two types: namely type 1, also known as insulin-dependent diabetes mellitus (IDDM) and type 2, also known as noninsulin-dependent diabetes mellitus (NIDDM). IDDM is caused by insulin deficiency due to immune-mediated destruction of the pancreatic β-cells that produce insulin. IDDM can be managed by insulin replacement therapy. The NIDDM is associated with compromised insulin secretion/signaling or peripheral resistance to the action of insulin due to either obesity or hereditary disorder. NIDDM is the most prevalent form of diabetes mellitus and usually manifest at about age 40. Since NIDDM is associated with lifestyle and genetics, it is usually managed through dietary adjustments, exercise, and use of hypoglycemic therapies.

Diabetes insipidus (DI) is associated with excessive thirst and passing of copious diluted urine that is not responsive to reduction of fluid intake by the individual. The most common type of DI is called central DI which is due to arginine vasopressin (antidiuretic hormone) deficiency. The other type of DI is called nephrogenic DI associated with the kidney's insensitivity to antidiuretic hormone.

Some common complications that a diabetic may likely have include hyperglycemia, diabetic ketoacidosis, hyperosmolar state, oxidative stress, impaired immune response with susceptibility to respiratory infections, and microvascular pathologies (such as retinopathy, neuropathy, and albuminuria).[21]

No adequate cure exists till date for diabetes. However, a number of management strategies have employed drugs (such as insulin therapy and oral hypoglycemic), and lifestyle/diet adjustments. Side effects from the use of orthodox therapeutics have stimulated research into the use of natural products including medicinal plants in the management of diabetes. Some of these side effects include lactic acidosis, peripheral edema, hypoglycemia, and gastrointestinal issues.[50] On the other hand, it is expected that the medicinal plants will elicit less side effects, be safer and cost effective. In Africa, there is an abundance of natural product especially of plant source that has shown potential in diabetes treatment. The local herbal medicine industry has helped to popularize a significant number of plants in the care for diabetes and other diseases. Consequently, a good number of these plants, and their potential active principle(s), have been studied using different animal models to investigate their antidiabetic properties.

This chapter focuses on promising studies with emphasis on the different animal models and potential application in humans. The chapter also highlights health benefits of African medicinal plants in the treatment of diabetes in the poor and less-technologically developed continent of Africa.

2.2 ANTIDIABETIC POTENTIAL OF AFRICAN MEDICINAL PLANTS IN DIFFERENT ANIMAL MODELS

Many of the research into the antidiabetic potentials of medicinal plants of African origin have been conducted in vivo using rats and mice. The experimental design of majority of the studies involves dividing the animals into at least four different treatment groups. At least two groups are composed of diabetic animals (usually induced by chemical agent, e.g., alloxan and streptozotocin; or genetic knockout animals) to have treated diabetic and untreated diabetics animals for comparison. The other two groups are usually untreated normal (nondiabetic) animal and treated normal (nondiabetic) animals. The duration of experiment ranged from 1 h to about 21 days. Many of the plants studied have been reported to contain phytochemicals that make them exhibit their antidiabetic effects. Such identified phytochemicals include tannins, flavonoids, saponins, cardiac glycosides, and alkaloids. Most of the investigations demonstrate a dose-dependent action of the administered plant materials. Of the several reviewed studies on antidiabetic investigation of medicinal plants across Africa, reports are predominantly from Nigeria, Cameroun, South Africa, Morocco, Egypt, and Tunisia.

The following are discussions of the reviewed studies according to the animal model used and the prospect for the application in human treatment of diabetes.

The murine studies have been predominantly in rats of the albino Wistar species and mice of the Swiss species. Some of these medicinal plants have also been reported to show benefits in attenuating secondary complications like retinopathy, nephropathy, dyslipidemia, neuropathy, and cardiovascular disease that are usually associated comorbidity of diabetes.[22,27,44,50] For example, some of the medicinal plants help to lessen the risk of secondary complications associated with diabetes by improving glucose tolerance, and restoring liver glycogen and antioxidant activity. Furthermore, they protect from the pathological damages like regeneration of the epithelium, expansion of glomeruli, clearance of hemorrhages, and cytoplasmic debris.[24,30]

2.3 POTENTIAL IMPACT OF AFRICAN ANTIDIABETIC MEDICINAL PLANTS ON THE DIABETES BURDEN

Data from WHO indicates that thousands of plant species with proven hypoglycemic activity are being used as traditional herbal therapies for the management of DM. The WHO, following laboratory investigations of these medicinal plants, has approved the use of herbal medicine for the treatment of certain ailments including diabetes.[9,24] Hence, a large proportion of the world's population use herbal remedies for the treatment of DM.[29,40] The use of plants in managing ailments and particularly DM is an ancient tradition in Africa that has been transmitted from one generation to another. The improvement in science and technology has helped in the experimental validation of these medicinal plants by indigenous researchers (Table 2.1).

The diabetes crisis in Africa is of serious concern as about 20 million people in Africa have been documented to be diabetic in 2013, which may increase to about 42 million by 2025.[43] Earlier, 10 million people in sub-Saharan Africa were reported to have diabetes in 2006 with the potential of rising to about 19 million by 2025.[26] This burden of DM is widespread in African countries with Nigeria having about 4 million people living with diabetes as of 2013,[48] and South Africa with about 500 thousand diabetic patients.[11] This incidence of DM explains the high rate of diseases such as cardiovascular diseases, impotence, and erectile dysfunction problems that are prominent in sub-Saharan Africa.[12] From the aforementioned, it can be said that the socioeconomic impact of DM cannot be overlooked and allowed to spiral into a pandemic that will overwhelm the already poor health care system.

TABLE 2.1 Antidiabetic Medicinal Plants of African Origin From Selected Experimental Diabetic Studies.

Plant	Plant part used	Animal model	Study design	Results	Country of study/ source of plant material
African mistletoes *(Viscum album (V. album))* (Loranthaceae)	Methanolic (ME) extract of whole plant[2] Leave extract[17]	Wistar albino rats[2]	The impact of extract on FBG of STZ-diabetic rats[2] Investigate stimulation of insulin secretion in normal and STZ-challenged rats[17]	Extract elicited significant hypoglycemic effects within 4 h of administration and anti-hyperlipidemia after daily administration for three consecutive weeks in STZ-induced diabetic rats.[2] Results suggest that extract from *V. album* leaves possesses antihyperglycemic, insulinotropic, and glucagonostatic properties.[17]	Nigeria[2,17]
Ageratum conyzoides	Aqueous extract of leaves	Wistar albino rats	Alloxan-induced diabetic animals were given 500 mg/kg bw aqueous extracts of the plant[7] Normoglycemic and STZ-induced diabetic rats given 100, 200, and 300 mg/kg aqueous extracts of the plant.[31]	FBG of animals reduced by 39.1%[7] 200 and 300 mg/ kg doses showed statistically significant hypoglycemic and antihyperglycemic activities.[31]	Nigeria[7] Cameron[31]
Argania spinosa (Sapotaceae)	Virgin argan oil (VAO) extracted from almond	Wistar rats	Investigate the prevention of development of diabetes by administering VAO to rats before and during alloxan administration.[8]	Treatment with VAO prior to alloxan treatment suppressed the body mass loss, caused significant drop of blood glucose and significant rise of liver glycogen compared with the untreated diabetic group.[8]	Morocco[8]
Artemisia campestris	Aqueous extract of leaves	Wistar rats	Investigate the effect of extract on alloxan challenged rats by assessing glycemia, lipid profile, lipid peroxidation, protein carbonyl content, and antioxidants.[47]	Treatment of diabetic rats with extract resulted in a significant reduction in glycemia, and improved antioxidant status of animals. Indication that plant may be active for controlling hyperglycemia and preventing diabetic complications [47].	Tunisia[47]

TABLE 2.1 *(Continued)*

Plant	Plant part used	Animal model	Study design	Results	Country of study/ source of plant material
Bauhinia thoningii	Aqueous extract of leaves	Wistar rats	Aqueous extract given orally to alloxan-challenged diabetic rats and fasting blood glucose examined spanning 7 days.[39]	81.37% drop in blood glucose in the experimental animal group over the 7 days period from 365 to 368 mg/dl.[39]	Nigeria[39]
Bryophyllum pinnatum (Crassulaceae)	Aqueous extract of leaves	Wistar rats	STZ-induced diabetes rats given plant extra 25–800 mg/kg i.p.[35]	The plant extract (25–800 mg/kg) caused significant hypoglycemia in rats.[35]	Nigeria[35]
Buchholzia coriacea (Capparaceae)	Ethanol extracts (EE) and butanol fraction (BF) of seeds	Mice and rats	The hypoglycemic properties in STZ-challenged mice and rats[6]	EE and BF significantly decreased fasting blood glucose in hyperglycemic mice and normoglycemic rats within 4 and 12 h, respectively, after administration. EE and BF also decreased hepatic injury induced by STZ.[6]	Nigeria[6]
Calamintha officinalis Moench. (Lamiaceae)	Water extract of aerial parts	Wistar rats	Investigate the effects of extract on plasma blood glucose concentrations and basal insulin levels in normal and STZ-induced diabetic rats.[25]	Plant extract demonstrated hypoglycemic properties in both normal and STZ-challenged rats. No changes were detected in basal plasma insulin levels of normal or STZ-challenged diabetic rats after treatment with extract. This implies that the underlying mechanism of action is independent of insulin secretion.[25]	Morocco[25]
Catharanthus roseus (Apocynaceae)	Methanolic leaf extract	Wistar albino rats	Effects of co-administration of leaf extract on the hypoglycemic activity of metformin and glibenclamide.[32]	Highest percentage reduction in blood glucose of 64.86% exhibited by metformin-extract combination, that is, leaf extract of *C. roseus* significantly increase hypoglycemic activity of metformin.[32]	Nigeria[32]

TABLE 2.1 *(Continued)*

Plant	Plant part used	Animal model	Study design	Results	Country of study/ source of plant material
Centaurium erythrea (Gentianaceae)	Aqueous extract of leaves	Wistar rats	Assess the protective potential of leaf extract in pancreas β-cell injury and antioxidant defense systems in STZ-challenged rats.[46]	Plant extract 200 mg/kg bw/day, i.p. for 30 days caused reduction in blood glucose and lipid peroxidation in STZ-treated rats. In addition degenerative changes of pancreatic β-cells of STZ-treated rats were restricted to near normal morphology by leaf extract as seen in histopathological examination.[46]	Morocco[46]
Cnestis ferruginea	Methanol (ME) and Ethylacetate (EE) extracts of leaves	Rats (Sprague Dawley) and Mice	250 mg/kg extract given orally for 10 successive days to STZ-challenged rats and 250 mg/kg lone extract dose were given to STZ-challenged mice. Then fasting blood glucose (FBG) was measured in both groups.[5]	ME and EE caused significant reduction in FBG within 4 h of administration. 10 days administration of ME and EE lowered significantly the FBG in STZ-induced diabetic rats respectively by 74% and 68%.[5]	Nigeria[5]
Commelina africana (Commelinaceae)	Aqueous extract of leaves	Wistar albino rats	Alloxan-induced diabetic animals were given 500 mg/kg bw aqueous extracts of the plant.[7]	FBG of experimental animals reduced by 78.0%[2]	Nigeria[7]
Cynanchum acutum L. (Asclepiadaceae)	Butanol fraction (BF) of methanolic extract of aerial parts	Swiss albino mice and albino rats	Assessment of antioxidant and antidiabetic properties of the BF in alloxan-challenged diabetic rats.[18]	The entire tested six isolates were very potent as antioxidants through. Three isolates 1Đ3 displayed potent antidiabetic property through an increase of insulin and a decrease of glucose levels.[18]	Egypt[18]
Diaphanathe bidens	Methanol extract of leaves	Wistar albino rats	Hyperglycemia induced by giving STZ i.p. 150, 300, and 600 mg/kg extract was administered orally.[41]	600 mg/kg extract reduced significantly blood glucose concentrations of rats between 120 and 480 min post administration in normal rats. Extract caused	Nigeria[41]

TABLE 2.1 *(Continued)*

Plant	Plant part used	Animal model	Study design	Results	Country of study/ source of plant material
				significant dose- and time- dependent reduction in the blood glucose levels in STZ- induced hyperglycemic rats.[41]	
Garcinia Kola (Guttiferae)	Seed extract	Wistar rats	Hypoglycemic and hypolipidemic effects of fractions from kolaviron (a *Garcinia kola* seed extract) and its fractions in normal and STZ-diabetic rats.[1]	Result affirms the anti-diabetic and hypolipidemic potentials of KV and two of its fractions, namely FI and FII (RF values of 0.48 and 0.71, respectively) in STZ-diabetic rats.[1]	Nigeria[1]
Hypoxis hemerocallidea (Hypoxidaceae)	Corm aqueous extract	Wistar rats	STZ-induced diabetes rats given plant extra 50 to 800 mg/kg p.o.[37]	Corm aqueous extract produced significant reductions in the blood glucose concentrations with maximal reductions at 800 mg/kg extract p.o.[37]	South Africa[37]
Lepidium sativum L. (Brassicaceae)	Aqueous extract of seed	Wistar rats	The hypoglycemic property of extract in normal and STZ-challenged rats.[16]	Lone dose 20 mg/kg acute or 15 daily recurrent chronic oral treatments caused significant decrease of blood glucose levels in STZ- diabetic and normal rats. Daily repeated oral administration of 20 mg/kg for 2 weeks normalized blood glucose concentrations.[16]	Morocco[16]
Momordica charantia (Cucurbitaceae)	Whole-plant aqueous extract	Wistar rats	Hypoglycemic properties of the plant extract were assessed using STZ-induced diabetes mellitus models.[38]	Acute oral administrations caused dose-dependent, significant hypoglycemia in normal and STZ-induced diabetic rats.[38]	South Africa[38]
Musa sapientum (Musaceae)	Methanolic (ME) and aqueous (AE) extracts of root	Wistar albino rats	Anti-hyperglycemic and antioxidant properties of extracts investigated in alloxan-induced diabetic rats[4]	Seven postinduction, blood sugar levels dropped in ME and AE-treated groups.[4]	Nigeria[4]

TABLE 2.1 *(Continued)*

Plant	Plant part used	Animal model	Study design	Results	Country of study/ source of plant material
Nigella sativa (Ranunculaceae)	Aqueous extract of seeds	Sprague Dawley rats	Investigate effects of extract of on intestinal glucose absorption in vitro using short-circuit current technique and in vivo using an oral glucose tolerance test.[28]	Aqueous extract elicited dose-dependent inhibition of sodium-dependent glucose transport across isolated rat jejunum. Oral glucose tolerance test indicates that *Nigella sativa* enhanced glucose tolerance also decreased body weight without any adverse effect.[28]	Morocco[28]
Psidium guajava (Myrtaceae)	Leaf aqueous extract	Wistar rats	Hypoglycemic effect of the extract was investigated in normal and STZ-induced diabetic rats.[34]	Acute oral administrations of 50–800 mg/kg, p.o. plant extract elicited dose-related, significant hypoglycemia in normoglycemic and STZ-challenged, diabetic rats.[34]	South Africa[34]
Sapium ellipticum (Hochst) Pax (Euphorbiaceaea)	Ethanol leaf extract	Wistar rats.	To study the impact of extract on hepatic glucokinase and glucose-6-phosphatase in STZ-induced diabetes.[23]	*S. ellipticum* extract significantly activated glucokinase activity by 40.31% and inhibited glucose-6-phosphatase activity by 37.29%. The extract also enhanced hepatic glycogen level.[23]	Nigeria[23]
Sclerocarya birrea (Anacardiaceae)	Stem-bark ethanolic extract[20]	Wistar rats	Nondiabetic and STZ-induced diabetic rats treated with plant extract[15,20]	Reduction in glucose blood levels in a dose-dependent fashion, associated with increased hepatic glycogen synthesis.[20]	South Africa[20]
	Stem-bark methylene chloride/methanol extract[15]			*Sclerocarya birrea* stem bark methanol/ methylene chloride extract reduced significantly blood glucose and significantly increased plasma insulin in diabetic rats.[15]	Cameroun[15]
Suthrelandia frutescens (Fabaceae)	Shoot aqueous extract.[33]	Wistar rats	Hypoglycemic effect of the extract was investigated in normal and STZ-induced diabetic rats.[33]	The 50–800 mg/kg extract p.o. or i.p. resulted in significant hypoglycemia in rats.[33]	South Africa[13,33]

TABLE 2.1 *(Continued)*

Plant	Plant part used	Animal model	Study design	Results	Country of study/ source of plant material
	Crushed leaves in drinking water.[13]		Investigated hypoglycemic and glucose uptake potential in prediabetic rats.[13]	Extract displayed normoinsulinaemic levels, after 8 weeks; with increase in muscle and adipose tissues glucose up takes, and decrease glucose up take by intestine. Thus extract is a promising type 2 antidiabetes candidate.[13]	
Telfaria occidentalis	Aqueous extract of leaves	Mice	The effect of oral administration of extract on blood glucose concentration studied in normal glycemic, glucose-induced hyper glycemic, and STZ-induced diabetic mice.[3]	Co-administration of extract and glucose, and preadministration of extra 60 min before glucose had antidiabetic activity in glucose-induced hyperglycemia group. Extract elicited drop in blood glucose level from day two of the administration in STZ-induced diabetic group.[3]	Nigeria[3]
Trema orientalis (Ulmaceae)	Aqueous stem bark extract	Wistar albino rats	Oral administration of extract to normal and STZ-induced diabetic rats[14]	75 mg/kg produced hypoglycemic activity; max effect of 29.67%, 5 h after administration STZ-induced diabetic rats[14]	Cameron[14]
Urtica dioica (Urticaceae)	Aqueous extract of aerial parts	Wistar rats and Swiss mice	Investigate the hypoglycemic effect of extract on hyperglycemia induced by oral glucose tolerance test and on alloxan-induced diabetic rats.[10]	Results indicate that plant extract had a significant antihyperglycemic effect in the oral glucose tolerance test. This may be associated with reduction in intestinal glucose absorption.[10]	Morocco[10]
Vemonia colorata (Wilid.) Drake (Asteraceae)	Aqueous extract of leaves	Wistar rats	Investigate antidiabetic properties in normoglycemic and alloxan intoxicated rats.[49]	The plant extract displayed both hypoglycemic and antidiabetic properties in normoglycemic and alloxan-challenged diabetic rats.[49]	Senegal[49]

TABLE 2.1 *(Continued)*

Plant	Plant part used	Animal model	Study design	Results	Country of study/ source of plant material
Vernonia amygdalina (Asteraceae)	Aqueous leaf extract	Sprague-Dawley rats	Hypoglycemic properties of the plant extract were assessed using alloxan-induced diabetes mellitus models.[40]	Administration of 500 mg/kg of body weight aqueous extract significantly decreased blood glucose in normal and alloxan-induced diabetic rats.[40]	Nigeria[40]
Zingiber officinale (roscoe) rhizomes (zingiberaceae)	Rhizomes ethanol extract	Wistar rats	STZ-induced diabetes rats given plant extra 50–800 mg/kg p.o.[36]	The 50–800 mg/kg plant extract produced a dose-dependent hypoglycemia in both normal and diabetic rats.[36]	Nigeria[36]

FGB, fasting blood glucose; STZ, streptozotocin; VAO, virgin argan oil.

2.4 SUMMARY

It is noteworthy to mention that deliberate investment by African governments in drug development program for the treatment of diabetes from these medicinal plants will, to a large extent, ease the health and socioeconomic burden occasioned by diabetes on the continent and make resources that would have been otherwise lost to the source available for other uses.

KEYWORDS

- **Africa**
- **antidiabetic**
- **diabetes**
- **hyperglycemia**
- **hypoglycemia**
- **medicinal plants**

REFERENCES

1. Adaramoye, O.; Adeyemi, E. Hypoglycaemic and Hypolipidaemic Effects of Fractions from Kolaviron, a Biflavonoid Complex from Garcinia Kola in Streptozotocin-Induced Diabetes Mellitus Rats. *J. Pharm. Pharmacol.* **2006,** *58*(1), 121–128.
2. Adaramoye, O.; Amanlou, M.; Habibi-Rezaei, M.; Pasalar, P.; Ali, M.-M. Methanolic Extract of African Mistletoe (Viscum Album) Improves Carbohydrate Metabolism and Hyperlipidemia in Streptozotocin-Induced Diabetic Rats. *Asian Pac. J. Trop. Med.* **2012,** *5*(6), 427–433.
3. Aderibigbe, A.; Lawal, B.; Oluwagbemi, J. The Antihyperglycamic Effect of Telfaria Occidentalis in Mice. *Afr. J Med. Med. Sci.* **1999,** *28*(3–4), 171–175.
4. Adewoye, E.; Taiwo, V.; Olayioye, F. Anti-Oxidant and Anti-Hyperglycemic Activities of Musa Sapientum Root Extracts in Alloxan-Induced Diabetic Rats. *Afr. J Med. Med. Sci.* **2009,** *38*(2), 109–117.
5. Adisa, R.; Choudhary, M.; Adewoye, E.; Olorunsogo, O. Hypoglycaemic and Biochemical Properties of Cnestis Ferruginea. *Afr. J. Tradit. Complement. Altern. Med.* **2010,** *7*(3), 185–194.
6. Adisa, R.A.; Choudhary, M.I.; Olorunsogo, O.O. Hypoglycemic Activity of Buchholzia Coriacea (Capparaceae) Seeds in Streptozotocin-Induced Diabetic Rats and Mice. *Exp. Toxicol. Pathol.* **2011,** *63*(7–8), 619–625.

7. Agunbiade, O.; Ojezele, O.; Ojezele, J.; Ajayi, A. Hypoglycaemic Activity of Commelina Africana and Ageratum Conyzoides in Relation to Their Mineral Composition. *Afr. Health Sci.* **2012,** *12*(2), 198–203.
8. Bellahcen, S.; Mekhfi, H.; Ziyyat, A.; Legssyer, A.; Hakkou, A.; Aziz, M.; Bnouham, M. Prevention of Chemically Induced Diabetes Mellitus in Experimental Animals by Virgin Argan Oil. *Phytother. Res.* **2012,** *26*(2), 180–185.
9. Bloom, A. Who Expert Committee on Diabetes Mellitus, 2nd Report: Technical Report Series 64. *J. R. Soc. Med.* **1980,** *73*(11), 834.
10. Bnouham, M.; Merhfour, F.-Z.; Ziyyat, A.; Mekhfi, H.; Aziz, M.; Legssyer, A. Anti-hyperglycemic Activity of the Aqueous Extract of Urtica Dioica. *Fitoterapia* **2003,** *74*(7–8), 677–681.
11. Board, W.G.o.t.N.D.A. Guidelines for the Management of Type Ii (Noninsulin-Dependent) Diabetes Mellitus at Primary Health Care Level in South Africa. *S. Afr. Med. J.* **1997,** *87*, 497–512.
12. Burke, J.P.; Jacobson, D.J.; McGree, M.E.; Nehra, A.; Roberts, R.O.; Girman, C.J.; Lieber, M.M.; Jacobsen, S.J. Diabetes and Sexual Dysfunction: Results from the Olmsted County Study of Urinary Symptoms and Health Status among Men. *J. Urol.* **2007,** *177*(4), 1438–1442.
13. Chadwick, W.A.; Roux, S.; van de Venter, M.; Louw, J.; Oelofsen, W. Anti-Diabetic Effects of Sutherlandia Frutescens in Wistar Rats Fed a Diabetogenic Diet. *J. Ethnopharmacol.* **2007,** *109*(1), 121–127.
14. Dimo, T.; Ngueguim, F.; Kamtchouing, P.; Dongo, E.; Tan, P. Glucose Lowering Efficacy of the Aqueous Stem Bark Extract of Trema Orientalis (Linn) Blume in Normal and Streptozotocin Diabetic Rats. *Pharmazie* **2006,** *61*(3), 233–236.
15. Dimo, T.; Rakotonirina, S.V.; Tan, P.V.; Azay, J.; Dongo, E.; Kamtchouing, P.; Cros, G. Effect of Sclerocarya Birrea (Anacardiaceae) Stem Bark Methylene Chloride/Methanol Extract on Streptozotocin-Diabetic Rats. *J. Ethnopharmacol.* **2007,** *110*(3), 434–438.
16. Eddouks, M.; Maghrani, M.; Zeggwagh, N.-A.; Michel, J. Study of the Hypoglycaemic Activity of Lepidium Sativum L. Aqueous Extract in Normal and Diabetic Rats. *J. Ethnopharmacol.* **2005,** *97*(2), 391–395.
17. Eno, A.; Ofem, O.; Nku, C.; Ani, E.; Itam, E. Stimulation of Insulin Secretion by Viscum Album (Mistletoe) Leaf Extract in Streptozotocin-Induced Diabetic Rats. *Afr. J Med. Med. Sci.* **2008,** *37*(2), 141–147.
18. Fawzy, G.A.; Abdallah, H.M.; Marzouk, M.S.; Soliman, F.M.; Sleem, A.A. Anti-diabetic and Antioxidant Activities of Major Flavonoids of Cynanchum Acutum L.(Asclepiadaceae) Growing in Egypt. *Z. Naturforsch. C* **2008,** *63*(9–10), 658–662.
19. Federation, I.D. Idf Diabetes Atlas 6th, 2015.
20. Gondwe, M.; Kamadyaapa, D.; Tufts, M.; Chuturgoon, A.; Musabayane, C. Sclerocarya Birrea [(A. Rich.) Hochst.][Anacardiaceae] Stem-Bark Ethanolic Extract (Sbe) Modulates Blood Glucose, Glomerular Filtration Rate (Gfr) and Mean Arterial Blood Pressure (Map) of Stz-Induced Diabetic Rats. *Phytomedicine* **2008,** *15*(9), 699–709.
21. Hasanpour, M.; Iranshahy, M.; Iranshahi, M. The Application of Metabolomics in Investigating Anti-Diabetic Activity of Medicinal Plants. *Biomed. Pharmacother.* **2020,** *128*, 110263.
22. Ighodaro, O.M. Molecular Pathways Associated with Oxidative Stress in Diabetes Mellitus. *Biomed. Pharmacother.* **2018,** *108*, 656–662.

23. Ighodaro, O.M.; Akinloye, O.A.; Ugbaja, R.N.; Omotainse, S.O. Sapium Ellipticum (Hochst) Pax Ethanol Leaf Extract Modulates Glucokinase and Glucose-6-Phosphatase Activities in Streptozotocin Induced Diabetic Rats. *Asian Pac. J. Trop. Biomed.* **2017,** *7*(6), 544–548.
24. Kayarohanam, S.; Kavimani, S. Current Trends of Plants Having Antidiabetic Activity: A Review. *J. Bioanal. Biomed.* **2015,** *7*(2), 55.
25. Lemhadri, A.; Zeggwagh, N.A.; Maghrani, M.; Jouad, H.; Michel, J.; Eddouks, M. Hypoglycaemic Effect of Calamintha Officinalis Moench. In Normal and Streptozotocin-Induced Diabetic Rats. *J. Pharm. Pharmacol.* **2004,** *56*(6), 795–799.
26. Levitt, N.S. Diabetes in Africa: Epidemiology, Management and Healthcare Challenges. *Heart* **2008,** *94*(11), 1376–1382.
27. Macdonald Ighodaro, O.; Mohammed Adeosun, A.; Adeboye Akinloye, O. Alloxan-Induced Diabetes, a Common Model for Evaluating the Glycemic-Control Potential of Therapeutic Compounds and Plants Extracts in Experimental Studies. *Medicina* **2017,** *53*(6), 365–374.
28. Meddah, B.; Ducroc, R.; Faouzi, M.E.A.; Eto, B.; Mahraoui, L.; Benhaddou-Andaloussi, A.; Martineau, L.C.; Cherrah, Y.; Haddad, P.S. Nigella Sativa Inhibits Intestinal Glucose Absorption and Improves Glucose Tolerance in Rats. *J. Ethnopharmacol.* **2009,** *121*(3), 419–424.
29. Ndip, R.N.; Tanih, N.F.; Kuete, V. Antidiabetes Activity of African Medicinal Plants. In *Medicinal Plant Research in Africa*; Elsevier: Amsterdam, Netherlands, 2013; pp 753–786.
30. Njan, A.A.; Adenuga, F.O.; Ajayi, A.M.; Sotunde, O.; Ologe, M.O.; Olaoye, S.O.; Erdogan, O.N.; Iwalewa, O.E. Neuroprotective and Memory-Enhancing Effects of Methanolic Leaf Extract of Peristrophe Bicalyculata in Rat Model of Type 2 Diabetes Mellitus. *Heliyon* **2020,** *6*(5), e04011.
31. Nyunaï, N.; Njikam, N.; Abdenneb, E.; Mbafor, J.; Lamnaouer, D. Hypoglycaemic and Antihyperglycaemic Activity of Ageratum Conyzoides L. In Rats. *Afr. J. Tradit. Complement. Altern. Med.* **2009,** *6*, 2.
32. Ohadoma, S.; Michael, H. Effects of Co—Administration of Methanol Leaf Extract of Catharanthus Roseus on the Hypoglycemic Activity of Metformin and Glibenclamide in Rats. *Asian Pac. J. Trop. Med.* **2011,** *4*(6), 475–477.
33. Ojewole, J. Analgesic, Antiinflammatory and Hypoglycemic Effects of Sutherlandia Frutescens R. Br.(Variety Incana E. Mey.)(Fabaceae) Shoot Aqueous Extract. *Methods Find. Exp. Clin. Pharmacol.* **2004,** *26*(6), 409–416.
34. Ojewole, J. Hypoglycemic and Hypotensive Effects of Psidium Guajava Linn.(Myrtaceae) Leaf Aqueous Extract. *Methods Find. Exp. Clin. Pharmacol.* **2005,** *27*(10), 689–696.
35. Ojewole, J.A. Antinociceptive, Anti-Inflammatory and Antidiabetic Effects of Bryophyllum Pinnatum (Crassulaceae) Leaf Aqueous Extract. *J. Ethnopharmacol.* **2005,** *99*(1), 13–19.
36. Ojewole, J.A. Analgesic, Antiinflammatory and Hypoglycaemic Effects of Ethanol Extract of Zingiber Officinale (Roscoe) Rhizomes (Zingiberaceae) in Mice and Rats. *Phytother. Res* **2006,** *20*(9), 764–772.
37. Ojewole, J.A. Antinociceptive, Anti-Inflammatory and Antidiabetic Properties of Hypoxis Hemerocallidea Fisch. & Ca Mey.(Hypoxidaceae) Corm ['African Potato'] Aqueous Extract in Mice and Rats. *J. Ethnopharmacol.* **2006,** *103*(1), 126–134.

38. Ojewole, J.A.; Adewole, S.O.; Olayiwola, G. Hypoglycaemic and Hypotensive Effects of Momordica Charantia Linn (Cucurbitaceae) Whole-Plant Aqueous Extract in Rats: Cardiovascular Topics. *Cardiovasc. J. South Afr.* **2006,** *17*(5), 227–232.
39. Ojezele, M.; Abatan, O. Hypoglycaemic and Coronary Risk Index Lowering Effects of Bauhinia Thoningii in Alloxan Induced Diabetic Rats. *Afr. Health Sci.* **2011,** *11*, 1.
40. Osinubi, A. Effects of Vernonia Amygdalina and Chlorpropamide on Blood Glucose. *Med. J Islamic World Acad. Sci.* **2007,** *16*(3), 115–119.
41. Ottah, A.A.; Augustine, O.; Obiora, I.O.; Maxwell, E. Antihyperglycemic Effects of the Methanol Leaf Extract of Diaphananthe Bidens in Normoglycemic and Streptozotocin–Induced Hyperglycemic Rats. *Asian Pac. J. Trop. Med.* **2012,** *5*(3), 192–196.
42. Patterson, C.C.; Karuranga, S.; Salpea, P.; Saeedi, P.; Dahlquist, G.; Soltesz, G.; Ogle, G.D. Worldwide Estimates of Incidence, Prevalence and Mortality of Type 1 Diabetes in Children and Adolescents: Results from the International Diabetes Federation Diabetes Atlas. *Diabetes Res. Clin. Pract.* **2019,** *157*, 107842.
43. Peer, N.; Kengne, A.-P.; Motala, A.A.; Mbanya, J.C. Diabetes in the Africa Region: An Update. *Diabetes Res. Clin. Pract.* **2014,** *103*(2), 197–205.
44. Roy, M.S.; Janal, M.N.; Crosby, J.; Donnelly, R. Markers of Endothelial Dysfunction and Inflammation Predict Progression of Diabetic Nephropathy in African Americans with Type 1 Diabetes. *Kidney Int.* **2015,** *87*(2), 427–433.
45. Saeedi, P.; Petersohn, I.; Salpea, P.; Malanda, B.; Karuranga, S.; Unwin, N.; Colagiuri, S.; Guariguata, L.; Motala, A.A.; Ogurtsova, K. Global and Regional Diabetes Prevalence Estimates for 2019 and Projections for 2030 and 2045: Results from the International Diabetes Federation Diabetes Atlas. *Diabetes Res. Clin. Pract.* **2019,** *157*, 107843.
46. Sefi, M.; Fetoui, H.; Lachkar, N.; Tahraoui, A.; Lyoussi, B.; Boudawara, T.; Zeghal, N. Centaurium Erythrea (Gentianaceae) Leaf Extract Alleviates Streptozotocin-Induced Oxidative Stress and B-Cell Damage in Rat Pancreas. *J. Ethnopharmacol.* **2011,** *135*(2), 243–250.
47. Sefi, M.; Fetoui, H.; Makni, M.; Zeghal, N. Mitigating Effects of Antioxidant Properties of Artemisia Campestris Leaf Extract on Hyperlipidemia, Advanced Glycation End Products and Oxidative Stress in Alloxan-Induced Diabetic Rats. *Food Chem. Toxicol.* **2010,** *48*(7), 1986–1993.
48. Shinkafi, T.S.; Bello, L.; Hassan, S.W.; Ali, S. An Ethnobotanical Survey of Antidiabetic Plants Used by Hausa–Fulani Tribes in Sokoto, Northwest Nigeria. *J. Ethnopharmacol.* **2015,** *172*, 91–99.
49. Sy, G.; Nongonierma, R.; Sarr, M.; Cisse, A.; Faye, B. Antidiabetic Activity of the Leaves of *Vemonia Colorata* (Wilid.) Drake (Composees) in Alloxan-Induced Diabetic Rats. *Dakar Med.* **2004,** *49*(1), 36–39.
50. Wannes, W.A.; Marzouk, B. Research Progress of Tunisian Medicinal Plants Used for Acute Diabetes. *J. Acute Dis.* **2016,** *5*(5), 357–363.
51. Williams, R.; Karuranga, S.; Malanda, B.; Saeedi, P.; Basit, A.; Besançon, S.; Bommer, C.; Esteghamati, A.; Ogurtsova, K.; Zhang, P. Global and Regional Estimates and Projections of Diabetes-Related Health Expenditure: Results from the International Diabetes Federation Diabetes Atlas. *Diabetes Res. Clin. Pract.* **2020,** *162*, E-article 108072.

CHAPTER 3

In Vitro Antioxidant and Nephroprotective Activities of *Carica papaya* Methanol Extracts Using Streptozotocin-Induced Diabetic Male Albino Rats

Modupe Olusola Adetayo, Oluwaseyi Adegoke Adetunji, Babafemi Tosin Ogunbiyi, Gogonte Hezekiah Amah, Omowunmi O. Akinnawo, Laide O. Olajide, and Toyosi S. Ajiboye

ABSTRACT

The antioxidant properties of ripe and unripe *Carica papaya* fruit were investigated on kidney parameters in diabetic streptozotocin-induced rats. Six groups of five male albino rats were designed. Groups 2, 3, 4, and 5 received 500 mg/kg body weight *C. papaya* for 21 days. Groups 1, 2, and 3 were normal and diabetic, respectively, and were treated with the unripe *C. papaya* extract, while the ripe *C. papaya* extract was used to treat groups 4 and 5 that were normal and diabetic, respectively. The diabetic control is group 6. Analysis was done for 1-1,diphenyl, 2-picryl hydrazine (DPPH) activity, ferric-reducing antioxidant power (FRAP), nitric oxide (NO) antioxidant potential, total antioxidant capacity (TAC) activities, and urea and creatinine activities. The DPPH activity, FRAP, and TAC of the unripe *C. papaya* were higher (0.498, 0.022, 0.01, 0.343) compared to ripe *C. papaya*, but the NO scavenging activity was higher (0.010) than that of unripe *C. papaya*. No remarkable effect was observed on the urea level, while the creatinine level of the treated animals was reduced significantly at $p<0.05$. Both unripe and ripe *C. papaya* possess antioxidant properties, but these are higher in unripe *C. Papaya*.

Curative and Preventive Properties of Medicinal Plants: Research on Disease Management and Animal Model Studies. Megh R. Goyal, Junaid Ahmad Malik, & Ademola Olabode Ayeleso (Eds.)

3.1 INTRODUCTION

Diabetes mellitus is a chronic, metabolic disorder associated with higher blood glucose amount leading to effective damage to the heart, blood vessels, kidneys, nerves, and eyes.[5] Diabetes is mainly classified into two forms: (1) Type 1 *diabetes mellitus* (insulin-dependent diabetes) is a chronic disease with little or no insulin being produced by the kidney; and (2) Type 2 *diabetes mellitus* (non-insulin-dependent diabetes), caused by the inability of cells to act properly to the effect of insulin. These two types of diabetes have been reported to affect persons of all age groups.[3]

Recently, around 150 million people are affected by *diabetes mellitus* and this data may be doubled by the year 2025.[12] The developing countries are expected to experience much of this increase as a result of population growth, unhealthy diet, ageing, obesity, and poor lifestyle. By the year 2025, most of the diabetic group in developed countries would be at the age of 65 years or older, while the affected age group in developing countries will be between the age 45 and 65 years. It was believed to be specific to a particular age bracket, but current research studies nullified age specificity and affirmed that all ages are affected. Symptomatic treatment is available, but presently there is no cure for the disease.

Disease conditions, such as arthritis and digestive disorders, have been treated using bioactive components (chymopapain and papain) of *C. papaya.*[19] The plant extracts of the ripe fruits are useful as antimalarial, anthelminthic, and antihypertensive medicine, while unripe fruits function as antidiabetic treatment. The unripe *C. papaya* fruit extract possesses antioxidant properties.[15]

Despite the wide consumption of *C. papaya* traditionally in the management of many illnesses, the ripe fruit antioxidant effect in the treatment of *diabetes mellitus* has not been reported. In view of this, it is the focus of this chapter to determine the free radicals scavenging properties and effect of ripe and unripe *C. papaya* methanol extract on kidney parameters in streptozotocin-induced diabetic albino rats.

Hyperglycemia brought about through the disruption of insulin metabolism is a major factor in *diabetes mellitus.* Between 1980 and 2014, people with incidences of diabetes grew from 108 million to 422 million.[2] The long-term hyperglycemia as seen in diabetes has been connected to chronic diseases and organs failure, as seen in the heart, kidney, eyes, and blood vessels.[2] Carbohydrate, fat, and protein metabolic abnormalities in diabetes are based on the deficient action of insulin, with uncontrolled diabetes resulting in life-threatening hyperglycemia, ketoacidosis, atherosclerotic cardiovascular, and cerebrovascular diseases.[1,21]

In *diabetes mellitus*, generation of reactive oxygen species resulting from glucose autoxidation and oxidative phosphorylation can be as a result of chronic hyperglycemia. This then results in oxidative stress when there is low activity of antioxidant enzymes, thereby resulting in β-cell dysfunction.[18] Diabetes mellitus currently has no cure, but good management can be achieved through diet modification and medications.

The study in this chapter intends to evaluate the in vitro free radical scavenging potential of ripe and unripe *C. papaya* methanol extract and its effect on the kidney function of streptozotocin-induced diabetic albino rats.

3.2 METHODS AND MATERIALS

3.2.1 EXPERIMENTAL ANIMALS

Male Wistar albino rats numbering 36 in total and weighing between 80 and 100 g were procured from Animal facility, Babcock University, Ilishan Remo, Ogun State. The rats were housed in clean, well-aerated, and well-illuminated battery cages. The animals were fed standard commercial rat chow and water ad libitum. The commercial rat chow was purchased from Kayus Animal Feeds, Iperu, Ogun State. The rats were acclimatized for a period of 2 weeks. Thereafter, seven groups having five animals each were set up:

- Group 1: Normal rats, untreated group (Normal control)
- Group 2: Unripe *C. papaya* methanol extract treated normal rats
- Group 3: Unripe *C. papaya* methanol extract Type 1 DM-induced rats treated group
- Group 4: Ripe *C. papaya* methanol extract treated normal rats
- Group 5: Ripe *C. papaya* methanol extract treated Type 1 DM-induced rats.
- Group 6: Untreated Type 1 DM-induced rats (Diabetic control).
- Group 7: Glibenclamide treated Type 1 DM-induced rats

3.2.2 INDUCTION OF DIABETES

Induction of diabetes was done intraperitoneally with 65 mg/kg body weight streptozotocin in groups 3, 5, and 6. Blood was extracted from the tail vein, after 48 h, for glucose analysis using the Accuchek™ Glucometer (model 35NZ, manufactured by Roche Diabetes Care, Mannheim, Germany). Diabetes was recorded in rats with fasting blood glucose of 200 mg/dL and above.

3.2.3 WEIGHING OF EXPERIMENTAL ANIMALS

The animals were weighed at the end of every week using an electronic scale. The average weight gain of the animals was gotten by subtracting the initial (starting weight) from the final weight after the whole experiment. This represented the average weight gain of each group at the end of the study.

3.2.4 SACRIFICING OF EXPERIMENTAL ANIMALS

After treatments for 3 weeks the animals were allowed to fast overnight for 12 h and then sacrificed. The experimental animals were bled via ocular puncture into sample bottles using a capillary tube after which the rats were sacrificed. The Urea and creatinine activity were assayed using the blood serum.

3.2.5 *IN VITRO* ANTIOXIDANT ASSAY

The DPPH activity of the ripe and unripe *C. papaya* extract was compared by adopting the procedure reported by McCune and Johns.[13] 3.0 mL of the reaction mixture will consist of 0.3 mM DPPH (1.0 mL), extract of the ripe (1.0 mL), and unripe (1.0 mL) *C. papaya* and methanol (1.0 mL). This was then incubated in the dark for 10 min. Absorbance was read at 517 nm. The percentage of antioxidant potential was calculated as follows:

$$\%\ \text{Antioxidant potential} = [(\text{Optical Density of control} - \text{Optical Density of sample}) / (\text{Optical Density of control})] \times 100 \quad (3.1)$$

The Benzie and Stain FRAP assay method was used.[6] Various concentrations of the ripe and unripe *C. papaya* fruit pulp extract were added to the FRAP reagent (2 mL) to form an intense blue Fe^{2+}-TPTZ complex. The absorbance was taken against blank at 593 nm after 2 min, using ascorbic acid as the standard. The potential to lower ferric ion of test sample was estimated as antioxidant capacity from mmol $FeSO_4$ equivalents standard curve per gram of sample.

Nitric oxide (NO) Griess reaction assay was employed using the Griess reagent, following the procedure of Green et al.[9] 10 mM sodium nitroprusside

(3.0 mL) was mixed to methanol extract (2.0 mL) of the *C. Papaya* ripe and unripe and ascorbic acid (20–100 g/mL). Incubation was done for the resulting solutions for 60 min at 25°C. This procedure was followed using ethanol as control. After incubation, 5 mL of Griess reagent was added and the absorbance formed was observed at 540 nm. The absorbance value of test and control preparation was calculated to derive the percentage inhibition of the nitrite oxide generated.

The total antioxidant capacity was measure by adopting the Prieto et al. method.[16] Molybdate solution (1 mL) was added to varying concentrations of the *C. papaya* ripe and unripe methanol extract. The molybdate solution was boiled for 90 min at 95°C. The resulting reaction solution's absorbance was taken at 695 nm after cooling at 25°C.

3.2.6 KIDNEY FUNCTION ANALYSIS

To determine the kidney function, creatinine level was estimated, which is better when compared to the evaluation of urea.[17] 0.1 mL of serum was added to the creatinine working reagent. The reaction absorbance was estimated and recorded as A1 after 30 s at 492 nm. Then, the absorbance was read again after 2 min and recorded as A2. The creatinine concentration in the serum was calculated as follows:

$$\left[\frac{\text{change in optical density of sample}}{\text{change in optical density of standard}} \text{X}\, \textit{standard concentration}]\right] \tag{3.2}$$

Determination of urea concentration[17]

For the quantification of urea concentration, 0.010 mL of serum, standard solution, and distilled water was measured into labeled sample, standard, and blank tubes, respectively. Reagent 1, R1 (0.5 mL) were measured into the experimental tubes, respectively, and were mixed carefully and at 37°C incubated for 10 min. 0.5 mL of R2 and R3 were added to all tubes and mixed gently. Incubation of the solution was followed at 37°C for 15 min, and absorbance reading was read at 546 nm.

$$\text{Concentration of Urea in mg/dl} =$$

$$\frac{\text{optical density of sample}}{\text{optical density of standard}} \times \textit{standard concentration} \tag{3.3}$$

3.2.7 *STATISTICAL ANALYSIS*

Data generated from this study was expressed as mean ± SD, whereas statistical comparison of the data was carried out by means of One-way Analysis of Variance (ANOVA) that was used to estimate level of significance among the study groups, using SPSS software, version 14.0. Duncan Multiple-Range Test (DMRT) was used to determine the post-hoc tests between the groups. The graphs and charts were constructed using Microsoft Excel © 2013. $p<0.05$ was considered statistical significance.

3.3 EXPERIMENTAL RESULTS

Figure 3.1 shows the DPPH radical scavenging activity of the extracts. It was shown that both ripe and unripe methanol extracts possessed DPPH radical scavenging activity, but they were found to be lower than those of the standard at $p> 0.05$. Figure 3.2 shows the ferric-reducing antioxidant power of the extracts.

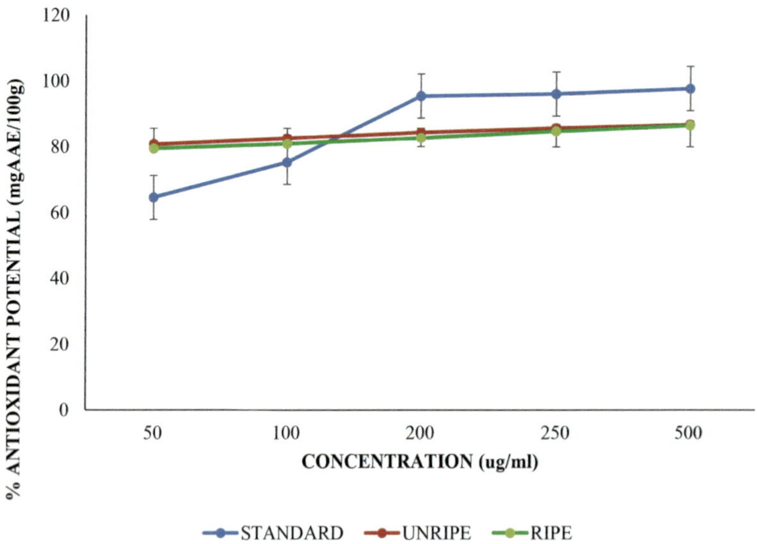

FIGURE 3.1 DPPH antioxidant activities of Unripe and ripe *C. papaya* methanol extract.

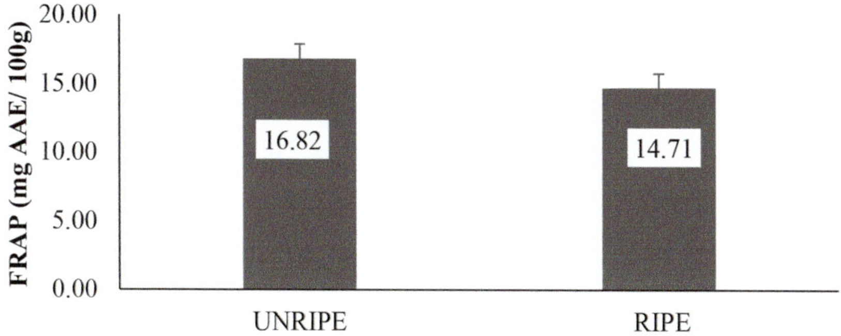

FIGURE 3.2 FRAP activities of unripe and ripe *C. papaya* methanol extract.

It was shown that the unripe *C. papaya* and ripe possessed ferric-reducing antioxidant power, but that of unripe *C. papaya* is higher than the ripe *C. papaya* and this was statistically significant difference at $p < 0.05$. The antioxidant nitric oxide activity of the extracts is shown in Figure 3.3. It was shown that the unripe *C. papaya* and ripe possessed nitric oxide antioxidant activity, but that of unripe *C. papaya* was higher than the ripe *C. papaya* and this was significant at $p < 0.05$. The total antioxidant capacity of the extracts is shown in Figure 3.4. It was shown that the unripe *C. papaya* and ripe possessed total antioxidant capacity, but that of unripe *C. papaya* was lower than the ripe *C. papaya* and this at $p > 0.05$ was not significant.

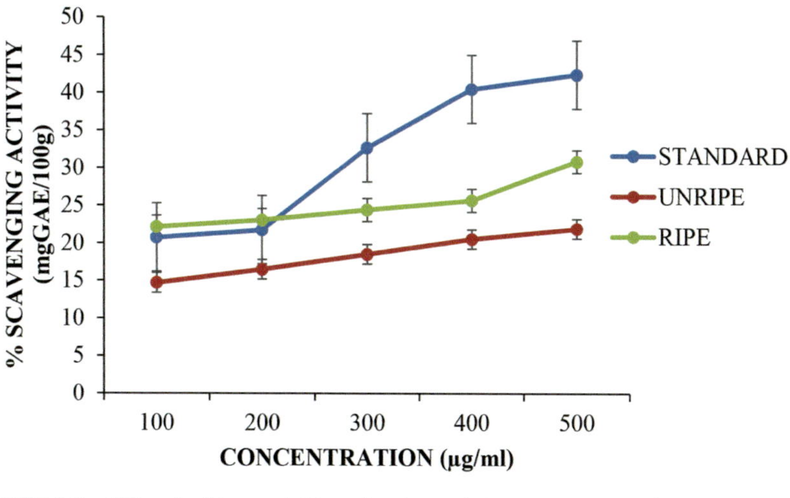

FIGURE 3.3 NO antioxidant activities of unripe and ripe *C. papaya* methanol extract.

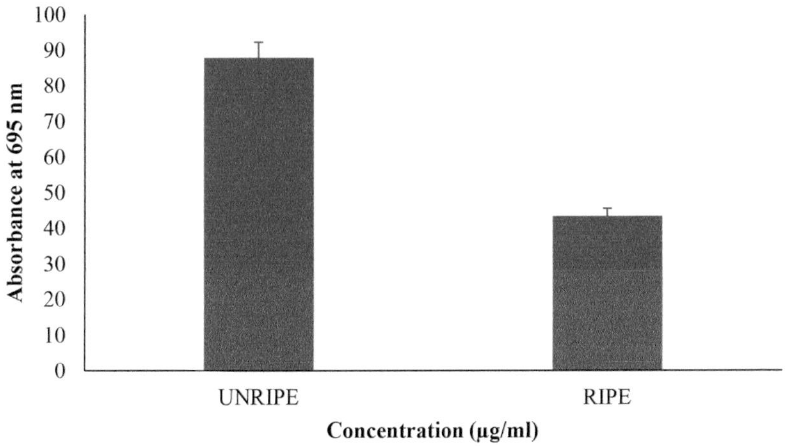

FIGURE 3.4 TAC of unripe and ripe *C. papaya* methanol extract.

Figure 3.5 shows the *C. papaya* extract effect on body weight change in normal and diabetic-treated animals**.** The results show that animals in group 4 (Normal rats treated with ripe *Carica papaya* extract) had the highest body weight gain, while animals in group 6 (diabetic control with no treatment) had the lowest body weight gain. The diabetic-treated animals' body weight was lower than that of the untreated animals, but the difference in weight at $p > 0.05$ was not significant. The serum urea activity of the experimental animals is shown in Figure 3.6. It was shown that group 6 (diabetic control) had the highest serum urea level, while the group one (normal control) showed lowest urea activity. The normal control group had lower urea activity in contrast to the normal group treated with extracts. The diabetic groups had lower urea activity in contrast to nondiabetic group. Figure 3.7 shows the serum creatinine levels of the experimental animals. It was shown that group 6 (diabetic control) had the highest serum creatinine level, while the group four (diabetic rats treated with methanol unripe pawpaw) had the lowest serum creatinine level. The groups treated with *C. papaya* had lower creatinine activity in contrast to the normal control. The diabetic groups had lower creatinine activity in contrast to the untreated diabetic group.

- Group 1: Normal rats with no treatment (Normal control).
- Group 2: Unripe *C. papaya* methanol extract treated normal rats.
- Group 3: Type 1 DM-induced rats treated with methanol extract of unripe *C. papaya.*
- Group 4: Ripe *C. papaya* methanol extract treated normal rats.

- Group 5: Type 1 DM-induced rats treated with methanol extract of ripe *C. papaya*.
- Group 6: Type 1 DM-induced rats with no treatment (Diabetic control).

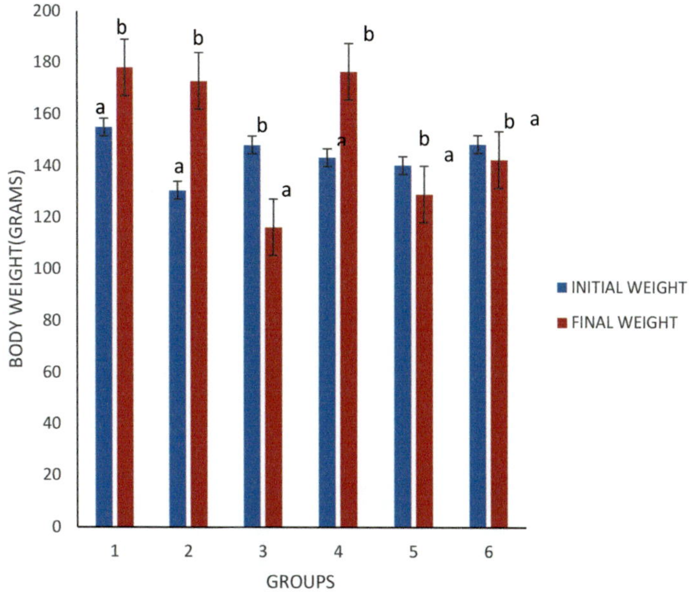

FIGURE 3.5 Effect of *Carica papaya* extract on body weight of experimental animals.

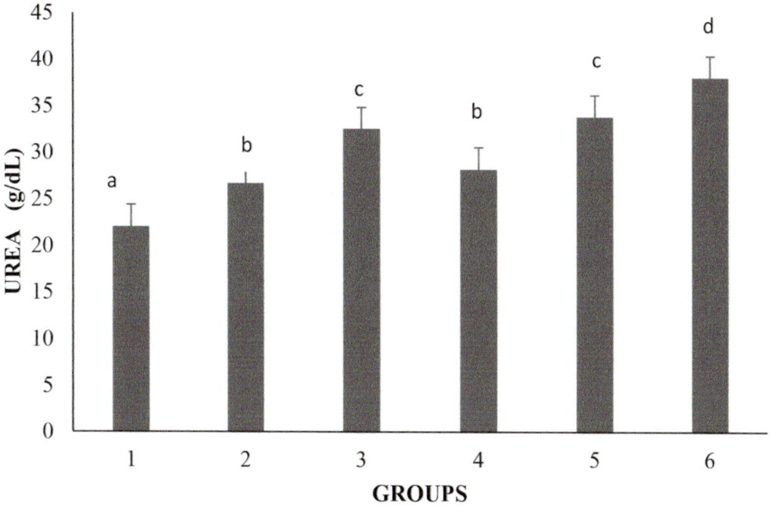

FIGURE 3.6 Serum urea levels of the experimental animals.

- Group 1: Normal rats with no treatment (Normal control).
- Group 2: Unripe *C. papaya* methanol extract treated normal rats.
- Group 3: Type 1 DM-induced rats treated with methanol extract of unripe *C. papaya.*
- Group 4: Normal rats treated with methanol extract of ripe *C. papaya.*
- Group 5: Ripe *C. papaya* methanol extract treated Type 1 DM-induced rats.
- Group 6: Type 1 DM-induced rats with no treatment (Diabetic control).

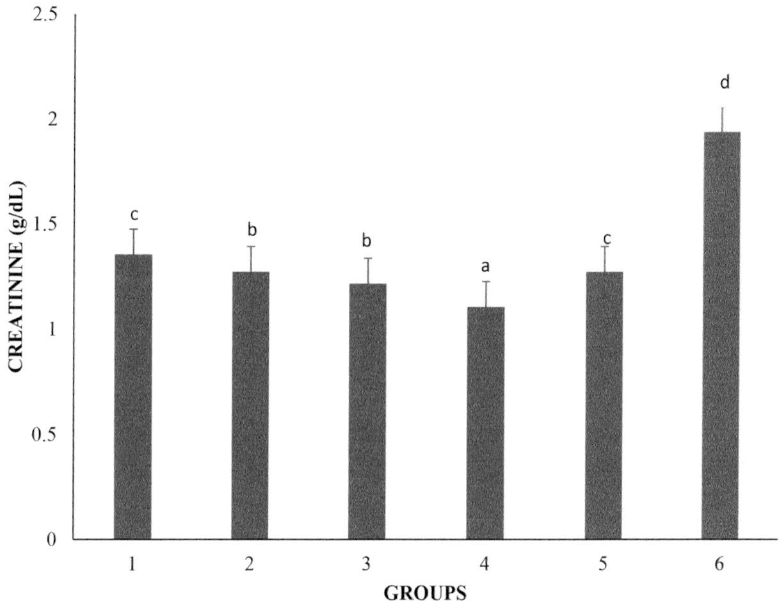

FIGURE 3.7 Serum creatinine levels of the experimental animals.

- Group 1: Normal rats with no treatment (Normal control).
- Group 2: Normal rats treated with methanol extract of unripe *C. papaya.*
- Group 3: Type 1 DM-induced rats treated with methanol extract of unripe *C. papaya.*
- Group 4: Normal rats treated with methanol extract of ripe *C. papaya.*
- Group 5: Type 1 DM-induced rats treated with methanol extract of ripe *C. papaya.*
- Group 6: Type 1 DM-induced rats with no treatment (Diabetic control).

3.4 DISCUSSION

Chronic kidney disease as seen in patients undergoing renal transplant[11] has been shown to be caused by diabetic nephropathy. This incidence is a major cause of cardiovascular mortality.[22] Diabetic nephropathy involves high production of albumin in urine (urinary albumin excretion—UAE) without the incidence of renal disease; it is broadly divided into microalbuminuria (UAE > 20 μg/min and ≤ 199 μg/min) and macroalbuminuria (UAE ≥ 200 μg/min).[11] Risk factors associated with this disease include genetic predispositions, hyperglycemia, and increased blood pressure.[8]

The prevalence of diabetic nephropathy has been reported among the Asians, Africans, and Native Americans than in Caucasians.[24] Statistical report indicated that diabetic nephropathy incidence doubled between 1991 and 2001 among patients undergoing renal treatment.[14] Luckily, the progression has decreased as a result of improved clinical diagnosis that helps to prevent diabetic nephropathy, thereby reducing advancement of kidney disease.[7]

Urea and creatinine levels are elevated in diabetic nephropathy with associated hyperglycemia in unregulated diabetes, as seen in advanced renal damage. Serum urea and serum creatinine are used as biomakers to test for evidence of kidney dysfunction in diabetes.[4]

Based on the findings from this study, the DPPH activity of the ripe *C. Papaya* was shown to have a low antioxidant capacity compared to unripe *C. papaya*. This finding is similar to that of Oke[15] which reported the DPPH radical scavenging activity of unripe *C. papaya*. For FRAP, the unripe *C. papaya* was shown to have a higher antioxidant capacity than the ripe *C. papaya*. Yee et al.,[23] reported similar findings as this study. Papaya peel and seed extracts demonstrated ferric-reducing antioxidant power. The ripe *C. papaya* was shown to have high nitric oxide radical scavenging capacity compared to unripe *C. papaya*. Srikanth et al.'s[20] investigations showed similar data that aqueous extract of *C. papaya* has nitric oxide radical scavenging. For TAC, the unripe *C. papaya* was shown to have a higher antioxidant capacity than the ripe *C. papaya*.

There was an increase in the body weight gained by rats in groups 1, 2, and 4. The ripe and unripe *C. papaya* methanol extracts helped to prevent excessive weight gain by regulating the weight gain based on what was observed in groups 2 and 4. There was a drop in the weight of group 3, 5, and 6 animals which was not significant. Group 5 that was treated with methanol ripe *C. papaya* extract did not have as much weight loss as compared to group 3 that was treated with methanol unripe *C. papaya* extract, while the diabetic control had the highest weight loss.

There was a decrease in the urea levels in the normal groups treated with ripe *C. Papaya* extract and unripe *C. Papaya* extract for the diabetic groups. Based on the results obtained from the study, the methanol ripe *C. Papaya* extract decreased urea levels for the diabetic rats compared to the methanol unripe *C. Papaya* extract. This is similar to that reported by Igbinovia et al.,[26] in which the *C. papaca* extract can potentially alter kidney function.

There was a reduction in the creatinine activity in the groups treated with both the methanol ripe *C. Papaya* extract and the methanol unripe *C. Papaya* extract for the diabetic groups. Based on the results obtained from the study, the methanol unripe *C. Papaya* extract decreased creatinine levels for the diabetic rats compared to the methanol ripe *C. Papaya* extract for the untreated groups. The diabetic control (group 6) had the highest serum level, while the lowest was observed in group 1 (Normal control). This is similar to that reported by Igbinovia et al.[17] in which the *C. papaca* extract can potentially alter kidney function.

It can therefore be concluded from the findings of this study that the methanol extract of both unripe and ripe *C. papaya* possesses antioxidant properties, but the unripe *C. papaya* showed a higher antioxidant potential. The *C. Papaya* extract has renal protective potential by regulating urea and creatinine levels. It is therefore recommended that further investigations should be employed to explore and recognize the active molecules of *C. papaya* which are essential in the control and treatment of diabetes. These bioactive compounds can be isolated and used in the production of drugs for the management of diabetes.

3.5 SUMMARY

The antioxidant properties of ripe and unripe *Carica papaya* fruit were investigated on kidney parameters in diabetic streptozotocin-induced rats. Six groups of five male albino rats were designed. Groups 2, 3, 4, and 5 received 500 mg/kg body weight *C. papaya* for 21 days. Groups 1, 2, and 3 were normal and diabetic, respectively, and were treated with the unripe *C. papaya* extract, while the ripe *C. papaya* extract was used to treat groups 4 and 5 that were normal and diabetic, respectively. The diabetic control is group 6. Analysis was done for 1-1,diphenyl, 2-picryl hydrazine (DPPH) activity, ferric-reducing antioxidant power (FRAP), nitric oxide (NO) anti-oxidant potential, total antioxidant capacity (TAC) activities, and urea and creatinine activities. The DPPH activity, FRAP, and TAC of the unripe *C. papaya* was higher (0.498, 0.022, 0.01, 0.343) compared to ripe *C. papaya*, but the NO scavenging activity was higher (0.010) than that of unripe *C. papaya*. No remarkable effect was observed on the urea level, while the creatinine level of the treated animals was reduced significantly at $p<0.05$. Both unripe and ripe *C. papaya* possess antioxidant properties, but these are higher in unripe *C. Papaya*.

KEYWORDS

- **antioxidants**
- ***Carica papaya***
- **diabetes mellitus**
- **nephroprotective**

REFERENCES

1. America Diabetes Association (ADA). Diagnosis and Classification of Diabetes Mellitus. *Diabetes Care* **2012,** *35*, S64–S71.
2. America Diabetes Association (ADA). Standards of Medical Care in Diabetes. *Diabetes Care* **2009,** *32*, S13–S61.
3. America Diabetes Association (ADA). Type-2 Diabetes in Children and Adolescents. *Diabetes Care* **2020,** *23*, 381–389.
4. Bamanikar, S.A.; Bamanikar, A.A.; Arora, A. Study of Serum Urea and Creatinine in Diabetic and Non-diabetic Patients in a Tertiary Teaching Hospital. *J. Med. Res.* **2016,** *2*(1), 12–15.
5. Baynes, H.W. Classification, Pathophysiology, Diagnosis and Management of Diabetes Mellitus. *J. Diabetes Metab.* **2015,** *6,* 541–549.
6. Benzie, I.F.; Strain, J.J. The Ferric Reducing Ability of Plasma (FRAP) as a Measure of "Antioxidant Power" The FRAP Assay. *Anal. Biochem.* **1996,** *239,* 70–76.
7. Craig, K.J.; Donovan, K.; Munnery, M.; Owens, D.R.; Williams, J.D.; Philips, A.O. Identification and Management of Diabetic Nephropathy in the Diabetes Clinic. *Diabetes Care* **2003,** *26*, 1806–1811.
8. Gall, M.A.; Hougaard, P.; Borch-Johnsen, K.; Parving, H.H. Risk Factors for Development of Incipient and Overt Diabetic Nephropathy in Patients With Non-Insulin Dependent Diabetes Mellitus: Prospective, Observational Study. *Br. Med. J.* **1997,** *314*, 783–788.
9. Green, L.; Wagner, D.; Glogowski, J.; Skipper, P.; Wishnok, J.; Tannenbaum, S. Analysis of Nitrate, Nitrite and 15N Nitrate in Biological Fluids. *Anal. Biochem.* **1982,** *126,* 131–138.

10. Igbinovia, E.N.S.; Isah, M.; Edebirin, O.E.; Uwuigbe, M.; Airhomwanbor, K.O.; Eghrevba, O. The Effect of Carica Papaya Seed Extracts on Urea, Creatinine and Uric Acid Levels in Wistar Rats. *Int. J. Herbs Pharm. Res.* **2015,** *4*, 4.
11. Jorge, L.; Gross, M.D.; Mirela, J.; Azavedo M.D.; Sandra, P.; Silveiro, M.D. Diabetic Nephropathy: Diagnosis, Prevention, and Treatment. *Diabeties Care* **2005,** 28, 1.
12. King, H.; Aubert, R.; Herman, W. Global Burden of Diabetes, 1995–2025, Prevalence, Numerical Estimates and Projections. *Diabetes Care* **1998,** *21*, 1414–1431.
13. McCune, L.; Johns, T. Antioxidant Activity in Medicinal Plants Associated With the Symptoms of Diabetes Mellitus Used by the Indigenous Peoples of the North American Boreal Forest. *J. Ethnopharmacol.* **2002,** *82*, 197–205.
14. U.S. Renal Data System, *USRDS Annual Data Report*: *Atlas of End-Stage Renal Disease in the United States*, National Institutes of Health, National Institute of Diabetes and Digestive and Kidney Diseases, Bethesda, MD, 2005**.**
15. Oke, J. Antidiabetic Potency of Pawpaw. *Afr. J. Biomed. Res.* **1998,** *1*, 31–34.
16. Prieto, R.L.; Pineda, M.; Aquilar, M. Spectrophotometric Quantitation of Antioxidant Capacity Through the Formation of a Phosphomolybdenum Complex: Specific Application to the Determination of Vitamin E. *Anal. Biochem.* **1999,** *269*, 337–341.
17. Reitman, S.; Frankel, S. A Colorimetric Method for the Determination of Serum Glutamic Oxalacetic and Glutamic Pyruvic Transaminases. *Am. J. Clin. Pathol.* **1957,** *28*, 56.
18. Robertson, R.P.; Harmon, J.; Tran, P.O.; Tanaka, Y.; Takahashi, H. Perspectives in Diabetes. Glucose Toxicity in b-cells: Type 2 Diabetes, Good Radicals Gone Bad, and the Glutathione Connection. *Diabetes* **2003**, *52*, 581–587.
19. Sofowa, A. Research on Medical Plants and Traditional Medicine in Africa. *J. Altern. Complement. Med.* **1996,** *2*(3), 365–372.
20. Srikanth, G.; Kavitha, C.H.N.; Vyaykumar, N.; Manohar, B.S.; Bhanoji, R.M.E.; Pradeep, C.H.; Studies on In-Vitro Antioxidant Activities of Carica Leaf Extract. *Res. J.Pharm. Biol. Chem. Sci.* **2010,** *1*(2), 59–65.
21. Steven, P.; Marso, M.D.; William, M.D. Peripheral Arterial Disease in Patients with Diabetes. *J. Am. Coll. Cardiol.* **2006,** *47*(7), 921–929.
22. Valmadrid, C. T.; Klein, R.; Moss, S. E.; Klein, B. E. The Risk of Cardiovascular Disease Mortality Associated With Microalbuminuria and Gross Proteinuria in Persons With Older-Onset Diabetes Mellitus. *Arch. Intern. Med.* **2000,** *160*(8), 1093–1100.
23. Yee, K.A.; Winne, C.M.; Hock, E.K.; Hip, S.Y. Antioxidant Potential of *Carica papaya* Peel and Seed. *Focusi. Modern Food Ind.* **2000,** *1*, 11.
24. Young, B.A.; Maynard, C.; Boyko, E.J. Racial Differences in Diabetic Nephropathy, Cardiovascular Disease, and Mortality in a National Population of Veterans. *Diabetes Care* **2003,** *26*, 2392–2399.

CHAPTER 4

Potential of *Moringa oleifera* for the Prevention of Cardiovascular Diseases in Animal Models

Olufemi Idowu Oluranti

ABSTRACT

Cardiovascular disease is a disease affecting the heart or blood vessels which is the main cause of mortality and is a major economic and health burden worldwide. Majority of the population use plants for their general health and well-being with aromatic and medicinal plants accounting for about 25% content of the integrated medications. Therefore, the use of natural products and herbal plants caught the attention of the populace bowing to their safety and efficacy. *Moringa oleifera* belong to monogenetic family Moringaceae and are known as drumstick, sahjana, and horseradish tree. It is originally from the foothills of the Himalaya and the dry tropical forest of north-west India. Its pharmacological potentials include anti-inflammatory, antimicrobial, neuroprotective, hepatoprotective, and cardioprotective. This chapter presents the therapeutic intervention of this plant in animal model of hypertension, cardiac and vascular dysfunction, and diabetic cardiomyopathy.

4.1 INTRODUCTION

Cardiovascular disease (CVD), a principal cause of mortality, is a disease affecting the heart or blood vessels that contribute a major burden in economic and health sector throughout the world. According to the World

Curative and Preventive Properties of Medicinal Plants: Research on Disease Management and Animal Model Studies. Megh R. Goyal, Junaid Ahmad Malik, & Ademola Olabode Ayeleso (Eds.)

Heart Federation, CVD accounts for 17.3 million deaths yearly, and probably 23.6 million deaths per year by 2030.[39] Regular risk factors for CVDs are diabetes mellitus, hypertension, hyperlipidemia, smoking, unhealthy diet, increased low-density lipoprotein cholesterol (LDL), and decreased high-density lipoprotein cholesterol (HDL) levels.[42]

Herbal medicines perform a significant part in health care for a large fraction of the world, which is also considered a part of various tribes' cultural legacy. Plants are the main source of conventional medicinal products used to treat various ailments. Of about 422,000 flowering plants reported all over the world, 50,000 plants are of medicinal importance for pharmaceutical purposes.[29] Approximately 80% of the populace depends on traditional remedies primarily for healthcare requirements.[41] Medicinal drug therapies are more commonly used as an alternative to allopathic drugs. Most of the commonly used current pharmaceutical products for the management of cardiovascular disorders are with side effects and alternative treatments with reduced side effects are urgently required.

The tree *Moringa oleifera* (MO), also called "drumstick tree," belongs to the genus Moringaceae, most commonly used and best known of the many species. It originates from the southern Himalayas to the northeastern parts of Afghanistan, Bangladesh, India, and Pakistan, and is now grown in America, Africa, and Asia, especially in their tropical and subtropical areas. It is a perennial tree that grows fast and shows great ecological plasticity as it can adapt to the most difficult soil conditions, precipitation, and temperature, becoming very drought-resistant.[15,17] The leaves and green pods are eaten as vegetables, while the seeds are ground to obtain flour along with the leaves in the preparation of soups, enhancing their nutritional quality.[6] The seed oil is used for solidifying and cooking because it has oxidative and thermal stability.[31]

Despite its bitter and a little astringent taste, MO leaf tea is increasingly consumed in Western communities.[14] The different part of the plant is a medicine commonly used in several different ways for multiple purposes in Africa and Asia.[7] Its therapeutic benefit includes hepatoprotective, neuroprotective, cardioprotective, antimicrobial, anti-asthmatic, antitumor, hypolipidemic, intestinal microbiota modulator, and antidiabetic. This is derived from the phytochemical elements in the plants, such as alkaloids, glycosides (saponins and glucosinolates), and phenolic compounds (coumarins, tannins, flavonoids), although some of these metabolites differ depending on the extraction method and geographical location.[23,25,13] MO leaves are abundant in antioxidants like vitamin E, beta-carotene, vitamin C, and polyphenols, and are a good source of natural antioxidants.[26]

Research laboratory animals are important for comprehending the pathophysiology and innovating therapeutic interventions for a disease; they are used in biomedical research for various reasons including feasibility, similarities to human, and drug safety.[24] The model of using animals to study cardiovascular disease has contributed substantially to disease pathogenesis understanding, giving rise to diagnostic techniques development as well as effectiveness of therapeutic procedures and preventive (whether interventional or pharmacological) verification.[11]

This chapter discusses the potential of *moringa oleifera* in ameliorating cardiovascular disease using the animal model.

4.2 SCIENTIFIC EVIDENCE OF EFFECTS OF *MORINGA OLEIFERA* ON CARDIOVASCULAR DISEASE USING ANIMAL MODELS

4.2.1 HYPERTENSION

Globally, hypertension is a public health problem, with about 9.4 million deaths and is a significant cause for other cardiovascular disorders like stroke, heart failure, coronary heart disease, left ventricular hypertrophy, and heart valvular disease.[35] Epidemiological studies support numerous medicinal plants' ability in preventing and treating cardiovascular diseases with limited side effects,[5] offering another source of medicines. Essential hypertension in human is a multifaceted disease with many factors that is affected by genetic and environmental conditions. In simulating hypertensive responses observed in humans, experimental hypertension of various models has been developed.[12]

4.2.1.1 N^{ω}-NITRO-L-ARGININE METHYL ESTER (L-NAME) MODEL

The L-NAME-mediated animal model of hypertension works primarily by initiating nitric oxide (NO) deficiency and oxidant/antioxidant imbalance. NO synthesis inhibition has several characteristics that reflect idiopathic hypertension in humans, such as reactive oxygen species-induced cardiovascular disorders, stimulation of sympathetic and renin–angiotensin nervous systems, inflammation, and cardiovascular remodeling.[2]

The effect of the MO hydroethanol extract on L-NAME-induced hypertension in Wistar rats revealed that 500 mg/kg BW of the extract orally administered for 14 days reduced the blood pressure caused by L-NAME significantly. Antihypertensive agents that stimulate the vascular smooth muscle tend to be effective to minimizing cardiovascular events and avoiding irreversible end-organ injury.[40] Continuous 50 mg/kg L-NAME administration for 21 days to male Wistar rats induced hypertension with resistance vessels destruction as indicated by endothelial relaxation impairment and increased contraction in mesenteric arterial beds mediated by the adrenergic.[2]

Moringa oleifera leaves (MO) aqueous extract mitigated high blood pressure, reduced adrenergic-mediated vasoconstriction, and endothelium-dependent vasorelaxation in the hypertensive animals. The therapy also decreased oxidative stress in hypertensive rats, in both vascular and systemic sections. This finding indicated that MO is antihypertensive, attributed to enhancing vasorelaxation based on endothelium and inhibiting vascular dysfunction induced by oxidative stress.[2] This study further investigated the mechanism of the action of MO aqueous extract to relax hypertensive rat mesenteric resistance arteries. It was revealed that MO induced vasorelaxation of hypertensive rats in isolated mesenteric arterial beds by acting on smooth muscle and endothelium. MO prompts active action on the endothelium by inducing hyperpolarization via a hyperpolarizing factor (EDHF) mechanism. It seems that another action of the extract is mediated by blocking extracellular calcium entry via voltage-regulated calcium channels (VOCCs) and receptor calcium channels (ROCCs), together with hindering the release of calcium from intracellular stores.[1] In the end, this capacity of MO can be exploited for use as a likely vasodilator in managing hypertension and other hyperactive vasoconstriction-associated conditions.

4.2.1.2 SPONTANEOUS HYPERTENSIVE RATS

Another animal model used for research purposes is spontaneous hypertensive rats (SHR), which are typified by an increased release of noradrenaline in circulation from the nervous system, leading to elevated blood pressure.[27] At 6 weeks old, the systolic and diastolic functions in the animals will be impaired owing to progressive development of left ventricular hypertrophy.[22]

Moringa oleifera (MO) seed powder was investigated in spontaneous hypertensive rats (SHR) as reported by Randriamboavonjy et al.[37] During the night-time active period, MO explicitly decreased SHR's heart rate (HR) without impact on diurnal arterial blood pressure and HR. They demonstrated the antihypertrophic function of MOI administration restored certain morphological parameters such as left ventricle posterior wall thickness on diastole (LVPWTd), interventricular septal thickness on diastole (IVSd), and relative wall thickness (RWT) similar to those reported in Wistar Kyoto (WKY), which are associated with left ventricle cardiac hypertrophy induced by pressure overload that is increased in SHR. In addition, cardiac fibrosis in the SHR control reduced substantially in the treated group. This action of MO was linked with increased expression of peroxisome proliferator-activated receptors (PPARα and δ), which are nuclear receptors and elevated levels of PPAR activator prostacyclin plasma. This indicated that MOI seed is beneficial against functional and structural cardiac changes associated with hypertension and scientifically supports its clinical in traditional Malagasy medicine to manage heart problems caused by overloading blood pressure.[37]

In another research, scientific examination of *Moringa oleifera* seed treatment on vascular oxidative and nitrosative stresses in SHR was revealed. The study showed that the oxidative and nitrosative stresses identified by fluorescent dye dihydroethidine in SHR aortas and protein nitrotyrosine staining in SHR aortas treated with MO were reduced. This was associated with a decline in circulating freely 8-isoprostane, vascular expressions p22phox and p47phox, and upregulation of SOD2. In addition, both circulating C-reactive protein and nitrites increased in SHR and decreased in MO-treated SHR. This was correlated with decreasing the expressions of NF-kB and iNOS protein after the treatment with MO. In functional studies, the relaxation caused by endothelium-dependent carbachol has been strengthened in SHR resistance arteries treated with MO. It can be said therefore that MO seeds possess anti-inflammatory, vascular antioxidant, and endothelial protective properties in the SHR model.[38]

4.2.2 VASCULAR DISEASE

Vascular aging leads to arterial wall functional and structural changes characterized by vascular stiffness and endothelial dysfunction. Decreased bioavailability of nitric oxide ensuing from increased oxidative stress is primarily responsible for endothelial dysfunction induced by aging. The aim of some antihypertensive therapy is to reduce total peripheral resistance by dilating the vessels.[21]

The effect of MO on endothelial dysfunction related to aging was investigated where 46-week-old males Wistar rats were fed 4 weeks with food either containing 750 mg/kg MO seed powder or not. MO therapy increased carbachol-stimulated relaxation in mesenteric arteries and aortas in the middle-aged rats. This has been associated with endothelial nitric oxide synthase activation increase and Akt signaling with arginase-1 downregulation in the aortas. The enhancement of the endothelial-determined relaxation in the mesenteric arteries was linked to an EDHF-related mechanism. The protective role of MO seeds against vascular malfunction in the resistance and conductance arteries during aging was indicated in these findings.[36]

4.2.3 MYOCARDIAL INFARCTION/INJURY

4.2.3.1 ISOPROTRENOL-INDUCED MODEL

Out of several well-known models the rat model of isoproterenol (ISO)-induced myocardial necrosis was mostly used to test many cardiac dysfunctions.[34]

Moringa oleifera polyphenolic oral fractions were administered for 4 weeks and afterward administered (s.c.) with 85 mg/kg ISO substantially decreased serum cardiac biomarkers and oxidative stress as a result of ISO which was administered for the last 2 days. In addition, on a histopathological examination, ISO-induced myocardium showed pronounced necrosis, disruption and degeneration of cardiac myofibers, and infiltration of inflammatory cell while ISO-induced rats pretreated with MOI showed normal cardiac fibers.[33]

The cardioprotective properties of an indole alkaloid from the *Moringa oleifera* leaves called N, α-L-rhamnopyranosyl vincosamide, in cardiotoxic rats induced by isoproterenol, were studied. Pretreatment with the alkaloid in rats treated with ISO significantly inhibited the ST-segment elevation and regulated the heart rate close to the control value. The cardio defense mechanism tends to include the prevention of cardiac myofibril damage, likely by reducing oxidative stress and thus enhancing contractile function of the heart. The mechanism of action of the compounds seems to be correct because it is understood that ISO creates oxidants and promotes lipid peroxidation, a contributing cause that damages myocardial tissues.[34]

In another research, a standardized lyophilized extract of *Moringa oleifera* protected against myocardial infarction in rats of Isoproterenol (ISO)-induced model. ISO administration results in significant rise in myocardial lipid peroxidation (MDA), a fall in antioxidant enzymes (GSH, CAT, and GPx) content, and an increase in myocardial creatinine kinase myocardial band (CK-MB isoenzyme) and lactate dehydrogenase (LDH) which indicates the incidence of myocardial injury. However, 200 mg/kg MO treatment enhances the antioxidant protection by significantly restoring myocardial antioxidant activities and prevents the leak of myocardial LDH and CK-MB enzymes and thus restores their function.[30] From the study it was observed that MO extract orally administered for a month protects the myocardium, prevents lipid peroxides accumulation, endogenous antioxidant resistance, and enhances cardiac function.

4.2.4 DIABETIC CARDIOMYOPATHY

The onset of cardiovascular disease is primarily responsible for the increase in type 2 diabetes prevalence.[16] Morbidity and mortality associated with diabetes is usually as a result of cardiovascular complications. Hyperglycemia with or without dyslipidemia produces oxidative stress that leads to different diabetic complications including vascular abnormality, destruction to membrane lipids, cellular proteins, and nucleic acids.[20,18] 300 mg/kg of MO leaves methanol extract administered to diabetic rats for 60 days effectively ameliorated hyperglycemia, oxidative stress, and related histological architecture in the heart of diabetic rats by reducing products of lipid peroxidation and increased both enzymatic and nonenzymatic antioxidants.[4]

4.2.5 CARDIOTOXICITY

Potassium bromate is an oxidizing agent, potent cardiotoxic as it stimulates lipid peroxidation with substantial decrease in heart antioxidant activities. It also initiated cardiac dysfunction as seen in increase in cardiac AST, ALT, and ALP. The study shows that potassium bromate in the studied tissue damaged the heart cells as demonstrated by decreased CAT and SOD activities. The study shows the chemopreventive advantage of the extract of MO leaf on cardiac oxidative damage in potassium bromate exposed rat as it significantly reduced the degree of antioxidant loss and restored cardiac dysfunction.[32] Intake of foods containing potassium bromate can cause heart trouble and should be avoided.

Aluminum phosphide (AlP) is an agrochemical pesticide used to increase productivity in agriculture.[8] AlP multisystem toxic involvement has been related to oxidative stress and phosphine gas.[28] Phosphine gas triggers oxidation stress by inhibiting enzymatic antioxidants like glutathione reductase, catalase, superoxide dismutase, and glutathione.[9] Investigation of the oral antioxidant activity of 100 mg/kg *Moringa oleifera* extract in neutralizing the oxidative stress caused by acute AlP toxic rat showed that the extract effectively protected the rats' heart by improving the antioxidants levels with a decrease in the production of lipid peroxidation compared with the untreated groups.[3]

Doxorubicin is a very active, well-known drug in multiple cancer treatments comprising small cell lung cancer, breast cancer, leukemia, solid tumors, soft tissue sarcoma, and esophageal cancer. Regardless of its anticancer potential, its clinical suitability is still limited due to its toxic effect on the heart tissue,[43] which could lead to temporary electrocardiographic irregularities and hypotension. Cheraghi et al.[10] evaluated the effect of N, α-L-rhamnopyranosyl vincosamide (VR), which was extracted from *Moringa oleifera* leaves in doxorubicin-stimulated cardio-toxic rats. The level of atrial natriuretic peptide, beta-major histocompatibility complex, and B-type natriuretic peptide expression decreased as the VR treatment concentration increases (200, 400, 800, 1000, and 2000 μg/mL) for 14 days every alternate day. After the administration of VR the reduced amount of GSH and SOD was increased. It also decreased the initially elevated level of lipid peroxidation in the cardiac tissue related to doxorubicin-induced generation of oxidative stress that can interact with membrane nucleic acids, lipids, and proteins preceding massive tissue injury.[19]

4.3 SUMMARY

Moringa oleifera is a rapidly growing tree in Africa, Asia, and America; its importance in traditional medicine cannot be over-emphasized which could be attributed to its phytochemical constituents. This chapter documents the therapeutic intervention of this plant in animal model of hypertension, cardiac, and vascular dysfunction and diabetic cardiomyopathy.

KEYWORDS

- **cardiovascular disease**
- **hypertension**
- **medicinal plant**
- ***Moringa oleifera***
- **vascular disease**

REFERENCES

1. Aekthammarat, D.; Pannangpetch, P.; Tangsucharit, P. *Moringa oleifera* Leaf Extract Induces Vasorelaxation Via Endothelium-Dependent Hyperpolarization and Calcium Channel Blockade in Mesenteric Arterial Beds Isolated From L-NAME Hypertensive Rats. *Clin. Exp. Hypertens.* **2020,** *42*(6), 490–501.
2. Aekthammarat, D.; Patchareewan, P.; Tangsucharit, P. *Moringa oleifera* Leaf Extract Lowers High Blood Pressure by Alleviating Vascular Dysfunction and Decreasing Oxidative Stress in L-NAME Hypertensive Rats. *Phytomedicine* **2019,** *54*, 9–16.
3. Ahmed, S. G.; El-Nabarawy, N. A.; Ibrahim S. F. *Moringa oleifera* Extract (Lam) Attenuates Aluminium Phosphide-Induced Acute Cardiac Toxicity in Rats. *Toxicol. Rep.* **2018,** *5*, 209–212.
4. Aju, B. Y.; Rajalakshmi, R.; Mini, S. Protective Role of *Moringa oleifera* Leaf Extract on Cardiac Antioxidant Status and Lipid Peroxidation in Streptozotocin Induced Diabetic Rats. *Heliyon* **2020,** *6*, e02935. https://doi.org/10.1016/j.heliyon.2019.e02935
5. Al Disi, S. S.; Anwar, M. A.; Eid, A. H. Anti-Hypertensive Herbs and Their Mechanisms of Action: Part I. *Front Pharmacol.* **2016,** *6*, 323–323.
6. Aluko, O.; Brai, M. R.; Adelore, A. O. Evaluation of Sensory Attributes of Snack From Maize-Moringa Seed Flour Blends. *Int. J. Nut. Food Eng.* **2013,** *7*, 944–946.
7. Anwar, F.; Latif, S.; Ashraf, M.; Gilani, A. H. *Moringa oleifera*: A Food Plant With Multiple Medicinal Uses. *Phytother. Res.* **2007,** *21*, 17–25.
8. Behera, C.; Krishna, K.; Bhardwaj, D. N.; Rautji, R.; Kumar, A. A Case of Accidental Fatal Aluminum Phosphide Poisoning Involving Humans and Dogs. *J. Forensic. Sci.* **2015,** *60*(3), 818–821.
9. Bumbrah, G. S.; Krishan, K.; Kanchan, T.; Sharma, M.; Sodhi, G. S. Phosphide Poisoning: A Review of Literature. *Forensic. Sci. Int.* **2012,** *214*(1–3), 1–6.
10. Cheraghi, M.; Namdari, M.; Daraee, H.; Negahdari, B.; Aiyelabegan, H. T. Cardioprotective Effect of Magnetic Hydrogel Nanocomposite Loaded N, α -L-rhamnopyranosyl Vincosamide Isolated From *Moringa oleifera* Leaves Against Doxorubicin-Induced Cardiac Toxicity in Rats: In Vitro and In Vivo Studies. *J. Microencapsul.* **2017,** *34*(4), 335–341.
11. Chorro, F. J.; Such-Belenguer, L.; López-Merino, V. Animal Models of Cardiovascular Disease. *Rev Esp Cardiol.* **2009,** *62*(1), 69–84.
12. Doggrell, S. A.; Brown, L. Rat Models of Hypertension, Cardiac Hypertrophy and Failure. *Cardiovasc. Res.* **1998,** *39*(1), 89–105.
13. Dou, Z.; Chen, C.; Fu, X. Bioaccessibility, Antioxidant Activity and Modulation Effect on Gut Microbiota of Bioactive Compounds From Moringa oleifera Lam. Leaves During Digestion and Fermentation In Vitro. *Food Funct.* **2019,** *10*, 5070–5079.
14. Fahey, J. W.; Wade, K.; Stephenson, K. K.; Shi, Y.; Liu, H.; Panjwani, A. A.; Warrick, C. R.; Olson, M. E. A Strategy to Deliver Precise Oral Doses of the Glucosinolates or Isothiocyanates From *Moringa oleifera* Leaves for Use in Clinical Studies. *Nutrients* **2019,** *11*, 1547.
15. Fejer, J.; Kron, I.; Pellizzeri, V.; Pl'uchtová, M.; Eliašová, A.; Campone, L.; Gervasi, T.; Bartolomeo, G.; Cicero, N.; Babejová, A. First Report on Evaluation of Basic Nutritional and Antioxidant Properties of *Moringa oleifera* Lam. From Caribbean Island of Saint Lucia. *Plants* **2019,** *8*, 537.
16. Feuvray, D. Cardiac Metabolism in the Diabetic Patient. *Heart Metab.* **2010,** *46*, 11–15.

17. Gandji, K.; Chadare, F.; Idohou, R.; Salako, V.; Assogbadjo, A.; Kakaï, R. G. Status and Utilisation of *Moringa oleifera* Lam: A review. *Afr. Crop. Sci. J.* **2018,** *26*, 137–156.
18. Giugliano, D.; Ceriello, A.; Paolisso, G. Diabetes Mellitus, Hypertension, and Cardiovascular Disease: Which Role for Oxidative Stress? *Metabolism* **1995,** *44*(3), 363–368.
19. Hamza, A.; Amin, A.; Daoud, S. The Protective Effect of a Purified Extract of *Withania somnifera* Against Doxorubicin-Induced Cardiac Toxicity in Rats. *Cell Biol. Toxicol.* **2008,** *24*(1), 63–73.
20. Johansen, J. S.; Harris, A. K.; Rychly, D. J.; Ergul, A.; Oxidative Stress and the Use of Antioxidants in Diabetes: Linking Basic Science to Clinical Practice. *Cardiovasc. Diabetol.* **2005,** *4*(1), 5–11.
21. Kjeldsen, S. E. Hypertension and Cardiovascular Risk: General Aspects. *Pharmacol. Res.* **2018,** *129*, 95–99.
22. Kokubo, M.; Uemura, A.; Matsubara, T.; Murohara, T. Noninvasive Evaluation of the Time Course of Change in Cardiac Function in Spontaneously Hypertensive Rats by Echocardiography. *Hypertens. Res.* **2005,** *28*, 601–609.
23. Kou, X.; Li, B.; Olayanju, J. B.; Drake, J. M.; Chena, N. Nutraceutical or Pharmacological Potential of *Moringa oleifera* Lam. *Nutrients* **2018,** *10*, 343–347.
24. Leong, X.; Ng, C.; Jaarin, K. Animal Models in Cardiovascular Research: Hypertension and Atherosclerosis. *BioMed. Res. Int.* **2015,** *2015*, Article ID 528757. http://dx.doi.org/10.1155/2015/528757
25. Linares, R. C.; Quiñones-Gálvez, J.; Pérez, M. A.; Carvajal, O. C.; Rivas, P. M.; Cid Valdéz, G. A.; Pérez, G. L.; La Rosa, G. S.; Capdesuñer, R. Y. K. Obtaining Phenolic Extracts From the Leaf of *Moringa oleifera* Lam by Using Different Extraction Methods. *Veg. Biotechnol.* **2018,** *18*, 47–56.
26. Mahmood, K. T.; Mugal, T.; Haq, I. U. *Moringa oleifera*: A Natural Gift—A Review. *J. Pharm. Sci. Res.* **2010,** *2*, 775–781.
27. McCarty, R.; Kopin, I. J. Alterations in Plasma Catecholamines and Behavior During Acute Stress in Spontaneously Hypertensive and Wistar-Kyoto Normotensive Rats. *Life Sci.* **1978,** *22*, 997–1005.
28. Mehrpour, O.; Jafarzadeh, M.; Abdollahi, M. A Systematic Review of Aluminium Phosphide Poisoning. *Arh. Hig. Rada Toksikol.* **2012,** *63*(1), 61–73.
29. Murad, W.; Azizullah, A.; Adnan, M.; Tariq, A.; Khan, K. U.; Waheed, S.; Ahmad, A. Ethnobotanical Assessment of Plant Resources of Banda Daud Shah, District Karak, Pakistan. *J. Ethnobiol. Ethnomed.* **2013,** *9*(1), 77.
30. Nandave, M.; Ojha, S. K.; Joshi, S.; Kumari, S.; Arya, D. S. *Moringa oleifera* Leaf Extract Prevents Isoproterenol-Induced Myocardial Damage in Rats: Evidence for an Antioxidant, Antiperoxidative, and Cardioprotective Intervention. *J. Med. Food* **2009,** *12*(1), 47–55.
31. Ogunsina, B.S.; Indira, T.N.; Bhatnagar, A.S.; Radha, C.; Debnath, S.; Krishna, A.G. Quality characteristics and stability of Moringa oleifera seed oil of Indian origin. *J. Food Sci. Technol.* **2014**, *51*, 503–510.
32. Oseni, O. A.; Ogunmoyole, T.; Idowu, K. A. Lipid Profile and Cardio-Protective Effects of Aqueous Extract of *Moringa oleifera* (lam) Leaf on Bromate- Induced Cardiotoxicity on Wistar Albino Rats. *Eur. J. Adv. Res. Biol. Life Sci.* **2015,** *3*(2), 52–66
33. Panda, S. Butanolic Fraction of *Moringa oleifera* Lam. (Moringaceae) Attenuates Isoprotrenol–Induced Cardiac Necrosis and Oxidative Stress in Rats: An EPR Study *EXCLI J.* **2015,** *14*, 64–74.

34. Panda, S.; Kar, A.; Sharma, P.; Sharma, A. Cardioprotective Potential of N,a-L-Rhamnopyranosyl Vincosamide, an Indole Alkaloid, Isolated From the Leaves of *Moringa oleifera* in Isoproterenol Induced Cardiotoxic Rats: In Vivo and In Vitro Studies. *Bioorg. Med. Chem. Lett.* **2013,** *23*(4), 959–962.
35. Pushpa, K.; Kanchana, R. Comparison of Waist-Hip Ratio, Prehypertension, and Hypertension in Young Male Bus Drivers and Non-Drivers of Bengaluru City. *Natl. J. Physiol. Pharm. Pharmacol.* **2019**, *9*, 90–94.
36. Randriamboavonjy, J. I.; Heurtebise, S.; Pacaud, P.; Loirand, G.; Tesse, A. *Moringa oleifera* Seeds Improve Aging-Related Endothelial Dysfunction in Wistar Rats. *Oxid. Med. Cell. Longev.* **2019,** *2019*, Article ID 2567198.
37. Randriamboavonjy, J. I.; Loirand, G.; Vaillant, N.; Lauzier, B.; Derbre, S.; Michalet, S.; Pacaud, P.; Tesse, A. Cardiac Protective Effects of *Moringa oleifera* Seeds in Spontaneous Hypertensive Rats. *Am. J. Hypertens.* **2016,** *45*(2), 153–156.
38. Randriamboavonjy, J. I.; Rio, M.; Pacaud, P.; Loirand, G.; Tesse, A.; *Moringa oleifera* Seeds Attenuate Vascular Oxidative and Nitrosative Stresses in Spontaneously Hypertensive Rats. *Oxid. Med. Cell. Longev.* **2017,** *2017*, Article ID 4129459.
39. Smith, S. C.; Collins, A.; Ferrari, R. Our Time: A Call to Save Preventable Death From Cardiovascular Disease (Heart Disease and Stroke). *J. Am. Coll. Cardiol.* **2012,** *60*, 2343–2348.
40. Touyz, R. M.; Alves-Lopes, R.; Rios, F. J.; Camargo, L. L.; Anagnostopoulou, A.; Arner, A.; Montezano, A. C.; Vascular Smooth Muscle Contraction in Hypertension. *Cardiovasc. Res.* **2018,** *114*, 529–539.
41. Ullah, R.; Iqbal, Z. H. Z.; Hussain, J.; Khan, F. U.; Khan, N.; Muhammad, Z.; Ayaz, S.; Ahmad, S.; Rehman, N. U.; Hussain, I. Traditional Uses of Medicinal Plants In Darra Adam Khel NWFP Pakistan. *J. Med. Plants Res.* **2010,** *4*(17), 1815–1821.
42. World Health Organization. Cardiovascular Diseases (CVDs). [Online] **2017**. https://www.who.int/news-room/factsheets/detail/cardiovascular-diseases-(cvds) (accessed Aug 26, 2020)
43. Zhou, S.; Palmeira, C. M.; Wallace K. B. Doxorubicin-Induced Persistent Oxidative Stress to Cardiac Myocytes. *Toxicol. Lett.* **2001,** *121*(3), 151–157.

CHAPTER 5

Blood Sugar Regulation and Proinflammatory Effects of *Desplatsia Dewevrei* in Streptozotocin-Induced Diabetic Wistar Rats

Oghale O. Godwin, and MacDonald Idu

ABSTRACT

In a pilot study, the blood sugar regulatory and proinflammatory effects of *Desplatsia dewevrei* were tested on streptozotocin-induced diabetic Wistar rats. Five groups with four subjects each were involved in this study. Groups 1 and 2 were used as negative and diabetic control. Group 3 received 30 mg of metformin/kg across the board; while groups 4 and 5 were administered 30 mg/kg of *D. dewevrei* aqueous and methanol extracts distinctively. Findings show that 30 mg/kg *D. dewevrei* aqueous extract was able to reverse blood fasting sugar in experimental subjects on the seventh day 7 of 14-days experimental period. Blood biochemical parameters and proinflammatory genes, IL-1β and MCP-1, were also moderated in treated groups against the control. In silico studies showed that gamma tocopherol with a binding score of −9.3 had the highest binding affinity with glucokinase of *D. dewevrei*. In conclusion, *D. dewevrei* may be considered a potent antidiabetic agent capable of checking inflammation in subjects suffering from type 1 *Diabetes mellitus* and eventually reversing the diabetic condition.

Curative and Preventive Properties of Medicinal Plants: Research on Disease Management and Animal Model Studies. Megh R. Goyal, Junaid Ahmad Malik, & Ademola Olabode Ayeleso (Eds.)

5.1 INTRODUCTION

About 3% of the world's mortality can be attributed to diabetes, which is a chronic metabolic disease with numerous complications. It is caused by either the destruction of pancreatic β-cells (type-1 diabetes) or unresponsiveness to insulin (type-2 diabetes).[9] Several laboratory studies have revealed that cells and tissues of the body's vital organs can be damaged by elevated reactive oxygen species thereby contributing to complications of *Diabetes mellitus* (DM).[2,23] Thus, the search for remedies to moderate hormonal imbalance or free-radical damage and up-regulate insulin deficiency has always been and is still a matter of continuous emphasis.[18] Understanding and improving medication is vital for the management of diabetes and its complications, and many of the available hypoglycemic drugs for the management of DM have undesired side effects.[10] Hence, in *(sic)* the scientific investigation to find alternative therapeutic strategies for diabetes treatment has become inevitable.

Burkil described and reported some ethnobotanical uses of *Desplatsia dewevrei* (a scarce species attributed useful in the management of many maladies). Although a fraction of the therapeutic claims of *D. dewevrei* has been verified by scientific experimentations,[4,13–16] yet many of such claims have not been verified.

This chapter explores the modulatory activities of leaves of *D. dewevrei* on blood glucose homoeostasis and inflammation resulting from diabetic complications using experimental animals.

5.2 MATERIALS AND METHODS

5.2.1 PREPARATION AND EXTRACTION OF PLANT SAMPLE

D. dewevrei leaves were harvested and air-dried after washing under running water. After 2 weeks of air-drying, the crisp leaves were crushed by hand and macerated in methanol and water separately. Thereafter, the filtrate was freeze-dried at −4°C.

5.2.2 ANIMAL STUDIES

Twenty Wistar rats of both sexes were involved in this study. All 20 animals were kept under a 12-h light/dark cycles, controlled temperature and humidity, with food and portable water readily available. All protocols were carried out following stated guidelines of the Animal Care and Ethics Committee of Adekunle Ajasin University, Akungba-Akoko, which conforms to the protocol for use of laboratory animals by the US National Institute of Health.

5.2.3 DIABETES INDUCTION

The 45 mg of streptozotocin (STZ) per kg was dissolved in acidified (pH 5.2) citrate buffer; thereafter, it was administered to the rats intraperitoneally after 12 h of fast. An Accu-Check Integra device (Roche Diagnostics Mannheim, Germany) was used to determine the blood glucose level from the blood sample obtained from Saphenous vein. After 3 days of administering STZ, a booster dose was given to animals that were not diabetic. After another 3 days, animals exhibiting blood glucose values of 200–300 mg/dL were presumed to be diabetic. The animals ($n = 20$) were randomized into five groups:

- group one served as basal control,
- group two—diabetic control,
- group three—standard (metformin of 30 mg/kg),
- group four was administered 30 mg/kg body weight of methanol extract of *D. dewevrei*, and
- group five was administered 30 mg/kg body weight of aqueous extract of *D. dewevrei*.

5.2.4 BLOOD FASTING SUGAR TEST

After a week of acclimatization, a 14-days treatment regimen was given, where animals received extracts orally on daily basis. For each group, dose concentrations of the extract were calculated based on the body weight of the animal. Both extracts (0.02 g) were dissolved in distilled water (14 mL). The 0.5 mL of resulting solution was administered orally across the treatment groups for 14 days. The control, however, received no treatment.

5.2.5 BIOCHEMICAL STUDIES

Based on the procedure by Ovuakporie-Uvo et al.,[12] some biochemical tests were conducted using blood (from the cardiac aorta) of the experimental rats.

5.2.6 ANTIOXIDANT STUDIES

In vivo antioxidant genes were used to measure the effect of *D. dewevrei* on oxidative stress mop-up. Genes under investigation included G6PDD, CAT, GSR, and GPx.

5.2.7 IN-SILICO STUDIES

Some phytoconstituents of *D. dewevrei* are characterized by GC–MS analyses as reported in an earlier study by Ovuakporie-Uvo et al.,[13] where plant samples were subjected individually to in silico studies in compliance to procedures adopted by Iyamah et al.[8] The 3D dimension of the phytocompounds were isolated, and binding affinity of each phytoconstituent to glucokinase was determined and correlation of each phytoconstituent was done using basic regression technique with Autovina 4.0 and mcule computer software.

5.2.8 STATISTICAL ANALYSIS

Data were expressed as mean ± standard error of mean. Statistical analyses were carried out using one-way analysis of variance (ANOVA). Multiple comparisons were done using Duncan multiple range tests (SPSS version 23), Tukey's multiple range tests (Graphpad prism 8), and Microsoft Excel software package. Significant levels were determined at $p < 0.05$.

5.3 RESULTS AND DISCUSSION

Administration of *D. dewevrei* aqueous and methanol extracts @ 30 mg/kg elicited antidiabetic effects on streptozotocin-induced rats. Blood fasting sugar was returned to normal in the diabetic-induced group of animals (that were administered 30 mg/kg of aqueous extract) after 7 days of oral administration of the aqueous leaf extract (Figure 5.1).

Results from serum biochemistry of Wistar rats in the antidiabetic study support the blood fasting sugar test as documented by cholesterol reduction, triglyceride, low-density lipoprotein, and an increased level of high-density lipoprotein (Figure 5.2).

Oxidative stress-reducing capacity of *D. dewevrei* was further subjected to gene expression studies using various oxidative and proinflammatory markers.

Downward regulation of CAT, glutathione reductase (GSR), glutathione peroxidase 1 (GPX-1), phosphoprotein tumor suppressing protein (P_{53}) further supported the antidiabetic activities of *D. dewevrei* extracts (Figure 5.3).

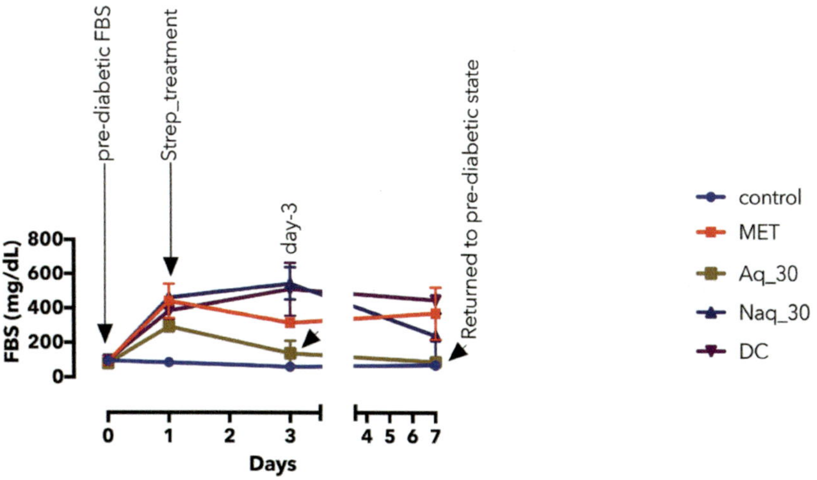

FIGURE 5.1 Blood fasting sugar of streptozotocin-induced Wistar rats treated with *D. dewevrei* extracts.

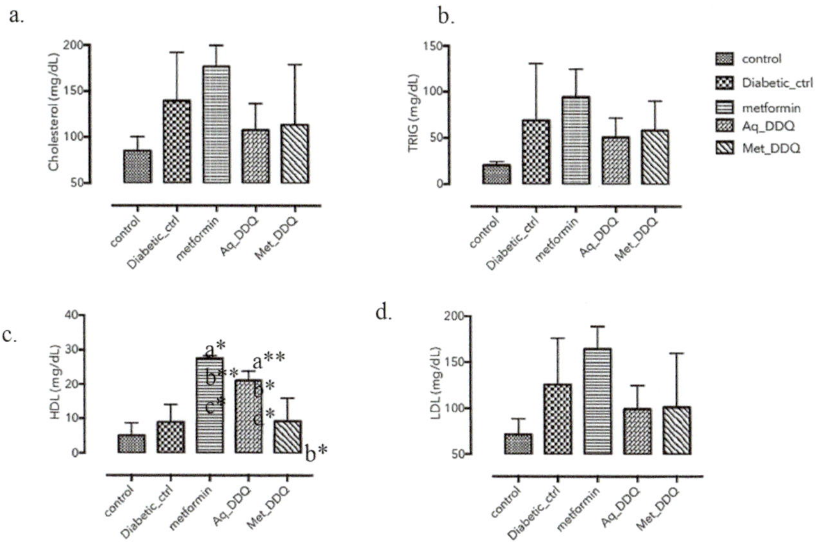

FIGURE 5.2 Serum biochemistry of streptozotocin-induced Wistar rats. Extract values in treatment groups are significantly different from control.

HDL, high-density lipoprotein; HDL Aq_DDQ, aqueous Desplatsia dewevrei extract; LDL, low-density lipoprotein; Met_DDQ, methanol Desplatsia dewevrei; TRIG, triglyceride.

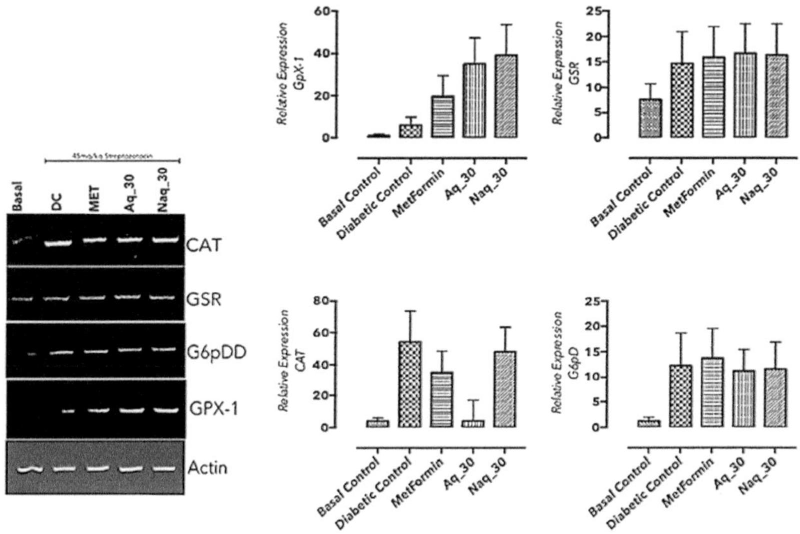

FIGURE 5.3(a–e) Oxidative stress-reducing capacity of *Desplatsia dewevrei* in streptozotocin-induced diabetic Wistar rats.

Aq_30, 30 mg/kg aqueous extract of Desplatsia dewevrei; Naq_30, 30 mg/kg of nonaqueous (Methanol) extract of Desplatsia dewevrei.

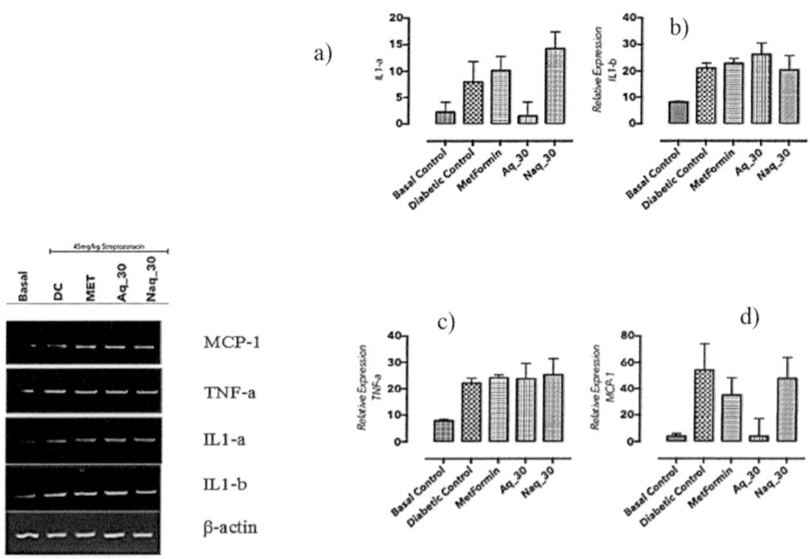

FIGURE 5.4(a–d) Blood vessel proinflammatory markers in diabetic rats.

Aq_30, 30 mg/kg aqueous extract of Desplatsia dewevrei; Naq_30, 30 mg/kg methanol extract of Desplatsia dewevrei.

The phytoconstituents in *D. dewevrei* methanol leaf extract were docked. The forward and backward sequences of all the genes queried in this study are presented in Table 5.1. The binding affinities of compounds having antidiabetic activities were documented, as shown in Table 5.2. Gamma tocopherol (−9.3) was found to the best fit and elicit agonistic response to target receptors that elicit antidiabetic effect. This was closely followed by tocopherol (−9.0) and gamma sistosterol (−8.7).

TABLE 5.1 Primer Sequences for Inflammatory and Oxidative Stress Studies in Streptozotocin-Induced Rats.

Gene used	Forward 5'-3'	Reverse 5'-3'
β-actin	ACACTTTCTACAATGAGCTGCG	ACCAGAGGCATACAGGACAAC
IL-1α	CTTCACATCCGCAGCTTTCC	GCGAGTGACTTAGGACGAGG
IL-1β	TTGAGTCTGCACAGTTCCCC	TCCTGGGGAAGGCATTAGGA
CAT	CCGACCAGGGCATCAAAA	GAGGCCATAATCCGGATCTTC
Gpx-1	AGTTCGGACATCAGGAGAATGGCA	TCACCATTCACCTCGCACTTCTCA
G6PDD	GCTATGCCCGTTCCATGCT	GCGTCCGTCATATCTGCCG
GSR	GGAAGTCAACGGGAAGAAGTT CACTG	CAATGTAACCGGCACCCACA ATAAC
MCP-1	TGCCAAGTAGCCACATCCAG	CACAGTGTGAGCAACTGGGA
TNF-α	CATCCGTTCTCTACCCAGCC	AATTCTGAGCCCGAGTTGG

The 3D pose of the phytoconstituents with antidiabetic potentials is reported in Figure 5.5. The 3D binding pose and in silico interaction of glucokinase with gamma tocopherol and the amino acids residues of gamma tocopherol pose are presented in Figure 5.6. The correlation curve of the gamma tocopherol gave a regression value of approximately 0.72 (Figure 5.7).

5.4 DISCUSSION

Hyperglycemia in its early stage causes vascular wall disorder involving increased oxidative stress, endothelial failure, platelet hyperactivity, and inflammation. Consequently, activation of these systems impairs endothelial function, augments vasoconstriction, increases inflammation, and promotes thrombosis.[21] In this chapter, *D. dewevrei* showed a tremendous antidiabetic effect on streptozotocin-induced diabetes after 7 days of daily oral administration as was evident from blood fasting sugar test (Figure 5.1). An elevation of HDL levels in the serum biochemistry in diabetic animals orally administered *D. dewevrei* aqueous and methanol leaf extracts further validated the blood fasting sugar test (Figure 5.2). According to Sadek et al.,[19] high doses of thiamine (50 mg/kg) revert fasting hyperglycemia in diabetic albino rats. This suggests that thiamine (B_1) content in *D. dewevrei* extract may be contributory to the experienced reversal of blood fasting sugar in this study.[12]

TABLE 5.2 Binding Affinity of Phytoconstituents of *Desplatsia dewvrei* With Antidiabetic Potentials.

Sl. No.	Compounds	Binding affinity (Kcal/mol)
1.	2-methylhexacosane	−6.6
2.	9-octadecenamide_Z	−6.8
3.	14-octadecenal.	−7.0
4.	Decanoic acid	−5.9
5.	Dodecanoic acid	−6.4
6.	Gamma-sitosterol	−8.7
7.	Gamma tocopherol	−9.3
8.	Hexadecanoic acid	−6.7
9.	Methyl stearate	−6.9
10.	Octanoic acid	−5.6
11.	Oleic acid	−6.8
12.	Pentadecanal	−6.4
13.	Tetradecanoic acid	−6.6
14.	Tocopherol	−9.0

According to Uruno et al.,[24] there is a close relationship between the Nrf2/Keap1 system and diabetes. Molecular insight enhances the thrust of the theory of oxidative stress, which is intimately linked to cellular energy balance. The ensuing detrimental biochemical pathway, MPs' release, changes in epigenetics, and miRs deregulation are resulted from the vascular accumulation of free radicals in diabetic patients, which adds to ROS generation and vascular inflammation in the patient.[17] Results in this chapter revealed that *D. dewevrei* potentiates oxidative stress-reducing capacity in streptozotocin-induced diabetic Wistar rats (Figure 5.3).

Antioxidant compounds also increase the number of β-pancreatic cells by enhancing the repair and restoration of these cells. It has been well documented

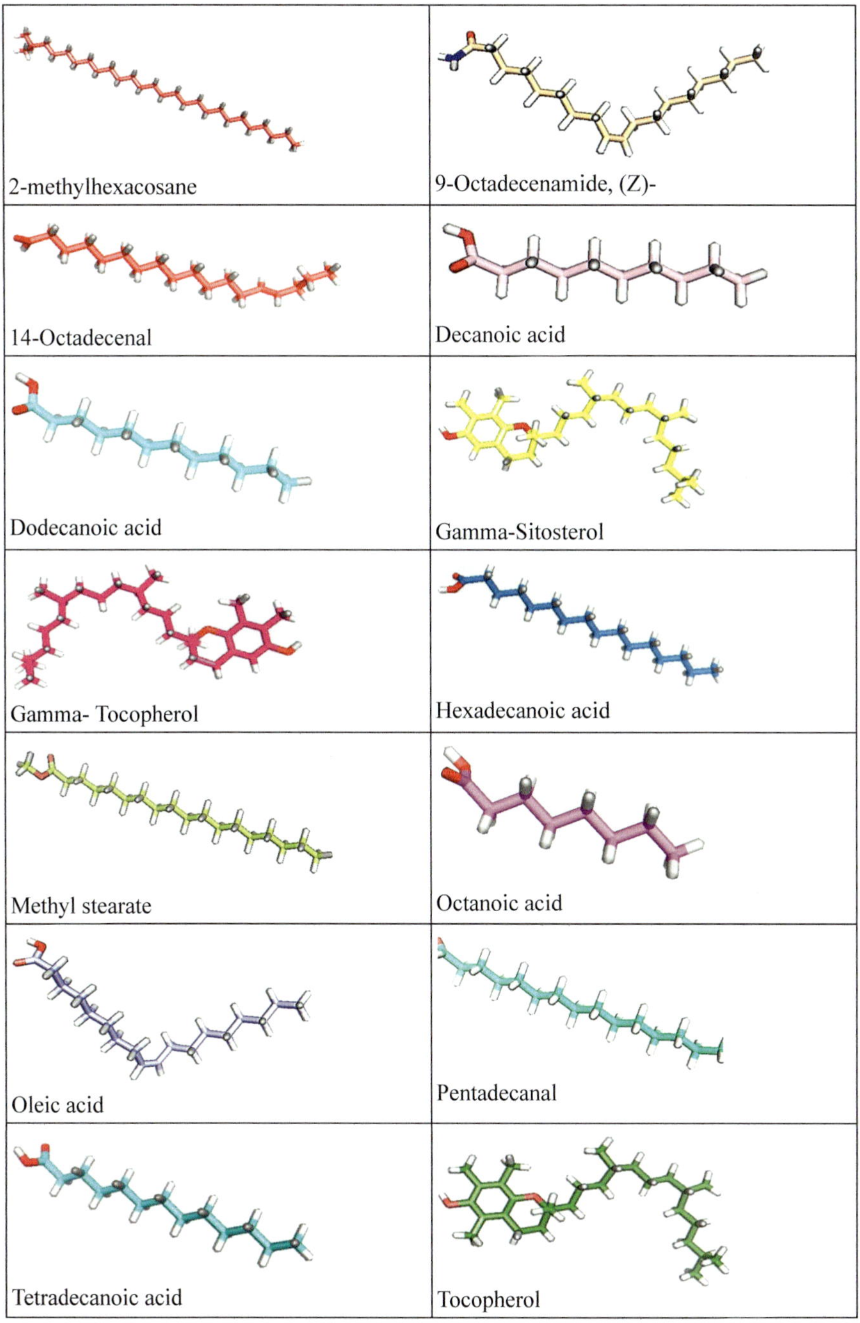

FIGURE 5.5 Three-dimensional structure of the phytochemical constituents present in *Desplatsia dewevrei* that were tested for their binding affinity to protein receptors that elicit antidiabetic effects.

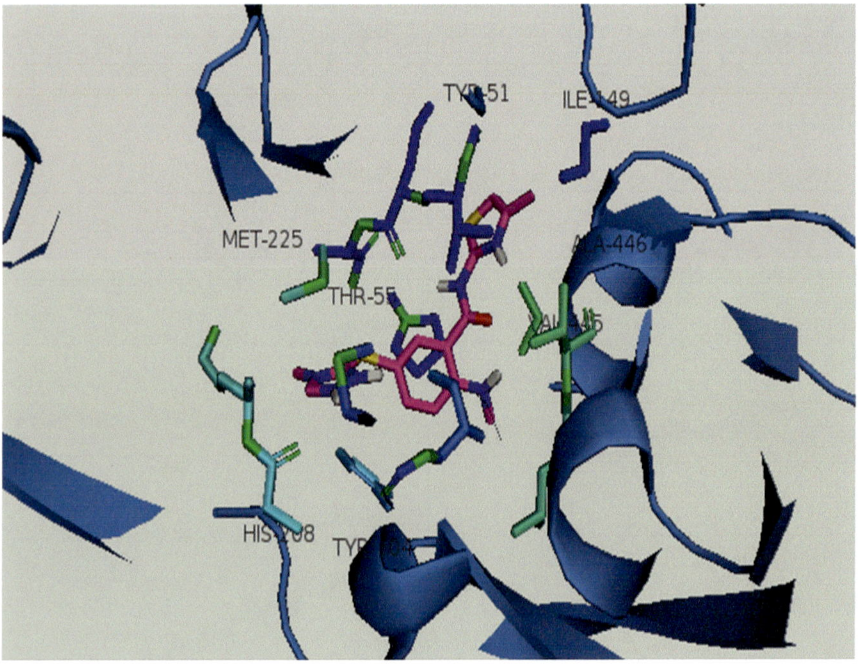

FIGURE 5.6 In silico analysis of the interaction of glucokinase with gamma tocopherol and the amino acid residues in the binding pose of gamma tocopherol.

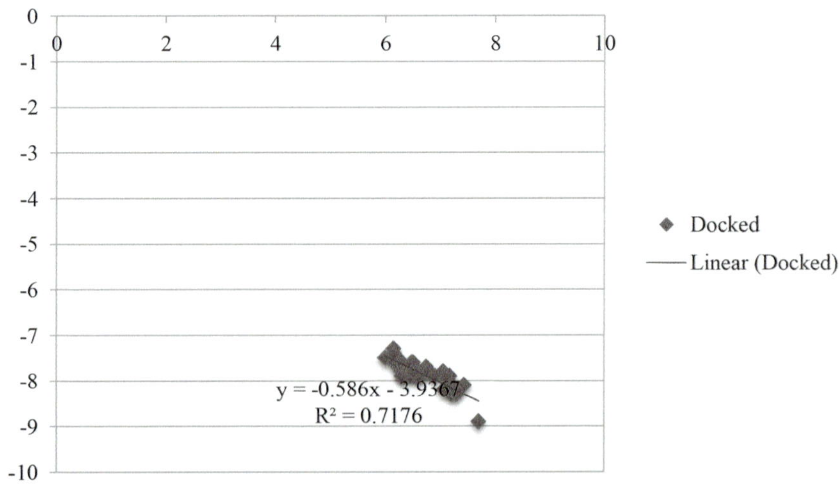

FIGURE 5.7 The coefficient of correlation (R^2) of the half maximal effective concentration (EC_{50}) of glucokinase.

that administration of antioxidants to diabetic rats significantly increases the number of β-cells.[3] In this chapter, however, *D. dewevrei* affects antioxidant activities as was evident in the downward regulation the expression of CAT, G6PDD, and amino phosphatase in the pancreas of diabetic-induced animals (Figure 5.3). According to Fontana and Klein,[6] catalase reduces the risk of transforming H_2O_2 into hydroxyl radicals through Fenton reaction catalyzed by chromium or ferrous ions. A downward regulation of CAT and G6PDD (Figure 5.3) is a double affirmative of the antioxidant activities of *D. dewevrei*. A reduction in cytokines responsible for proinflammatory activity in the presence of any disease state (such as IL1-α and MCP-1) was observed to be down-regulated in the present study. This is a further attestation to the cholesterol-lowering effect of *D. dewevrei*.

Inhibition of MCP-1 is an expression of protein restriction while IL1-α plays a major function of maintaining inflammation by utilizing autocrine and paracrine induction of complementary and alternative medicines, chemotactic factors, and cytokines.[11,22] A down-regulation of IL-1α and MCP-1 following the administration of 30 mg/kg of *D. dewevrei* aqueous leaf extract (Figure 5.4) implies that the plant has the ability to inhibit inflammatory response of blood vessels in a diabetic condition. This is perhaps the mechanism that out-played in the blood sugar fasting test earlier discussed (Figure 5.1). Results in this study concur with the earlier reports by Faubel et al.[5]

The 3D structures of the phytocompounds got from an earlier report on *D. dewevrei* showed that some compounds in the plant have relatively high binding affinity to glucokinase responsible for eliciting antidiabetic effects (Figure 5.5). The phytocompound that had the best binding affinity in the antidiabetic study was Gamma tocopherol with a binding score of 9.3. Gamma tocopherol has been implicated to be useful as antioxidative and anti-inflammatory agent.[1]

In vivo, an ability shown by gamma tocopherol is to inhibit ascorbate oxidation and protein nitration caused by inflammation. When compared to alpha-tocopherol, it also decreases the proinflammatory prostaglandins and leukotrienes better.[7] Tocopherol, which was next to gamma tocopherol, had a binding affinity of −9.0 (Table 5.2). Tocopherols are abundant in lipid-rich areas of cells (e.g., mitochondrial membranes), fat depots, and lipoproteins, such as, high-density lipoprotein cholesterol.

Health benefits of tocopherols may be explained by regulation of gene expression, signal transduction, and modulation of cell functions. Other possible health benefits of tocols include prevention of certain types of cancer, heart disease, and other chronic ailments.[20] From the pictorial representation of the binding pose of gamma tocopherol, interaction with amino acid residues (Figure 5.6) and the coefficient of correlation of half-maximal showing the effective concentration shows a correlation of R^2= 0.7176 (Figure 5.7), we can conclude that *D. dewevrei* leaves are hypoglycemic and may be useful in reducing inflammation in diabetic subjects.

5.5 SUMMARY

D. dewevrei is a scarce medicinal plant species in Africa with reported high nutritional contents and phytoconstituents that can be harnessed in the management of diabetes and complication from diabetes specifically inflammation. This study verifies assertions from a pilot toxicity study confirming the antidiabetic effect of *D. dewevrei* using blood sugar test and biochemical profiles. Gene expression and in silico studies further supports that *D. dewevrei* contains cytokines that affect antidiabetic effects.

ACKNOWLEDGMENTS

The authors acknowledge Dr. O. I. Omotuyi of Biochemistry Department, Adekunle Ajasin University Akoko-Akungba; Dr. Gerald Eze of Anatomy Department, University of Benin, and the entire management and staff of the Centre for Bio-computing and Drug Development, Adekunle Ajasin University, Akungba Akoko, Ondo State, Nigeria for their selfless services and professional support during this research. This research in this chapter was solely funded by the authors without conflict of interest.

KEYWORDS

- ***Desplatsia dewevrei***
- **diabetics**
- **inflammation**
- **medicinal plant**

REFERENCES

1. Agarwal, T.; Gupta, P.; Asthana, S.; Khursheed, A. *In Silico* Analysis to Access the Antibacterial Effect of Thiazides on Pdfs: Molecular Docking Approach. *Int. J. Pharm. Pharm. Sci.* **2014,** *6*(6), 387–391.
2. Ben Salem, M.; Kolsi, R. B. A.; Dhouibi, R.; Ksouda, K.; Charfi, Slim.; Yaich, M.; Hammami, S.; Sahnoun, Z., Zeghal, K. M.; Jamoussi, K.; Affes, H. Protective Effects of *Cynara scolymus* Leaves Extract on Metabolic Disorders and Oxidative Stress in Alloxan-Diabetic Rats. *BMC Complement. Altern. Med.* **2017,** *17*, 328–334. DOI: 10.1186/s12906-017-1835-8.
3. Bharti, S. K.; Kumar, A. Sharma, N. K.; Prakash, O.; Jaiswal, S. K.; Krishnan, S.; Gupta, A. K.; Kumar, A. Tocopherol From Seeds of *Cucurbita pepo* Against Diabetes: Validation by *In Vivo* Experiments Supported by Computational Docking. *J. Formos. Med. Assoc.* **2013,** *112*, 676–690.
4. Burkill, H. M. (Ed.) In *The Useful Plants of West Tropical Africa*, 2nd ed.; Families A–D; Royal Botanic Gardens: Kew, United Kingdom, 1985; vol 1, p 462.
5. Faubel, S.; Lewis, E. C.; Reznikov, L.; Liubanovic, D.; Hoke, T. S.; Somerset, H.; Oh, D. J.; Lu, L.; Klein, C. L.; Dinarello, C. A.; Edelstain, C. L. Cisplatin-Induced Acute Renal Failure is Associated With an Increase in the Cytokines Interleukin (IL)-1Beta, IL-18, IL-6, and Neutrophil Infiltration in the Kidney. *J. Pharmacol. Exp. Ther.* **2007,** *322*, 8–15.
6. Fontana, L.; Klein, S. Aging, Adiposity, and Calorie Restriction. *J. Am. Med. Assoc.* **2007,** *297*, 9–11.
7. Himmelfarb, J.; Phinney, S.; Alplkizler, T.; Kane J.; McMonagle, E.; Miller, G. Gamma-Tocopherol and Docosahexaenoic Acid Decrease Inflammation in Dialysis Patients. *J. Ren. Nutr.* **2007,** *17*(5), 296–304.
8. Iyamah, P.; Famuti, A.; Idu, M. GC-MS and Molecular Docking Studies for Identification of Anti-malarial Compounds in Agbo-Iba PMII-a Polyherbal Formulation. *Chem. Res. J.* **2017,** *2*(1), 46–56.
9. Kang, O. H.; Shon, M. Y.; Kong, R.; Seo, Y. S.; Zhou, T.; Kim, D. Y.; Kim, Y. S.; Kwon, D. Y. Anti-Diabetic Effect of Black Ginseng Extract by Augmentation of AMPK Protein Activity and Upregulation of GLUT2 and GLUT4 Expression in db/db Mice. *BMC Complement. Altern. Med.* **2017,** *29*(17), 341–348. https://doi.org/10.1186/s12906-017-1839-4
10. Kasetti, R. B.; Nabi, S. A.; Swapna, S.; Apparao, C. Cinnamic Acid as One of the Anti-diabetic Active Principle(s) From the Seeds of *Syzygium alternifolium*. *Food Chemo. Toxicol.* **2012,** *50*, 1425–1431.
11. Netea, M. G.; Kullberg, B. J.; Boerman, O. C.; Verschueren, I.; Dinarello, C. A; Jos, W. M. V. Soluble Murine IL1 Receptor Type 1 Induces Release of ConstitutiveIL1-α. *J. Immunol.* **1999,** *162*, 4876–4881.
12. Ovuakporie-Uvo, O.; Idu, M.; Eze, G. O.; Ozolua, R. I. Toxicological Studies of *Anchomanes difformis* Blume Using Rats and Mice. *Int. J. Basic Clin. Pharmacol.* **2015,** *4*, 1228–1234.
13. Ovuakporie-Uvo, O.; Idu, M.; Itemire, O. A. Phytochemistry, Gas Chromatography-Mass Spectrometry Analysis and *In Vitro* Anti-Bacterial Activities of *Desplatsia dewevrei*. *Braz. J. Biol. Sci.* **2018,** *5*(10), 373–404.
14. Ovuakporie-Uvo, O.; Idu, M.; Obarisiagbon, P.; Abode, C. Analgesic, Pro and Antiinflammatory Activities of *Desplatsia dewevrei*: Cytokine Gene Expression Using Wistar Rats and Mice. *J. Phytopharmacol.* **2018,** *7*(2), 185–190.

15. Ovuakporie-Uvo, O.; Idu, M.; Omoregie, E. S. Nutrients and Chemical Composition of *Desplatsia dewevrei*. *Food Sci. Nutr.* **2019,** *2019*, 1–10. https://doi.org/10.1002/fsn3.1019.
16. Ovuakporie-Uvo, O.; Idu, M.; Omotuyi, I. O. Toxico-Genetic Studies of *Desplatsia dewevrei* Using Gene Expression of Blood, Pancreatic and Intestinal Genes in Wistar rats. *Can. J. Biotechnol.* **2019,** *2*(4), 124–131.
17. Paneni, F.; Beckman, J. A.; Creager, M. A.; Cosentino, F. Diabetes and Vascular Disease: Pathophysiology, Clinical Consequences, and Medical Therapy: Part I. *Eur. Heart J.* **2013,** *34*, 2436–2446.
18. Pia, V. R.; Bingbing, W.; Yixian, L.; Weiping, H. Pancreatic Regulation of Glucose Homeostasis. *Exp. Mol. Med.* **2016,** *48*(3), 219–225. DOI: 10.1038/emm.2016.6
19. Sadek, K. M.; Taha, N.; Korshom, M.; Mandour, A. Thiamine Ameliorates Hepatic, Renal Dysfunction and Dyslipidemia in Diabetic Rats. *J. Curr. Res. Sci.* **2013,** *1*(1), 35–39.
20. Sena, C. M.; Pereira, A. M.; Seiça, R. Endothelial Dysfunction: A Major Mediator of Diabetic Vascular Disease. *Biochim. Biophys. Acta Mol. Basis Dis.***2019,** *1832*(12), 2216–2231.
21. Shahidi, F.; de Camargo, A. C. Tocopherols and Tocotrienols in Common and Emerging Dietary Sources: Occurrence, Applications, and Health Benefits. *Int. J. Mol. Sci.* **2016,** *17*(1745), 1–29. DOI: 10.3390/ijms17101745
22. Taal, M. W.; Zandi-Nejad, K.; Weening, B.; Shahsafaei, A.; Kato, S. H.; Lee, K. W.; Ziai, F.; Jiang, T.; Brenner, B. M.; Mackenzie, H. A. S. Proinflammatory Gene Expression and Macrophage Recruitment in the Rat Remnant Kidney. *Kidney Int.* **2000,** *58*, 1664–1676.
23. Tagang, A.; Joseph, O. A.; Alkali, K.; Olusola, O. O. Amelioration of Hyperglycaemia, Oxidative Stress and Dyslipidemia in Alloxan-Induced Diabetic Wistar Rats Treated With Probiotic and Vitamin C. *Nutrients* **2016,** *8*, 151–154.
24. Uruno, A.; Yagishita, Y.; Yamamoto, M. The Keap1-Nrf2 System and *Diabetes Mellitus*. *Arch Biochem. Biophys.* **2015,** *566*, 76–84.

PART II

POTENTIAL OF MEDICINAL PLANTS IN NEURODEGENERATIVE DISEASES

CHAPTER 6

Plant-Based Antioxidants in the Prevention and Treatment of Neurodegenerative Diseases: Pros and Cons

Folake Olubukola Asejeje and Gbolahan Iyiola Asejeje

ABSTRACT

The research in the area of medicinal plants for the prevention and treatment of human diseases has yielded very positive results with the discovery of efficacious plant-sourced compounds that have fewer or milder side effects. However, there is a need to strike a balance between the beneficial and harmful effects of these plant-sourced compounds before recommending their usage in the treatment of human diseases. Several animal models have been used to investigate the potentials of the different plant-derived antioxidants in the treatment of human diseases. This chapter presents a discussion of both the potential benefits and limitations of plant-sourced antioxidants in the prevention and treatment of human neurodegenerative diseases from a review of research reports that have used different animal models to investigate the potentials of medicinal plants in the treatment of human diseases. The aim of this chapter is to discuss the potentials of medicinal plants in the treatment of neurodegenerative diseases with a focus on plant-derived antioxidants as therapeutics.

Curative and Preventive Properties of Medicinal Plants: Research on Disease Management and Animal Model Studies. Megh R. Goyal, Junaid Ahmad Malik, & Ademola Olabode Ayeleso (Eds.)

6.1 INTRODUCTION

Lately, research in the area of medicinal plants in the prevention and treatment of human diseases, including neurodegenerative disease, has yielded very positive results with the discovery of efficacious plant-sourced compounds that have fewer or milder side effects.[113,136] However, there is a need to strike a balance between the beneficial and harmful effects of these plant-sourced bioactive compounds before recommending their usage in the treatment of human diseases. Several animal models have been used to investigate the potentials of different plant-derived antioxidants in the treatment of human diseases. It is important to consider dosage and bioavailability, the animal model used, and the means or method of induction of the disease condition in the animal model used. This is imperative because some antioxidants, under certain conditions, have been shown to act as prooxidants with potential for cellular toxicity.[17,73]

This chapter aims to highlight previous research reports on investigation of the potentials of medicinal plants or plant-derived compounds in the treatment of human neurodegenerative diseases. In this chapter, we present a discussion of both the potential benefits and limitations of plant-sourced antioxidants in the prevention and treatment of human neurodegenerative diseases from a review of research reports by several researchers that used various animal models to investigate the potentials of medicinal plants in the treatment of human diseases.

6.2 PLANT-SOURCED ANTIOXIDANTS

It is established that oxidative stress, characterized by elevated oxidant levels in the living system resulting in structural and functional damages, is involved in the onset of several human diseases.[82,111] This knowledge has spurred research in the area of antioxidants for the treatment of human diseases.[53,54,82] Plant-sourced active principles are useful in the prevention and treatment of certain human diseases.[106,120] These active principles are usually phenolic antioxidant compounds present as secondary metabolites in plants.[22,56,90] The chemical structures of these compounds, especially the number and position of the hydroxyl group, are largely responsible for their antioxidant properties.[56,119] These plant-derived antioxidants are classified into three broad groups (Figure 6.3), such as flavonoids, nonflavonoids, and conjugates. [54,132] The flavonoids are a major and well-known class of the biophenols.[39] They have the diphenyl propane (C_6–C_3–C_6) structure (the flavan nucleus) composed of two aromatic rings (A and B) connected through an oxygenated heterocycle (C) formed by three carbon atoms (Figure 6.1). The predominantly occurring plant flavonoids are flavones, flavonols, flavanols, isoflavonoids, flavanonoes, and anthocyanins.[35,77]

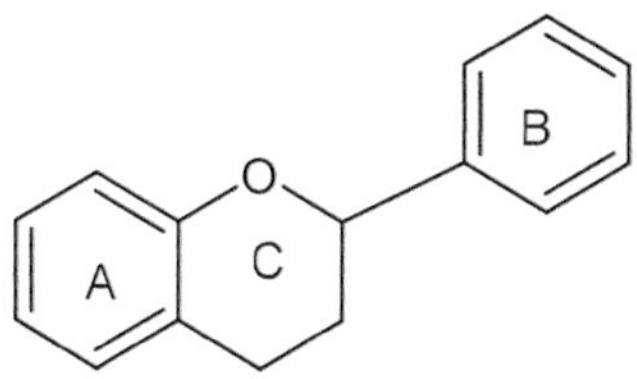

FIGURE 6.1 The flavan nucleus.

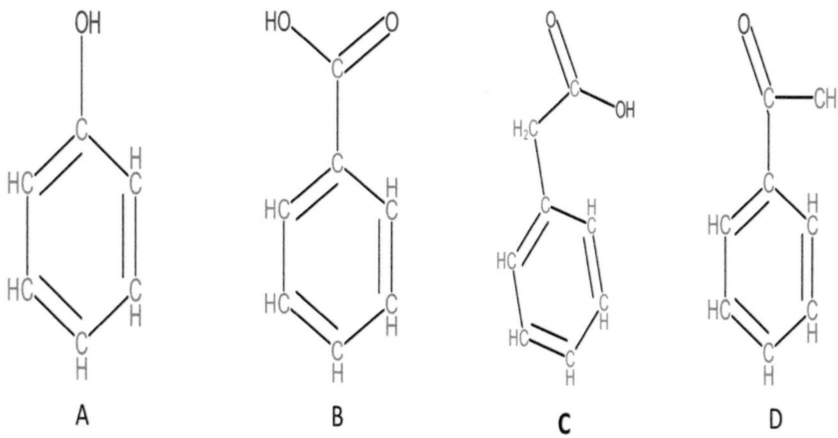

FIGURE 6.2 The nonflavonoids group.

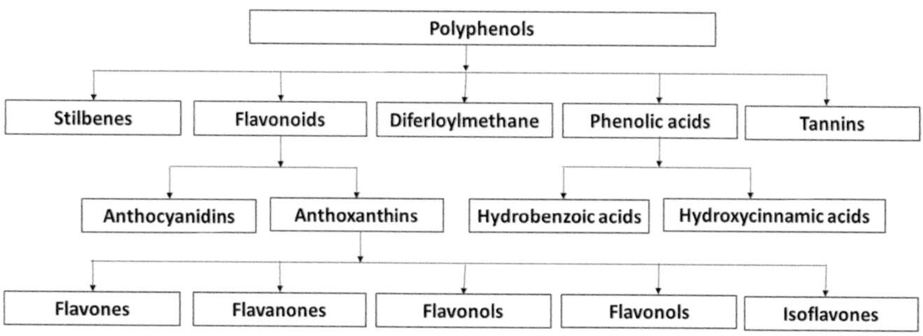

FIGURE 6.3 Classification of polyphenols.

The nonflavonoids group comprises the simple phenols with a C_6 aromatic ring substituted by an alcohol in one or more positions (Figure 6.2A), phenolic acids with a C_6–C_1 having a carboxyl attached to the benzene structure (Figure 6.2B), and the less common phenylacetic (Figure 6.2C) and acetophenones with the C_6–C_2 structure (Figure 6.2D).[54]

Several of these naturally occurring phenolic compounds have been isolated and shown to be efficacious in the prevention and treatment of different human diseases including neurodegenerative illnesses, with minimal side effects relative to the conventional therapeutic drugs.[90] For example, the intake of natural phenolic-rich foods such as vegetables, cereals, and olive oil has been linked with a lower incidence of dementia and Alzheimer's disease (AD).[118] Similarly, consumption of turmeric, a plant rich in the biophenol curcumin, has been associated with a markedly lower frequency of AD in the Indian population compared with the US population.[45] In addition, there is evidence that prolong consumption of tea may reduce the incidence of Parkinson's disease (PD).[10] Also, the ability of some polyphenols or polyphenol metabolites to cross the brain–blood barrier in animal models, albeit in small concentrations; further speaks to the potentials of plant-derived antioxidants in the prevention of neurodegenerative diseases.[22,41,60,61,94,125]

6.3 OXIDATIVE STRESS AND NEURODEGENERATIVE DISEASES

Neurodegenerative diseases involve chronic loss of function of the central nervous system (CNS), usually accompanied by a partial loss of function and the attendant mental disorders.[23] Prominent among the neurodegenerative diseases are AD, PD, dementia, and multiple sclerosis. It is now known that oxidative stress, neuroinflammation, certain viruses,[23] protein misfolding, cellular energy failure, and glutamate release[32] are involved in the mechanisms of neurodegenerative diseases.

Oxidative stress-mediated cellular damages in the CNS are involved in the pathogenesis of the neurodegenerative diseases. It is also worthy of note to state here that the brain is highly susceptible to oxidative stress because of the several metabolic activities in the brain. For instance, the brain is about 2% of the body weight but it utilizes about one-fifth of inspired oxygen for metabolism.[30] Furthermore, the brain has a low antioxidant capacity, for example, poor catalase and superoxide dismutase activities and lower levels of glutathione peroxidase and vitamin E compared to the liver.[30] Oxidative stress has been shown to be a trend in animal models of neurodegenerative illnesses.[20,22,54,65,102] On the other hand, antioxidants mitigate these conditions by different mechanisms that have been elucidated by several scientific investigators.[33,37,51,62,88] For example, antioxidant compounds of plant origin elicit their protective properties in the biological system by inhibiting oxidants (reactive species) and thus delay or prevent oxidation of biomolecules like DNA, proteins, and lipids.[3,39]

6.3.1 ALZHEIMER'S DISEASE (AD)

AD is the main cause of dementia which is attendant mainly with memory loss, and other impairments of normal living in humans.[22,98] AD occurrence increases with advancement in age. It has been demonstrated with data that AD is more likely to occur from age 65 years and above.[22,57,103] The disease presents in two forms including the sporadic type which is the commonest and accounts for about 90% of all AD occurrences; and the less common hereditary type which accounts for about less than 5% of the disease occurrence.[129]

AD is characterized by imbalances in cortical levels of neurotransmitters and toxic effects of overexposure of the cortical neurons to glutamate.[11,49] There are four major hypotheses put forward to explain the pathophysiology of AD. These are Amyloid-β (Aβ) or Amyloid cascade hypothesis, tau hypothesis, cholinergic hypothesis, and oxidative hypothesis. The Aβ theory suggests that the deposition of Aβ peptides results in the formation of senile plaques or amyloid plaques.[19,22,55,82,133] On the other hand, the tau hypothesis suggests that abnormal phosphorylation of tau proteins results in the formation of neurofibrillary tangles, loss of neurons, and loss of synapses.[68,114] The Aβ and tau proteins are important in the pathogenesis of AD; being the constituents of the usual histological features of the disease. The former is the constituent of the senile plagues, while the latter is the constituent of the neurofibrillary tangles.[18] The cholinergic hypothesis involves the loss of cholinergic neurotransmission due to the degeneration of cholinergic neurons in the forebrain believed to be mediated by upregulated activities of acetylcholinesterase (AChE) and butylcholinesterase (BChE); enzymes responsible for normal physiological maintenance of choline levels. Three main paths have been elucidated to show the contribution of oxidative stress to the progression of AD. These include Aβ-metal ion (copper, zinc, etc.) redox potential, macromolecule (DNA, lipid, protein, etc.) oxidation, and mitochondrial dysfunction.[22]

There is no approved drug or treatment for the cure of AD.[22,25,90] However, the US Food and Drug Administration has approved some drugs for only symptomatic treatments. These include the cholinesterase inhibitors (ChEIs) rivastigmine and donepezil; and the galantamine/N-methyl-D-aspartate (NMDA) antagonist memantine.[40] The ChEIs act by inhibiting AChE and BChE to improve the levels of the neurotransmitter acetylcholine (ACh) and butylcholine (BCh)[11]; while on the other hand, memantine ameliorate the effect of overexpression of glutamate.[49] It is noteworthy to mention that in addition to the fact that these drugs do not cure AD, they also elicit substantial negative side effects which include symptoms such as tremor, nausea/vomiting, diarrhea, muscle weakness, hepatotoxicity, hallucination, confusion, fatigue, dizziness, headache, and constipation.[5,124] These downsides in the pharmacotherapy of AD have motivated interest in finding alternative treatments of AD. Some of the alternative disease treatment or management options include lifestyle and dietary habit modifications, and investigation of plant-sourced active principles in the treatment of AD.

6.3.2 PARKINSON'S DISEASE (PD)

PD was first described by James Parkinson over 200 years ago.[59] PD ranks second after AD in the prevalence of human neurodegenerative diseases.[116] The familial form of PD accounts for about 10% of the patients; while the sporadic for of the disease in the most common.[36] PD risk factors for humans include exposure to environmental toxins like paraquat and rotenone.[8,21,29,48,85] There is documentary evidence that the incidence and progression of PD are higher in men than women.[7,99,105,117,130] It has been postulated that these differences may be sex hormone related.

PD is characterized by the following symptoms: difficulties in movement, rigidity of the limbs, resting tremor, and instability of posture.[36,83] The disease involves the impairment of the midbrain's *substantia nigra pars compacta* (SNpc) due to the loss of dopaminergic neurons. PD is essentially the apoptosis of dopaminergic neurons in the *substantia nigra* mainly via oxidative stress.[58] It has also been shown that insoluble Lewy-body inclusions, which are rich in fibrillar forms of the α-synuclein (α Syn) protein, are usually present in PD. α-synuclein is a soluble presynaptic protein whose biological function is not well elucidated yet. They are thought to be involved with the recycling and storage of neurotransmitters.[71] The aggregation of αSyn is thought to precipitate the formation of Lewy-body in PD.[59,123] Oxidative stress, mitochondrial damage, neuroinflammation, and αSyn aggregation have been implicated in the onset and progression of PD.[9,13,15,112,123,134,139]

Similar to AD, the available pharmacotherapies for PD do not target or reverse the underlying neurodegeneration condition but only provide temporary relief of the symptoms.[59] An outstanding therapy amongst the current management options available for PD is dopamine replacement therapy. This involves the use of medications such as L-DOPA and dopamine receptor agonists that help to improve the levels of the neurotransmitter dopamine. However, these therapies have their downsides which include such deleterious side effects as dyskinesia, aggression, and insomnia.[72]

6.4 PLANT-BASED ANTIOXIDANTS VERSUS HUMAN NEURODEGENERATIVE ILLNESS

In the meantime, there are no approved therapies for cure of the human neurodegenerative diseases. All of the currently available pharmacotherapies only at best assuage the disease symptoms but do not stop or reverse the neurodegeneration condition. This limitation has opened a vista of research into better elucidation of the mechanisms of the neurodegenerative disorders and the development of pharmacotherapies that can address the underlying degeneration condition.[36] As a fallout of these efforts, there has been an increased research focus on dietary-derived materials; especially plant-sourced, for the prevention of and/or slowdown of the onset of the neurodegenerative diseases.[31,46,75,78,93,106,120] As a result, several investigators have demonstrated that plant-based phenolic compounds have potent antioxidant effects against neurodegenerative disorders in humans and thus may improve the outcomes of neurological disorders such as AD, PD, and dementia.[115,128] The conclusion of these studies has suggested the possibility of plant-sourced dietary supplementation as an effective prevention for human neurodegenerative diseases.

Lately, research in the area of plant-sourced antioxidants for the prevention and treatment of human neurodegenerative disorders has yielded very positive results with the discovery of efficacious compounds with fewer or milder side effects and better tolerability. Different animal models have been used to demonstrate the potential benefits of plant-sourced antioxidants in the treatment of human neurodegenerative diseases. The plant-sourced antioxidants have multiple biological properties that give them an advantage as good candidates for the treatment of human neurodegenerative disorders.[67]

Several studies have demonstrated that these antioxidants have in addition to their antioxidant capacity; free radical scavenging property, enzyme-modulating ability, and metal ion chelating ability.[6,97,110] Conversely, some scientific researchers have also reported that certain plant-derived principles can enhance oxidative stress (i.e., act as pro-oxidants) and thus may exacerbate neurodegeneration.[84] This suggests that dose- and time-dependent considerations are important when studying any plant-sourced antioxidant as potential therapeutic agent in the treatment of neurodegenerative disorders. In addition, some research reports have also shown no relationship between antioxidant intakes and some human neurodegenerative diseases.[75,79]

Luchsinger et al.[80] reported that they found no relationship between the incident of AD and antioxidant in the Washington Heights Aging Project. Other issues that make the benefits offered by antioxidant molecules derived from medicinal plants in the treatment and prevention of neurodegenerative diseases contestable include factors such as poor in vivo bioavailability, rapid systemic elimination, and selective permeability of the blood–brain barrier in the human system.[94]

Therefore, the authors of this chapter have presented this review of several research reports on therapeutic potential of plant-based antioxidants in the human neurodegenerative diseases using different animal models, based on therapeutic targets or mode of action. Table 6.1 is a summary of the discussed research reports.

6.4.1 α-SYNUCLEIN AND AMYLOID-B FIBRIL TARGETS

The Aβ and tau proteins are important in the pathogenesis of AD; being the constituents of the usual histological features of the disease. The former is the constituent of the senile plagues, while the latter is the constituent of the neurofibrillary tangles.[17] It has been shown by several researchers that polyphenols interfere with Aβ formation, α-synuclein aggregation, and hyperphosphorylation of tau proteins.[28]

Bieschke et al. using PC12 cells (rat pheochromocytoma, American Type Culture Collection) demonstrated that epigallocatechin gallate (EGCG), a polyphenol present in green tea,[64,131] inhibited α-synuclein, and amyloid-β fibril formation. Their work showed that EGCG by directly binding to large α-synuclein and amyloid-β fibrils and transform them into smaller amorphous forms that are nontoxic to mammalian cells.[14] This ability of EGCG to alter the structure of mature amyloid-β sheets and thus prevent the characteristic aggregation of Aβ peptides implicated in AD; provides a therapeutic target in the treatment of the disease. However, it is important to elucidate whether this ability of EGCG to change the conformation of amyloid fibrils, will not increase cellular amyloid toxicity because of increased the amount of oligomeric amyloid proteins in the cell.[52] This may be a downside for EGCG

TABLE 6.1 Summary of Potential Neuroprotective Properties of Some Plant-Derived Antioxidants.

Plant source	Active compound	Animal/disease model	Mode of action or Target	Reference
Green tea	EGCG	Rat pheochromocytoma/AD	α-synuclein and amyloid-β fibril	14
		AD Transgenic mice	α-secretase activity	108
		MPTP-induced AD in mice	Fe and α-synuclein accumulation	81
		MPTP-induced PD in C57B6 mice	nNOS activity	24
Salvia lavandulifolia	Essential oil	H_2O_2 treated PC12 cells	Cell death/ caspase-3 activity	101
Panax ginseng	Ginseng saponins	AD induced in Sprague–Dawley rats	RAGE and NF-κB	126
		MPTP-treated C57BL/6J mice	Wnt/β-catenin signaling pathway	140
Curcuma longa (turmeric)	Curcumin	PC12 rat cells	Aβ peptides and tau proteins	95
		AD Tg2576 mice brain sections	Aβ fibrils	91,138
		Hippocampal HT22 rat cell lines	Formation of fibrils	66
		Sprague–Dawley rats hippocampus	Synaptic plasticity	1
Alchemilla vulgaris and *Filipendula ulmaria*	Plant extracts	In vivo system	Inhibition of AChe and tyrosinase	87
Allanblackia onticola fruits	Garcinol	In vitro system	Inhibition of AChe and BChe	74
Asteraceae family	*Pulicaria incisa,* infusions	LPS-treated microglial cell culture	IL-1β and IL-6 transcription	38
Pine trees/cocoa bushes	Resveratrol	Colchicine-induce AD rats	LPO and cognitive functions	69
Several plants	Catechin, morin, and quercetin	MG treated P5 C57/BL6 mice	MG and GSH, glo 1 activity	42
Several Plants	Chrysin	6-OHDA-induced PD in mouse	IL-1β, TNFα, NF-κB, and IL-10	47
Silybum marianum plant	Silymarin	Mouse model of MPTP-induced PD	Rescue striatal dopamine loss	96

as a treatment for neurodegenerative disorders as shown for other similar molecules that also inhibit amyloid formation.[27,121]

Rezai-Zadeh et al.[108] reported that EGCG reduced the generation of Aβ peptides in Alzheimer's transgenic mice. This work demonstrated that green tea EGCG elevated α-secretase activity, thus promoting the cleavage of amyloid precursor protein (APP), and consequently reducing Aβ levels and Aβ plaques in the brain.[108] The overexpression of APP and its subsequent proteolysis is key for the production of Aβ peptide which is implicated in AD.[44]

Mandel et al.[81] using 1-methyl-4-phenyl-1,2,3,6-tetrahydropyridine (MPTP)-treated mice reported that oral administration of EGCG prevented iron and α-synuclein accumulation via its antioxidant and iron-chelating properties. Their conclusion suggested that EGCG may offer neuroprotection through its radical scavenging and iron chelating properties. Although this iron chelating ability of EGCG may be a prospective therapeutic approach in neuroprotection, it is desirable to investigate that if there are any additional effects on the cell-signaling or cell-survival pathways to ensure that it does not cause any deleterious imbalance of brain iron metabolism.

Another well-studied plant antioxidant, curcumin; a polyphenolic derived from the rhizome of *Curcuma longa* or the spice turmeric, prevented Aβ toxicity and ameliorated the hyperphosphorylation of tau proteins in PC12 rat cells and umbilical endothelial cells.[95] Curcumin also destabilized preformed Aβ fibrils in a dose-dependent manner in Tg2576 mice brain sections[91,138], and slowed down the formation of fibrils in HT22 cells (a rat cell line from the hippocampus) in an analog-dependent manner.[66] In addition, curcumin restored synaptic plasticity in Aβ-peptide-exposed hippocampal slices from male Sprague–Dawley rats.[1]

Furthermore, curcumin has been found to have neuroprotective effect in different models of neurotoxicity; for example, in 3-Nitropropionic acid (3-NP)-induced cognitive impairment in rats,[70] in human immunodeficiency virus type 1 glycoprotein 120 V3 loop peptide-induced spatial memory impairment in rats,[127] and against glutamate excitotoxicity in rat cerebral cortex.[135] These findings have highlighted the above findings and many other authors have suggested that curcumin may be of value in the pharmacotherapy of AD and PD. However, there is a downside to this potential of curcumin due to its poor bioavailability because it is poorly absorbed and easily undergoes glucuronidation in the intestinal wall and the liver.[18,109] This poor bioavailability of curcumin suggests that larger oral dose or prolong time of administration may be necessary to derive its full benefit in the treatment of the disease. Several research studies have been conducted for the prospects of augmented delivery of curcumin to improve its bioavailability.[22]

Some delivery systems that have been explored to improve bioavailability of curcumin include nanoparticles, liposomes, and micelles.[4,107]

6.4.2 OXIDATIVE STRESS AND IMMUNOMODULATORY TARGETS

Oxidative stress has been shown to be an important contributor to the development of the human neurodegenerative diseases.[22] Since oxidative stress-mediated cellular damages are involved in the pathogenesis of the neurodegenerative diseases,[17,45,57,83] it is logical to devise treatment strategies that target the prevention of oxidative stress conditions in neurodegenerative illnesses.

The green tea containing EGCG improved MPTP-induced PD by inhibiting neuronal nitric oxide synthase (nNOS) activity in male C57B6 mice as reported by Choi et al.[24] The expression of nNOS was decreased in the *substantia nigra* an indication that EGCG may offer neuroprotection via an antioxidant mechanism. This finding by Choi et al.[24] explains the mechanism of action involved in the previously reported work of Levities et al.[76] in which EGCG mitigated MPTP-induced loss of dopamine neuron in the *substantia nigra.* It is worthy of note to state here that NO has established anti-inflammatory and anticancer properties.[24] Therefore, the regular consumption of green tea may lead to chronic inhibition of NOS and NO production which may impair cellular anti-inflammatory functions with potential for inflammatory disorders and incidence of cancers; albeit minimal due to the anti-inflammatory properties of green tea EGCG itself.[63,86]

Furthermore, essential oils obtained from *Salvia lavandulifolia*, a perennial herb indigenous to the mountainous areas of East Spain, North Africa, and South France have been reported to possess neuroprotective properties against oxidative stress-mediated neuronal injury using H_2O_2 treated PC12 cells as an experimental model.[101] In this study, essential oils from *S. lavandulifolia* elicited neuroprotection by preventing cell death through upregulation of antioxidant activity, caspase-3 activity, and decrease in lipid peroxidation.[101] However, the efficacy of *S. lavandulifolia* is limited with the phenological stage of the plant, as demonstrated by the fact that better results were obtained when the plant used was at the flowering stage or obtained from high-density cultivation of the plant.[100,101]

Panax ginseng, a species of the popular perennial medicinal plant Ginseng have also been shown to have a beneficial effect in the prevention of human neurodegenerative illness mainly through the antioxidative and immunomodulatory properties of it active principles—the ginsenosides or ginseng saponins.[67]

P. ginseng mitigated AD pathophysiological changes in male Sprague–Dawley rats, by down-regulating expressions of receptors for advanced glycation end products (RAGE), and nuclear factor-kappa-light-chain-enhancer of activated B cell (NF-κB).[126] Similarly, water extract of *P. ginseng* provided neuroprotection in MPTP-induced PD C57BL/6J mice model through the Wnt/β-catenin signaling pathway including Wnt-1, β-catenin, GSK-3β, and p-GSK-3β.[140] Although there are abounding evidence supporting the benefit of *P. ginseng* treatment of neurodegenerative diseases, it has also been reported that ginseng may interfere with blood hemostasis and anticoagulation.[89]

Furthermore, the flavonoids—resveratrol, morin, rutin, quercetin, and myricetin are other plant-derived compounds that have received research attention in the treatment of the human neurodegenerative disorders. Resveratrol, a flavonoid from pine trees and cocoa bushes has been shown to elicit neuroprotection by reducing lipid peroxidation and enhancing cognitive functions when orally administered to colchicine-induce AD rats.[69] Similarly, myricetin, quercetin, morin, rutin, and apigenin all possess antiamyloidogenic and Aβ fibril-destabilization properties in in vitro models of AD.[92] These effects are linked to their reactive oxygen species (ROS) scavenging capacities. However, there are cons in the use of these flavonoids for the treatment of AD because it has been suggested that their potential usefulness may be limited to the early stage of the disease (limiting their therapeutic window).[22,34] They become cytotoxic at higher concentrations due to their autoxidation to toxic quinones,[32] and they have issues of bioavailability in vivo necessitating prolonged use or supplementation to produce the desired neuroprotection.[32]

Another important antioxidant pathway and potential drug target in human neurodegenerative disorder are the glyoxalase pathway (Figure 6.4).[43] The glyoxalase pathway function to control cellular dicarbonyl stress occasioned by an accumulation of reactive dicarbonyl metabolites leading to DNA and protein modifications.[104] The glyoxalase pathway is present in all body cells and provides antioxidant defense by eliminating highly reactive dicarbonyl compounds, for example, methylglyoxal (MG); that cause cellular inflammation and oxidative stress.[2,42,104,122] MG is a cytotoxic membrane-permeable reactive carbonyl species produced as a glycolytic pathway by-product.[2] The glyoxalase pathway involves two reactions catalyzed by glyoxalase 1 (glo 1) and glyoxalase 2 (glo 2) to neutralize MG through its reaction with reduced glutathione (GSH) to produce D-lactate.[12,42,50] The glyoxalase pathway is very important in the brain because the brain has a very high rate of metabolism, low antioxidant capacity (e.g., poor catalase and superoxide dismutase activities), and high concentration of easily oxidizable biomolecules (e.g., long-chain polyunsaturated fatty acids) that can be attacked by reactive carbonyls species.[30,50,51,137] The accumulation of methylglyoxal (MG) due to glyoxalase pathway dysfunction has been implicated in some human neurodegenerative diseases.[43,104,122]

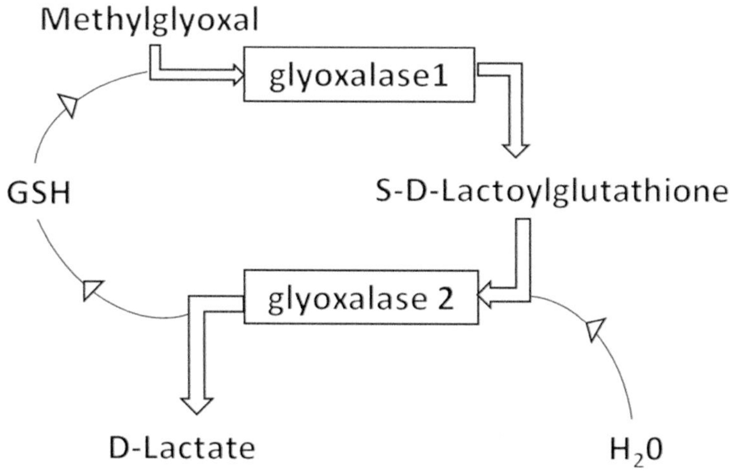

FIGURE 6.4 The glyoxalase pathway.

Hence, plant active principles that have the capacity to enhance the neuronal glyoxalase pathway can be good neuroprotection candidates in drug development for the treatment of neurodegenerative diseases. A number of flavonoids have been reported to be effective in the modulation of the glyoxalase pathway.[43] Catechin, morin, and quercetin attenuated the effects of MG in cerebellar neurons from P5 C57/BL6 mice by reducing the concentration of MG, increasing the activity of glyoxalase 1, and increasing the concentration of GSH.[42]

Another therapeutic strategy in combating neurodegenerative disease is the inhibition of microglial activation and the suppression of the neuroinflammatory response it mediates.[38] Activation of microglial cells results in the secretion of immune modulators such as inflammatory cytokines tumor necrosis factor alpha (TNFα) and interleukin 1β (IL-1β), and neurotoxins such as glutamate, nitric oxide (NO), and ROS.[16] Therefore, plant-derived active principles that have anti-inflammatory/antioxidant properties may have prospects for the treatment and prevention of neurodegenerative diseases by targeting microglial cell activation. For example, infusion of the medicinal plant *Pulicaria incisa,* suppressed IL-1β transcription and secretion, and IL-6 secretion when added to lipopolysaccharide (LPS)-treated microglial cell culture. Similarly, infusion of *P. incisa* ameliorated NO toxicity induced by activation of microglial cells by LPS.[38] This study[38] demonstrated that *P. incisa* infusion has anti-inflammatory effects on activated microglial cell and thus maybe relevant in development of pharmacotherapy for neurodegenerative diseases. *P. incisa* is a medicinal plant that belongs to the *Asteraceae* family.

Chrysin, a natural flavonoid present in various plants, have also been shown to be neuroprotective through its anti-inflammatory properties. In a mouse model of 6- hydroxidopamine (6-OHDA)-induced chronic PD, oral administration of chrysin over a 28-day period reversed the 6-OHDA-induced elevated striatal proinflammatory IL-1β, TNFα, and decreased anti-inflammatory IL-10 levels. In addition, chrysin reduced the increase of NF-κB levels caused by 6-OHDA.[47]

6.4.3 CHOLINESTERASE AND CHOLINESTERASE INHIBITORS TARGETS

A major treatment option of the human neurodegeneration diseases, specifically AD, is to target the neurotransmitter of the cholinergic system. This may involve introduction of choline (Ch) precursors, Ch-releasing agents, and ChEIs to improve the concentration of choline and thereby restore cholinergic neurotransmission functions.[67,74] Several investigators have reported the potentials of some plant-sourced natural active principles as efficient ChEIs with the ability to inhibit the ChEs and thus improve the levels of respective neurotransmitters. In this regard, acetylcholinesterase (AChe) and butylcholinesterase (BChe) are attractive targets in the research and development of pharmacotherapies for neurodegenerative diseases specifically AD.

Lenta et al.[74] reported that garcinol, a benzophenone isolated from the Cameroonian *Allanblackia monticola* fruits, showed potent in vitro ChEI properties on AChe and BChe. They reported that garcinol was active against AChe as about the activity of galathamine, the reference drug used for the study.

Silymarin, a complex of flavonolignans present in seeds of the *Silybum marianum* plant, have also been shown to demonstrate neuroprotective properties. Silymarin in a dose-dependent manner rescued striatal dopamine loss in a mouse model of MPTP-induced PD.[96] However, the mechanism by which Silymarin elicit it neuroprotective effect has not been fully elucidated. Also, in the work of Pérez-H et al.,[96] Silymarin was used intraperitoneally which may limit its potential in human use. Studies involving oral route of administration are necessary to determine the potential clinical use of Silymarin in humans.

Furthermore, two Romanian medicinal plants, *Alchemilla vulgaris* and *Filipendula ulmaria*, have been reported to possess neuroprotective potentials. Extracts from these plants exhibited AChe inhibitory and tyrosinase inhibitory properties in vivo.[87] This observation suggests that these plants have potential therapeutic value in the treatment of both AD and PD.

6.5 SUMMARY

At present the drugs available for the management of human neurodegenerative diseases do not actually target or reverse the degeneration condition. Therefore, alternative therapies that target or reverse the degeneration condition and not just only assuage the disease symptoms are desirable. Hence, the trend toward research into the potentials of plant-sourced agents for the prevention and treatment of human degenerative diseases. From our preceding discussions, it can be concluded that plant-sourced antioxidants have potential neuroprotective properties in various animal models of the human neurodegenerative disorders. This natural approach may help in addressing the limitations associated with the current synthetic drugs.

Hence, the health-promoting effects and potentials of medicinal plants for the prevention and treatment of neurodegenerative diseases are no longer a new knowledge. The new challenge is to develop acceptable standards for the testing of the several known beneficial medicinal plants to determine their safe dosage range and define their therapeutic usage in the treatment of human diseases. Therefore, future research is desirable into the safe use of the active principles; especially antioxidants from medicinal plants as neuroprotective agents in the treatment of human neurodegenerative diseases. That is, the plant-sourced agents should be effective in targeting/reversing the mechanisms of neurodegenerative diseases based on the European Medicines Agency's definition of neuroprotection,[26] and at the same time safe to the human. The latter is important because doses usually used in the in vitro studies are generally high, and some of the studies have shown that high concentrations of the plant-sourced active principle are inimical to cell viability or produce other toxic effects. This prooxidant effect of plant-sourced antioxidants may compromise their neuroprotective roles in the management of human neurodegenerative diseases.[22,84] Therefore, a better understanding of their mechanisms of action, optimal dose, most appropriate stage of the disease to administer the antioxidant agents, and effective delivery or bioavailability in vivo a very important for the application of plant-sourced molecules in the treatment of neurodegenerative diseases. Research efforts geared toward the advancement of in vivo bioavailability of identified beneficial plant-sourced active principles will help in drug discovery and therapeutic viability of medicinal plants in the treatment and prevention of neurodegenerative diseases.

In conclusion, it is thought that the development of better experimental models that sufficiently mimic the initiation/progression of human neurodegenerative diseases on the one hand; and on the other hand, the understanding of metabolism and bioavailability of the active principle of medicinal plants will help in the investigation and identification of medicinal plants that are safe and readily available neuroprotective agents against human neurodegenerative diseases.

KEYWORDS

- **antioxidants**
- **medicinal plants**
- **neurodegenerative diseases**
- **neuroprotection**
- **oxidative stress**

REFERENCES

1. Ahmed, T.; Gilani, A. H.; Hosseinmardi, N.; Semnanian, S.; Enam, S. A.; Fathollahi, Y. Curcuminoids Rescue Long-Term Potentiation Impaired by Amyloid Peptide in Rat Hippocampal Slices. *Synapse* **2011**, *65*(7), 572–582.
2. Allaman, I.; Bélanger, M.; Magistretti, P. J. Methylglyoxal, the Dark Side of Glycolysis. *Front. Neurosci.* **2015,** *9*, 23–30.
3. Amarowicz, R.; Shahidi, F. Antioxidant Activity of Broad Bean Seed Extract and Its Phenolic Composition. *J. Funct. Foods* **2017,** *38*, 656–662.
4. Anand, P.; Kunnumakkara, A. B.; Newman, R. A.; Aggarwal, B. B. Bioavailability of Curcumin: Problems and Promises. *Mol. Pharm.* **2007,** *4*(6), 807–818.
5. Arrieta, J. L.; Artalejo, F. R. Methodology, Results and Quality of Clinical Trials of Tacrine in the Treatment of Alzheimer's Disease: A Systematic Review of the Literature. *Age Ageing* **1998,** *27*(2), 161–179.

6. Assunção, M.; Santos-Marques, M. J.; Carvalho, F.; Andrade, J. P. Green Tea Averts Age-Dependent Decline of Hippocampal Signaling Systems Related to Antioxidant Defenses and Survival. *Free Radic. Biol. Med.* **2010,** *48*(6), 831–838.
7. Baldereschi, M.; Di Carlo, A.; Rocca, W.; Vanni, P.; Maggi, S.; Perissinotto, E.; Grigoletto, F.; Amaducci, L.; Inzitari, D. Parkinson's Disease and Parkinsonism in a Longitudinal Study: Two-Fold Higher Incidence in Men. *Neurology* **2000,** *55*(9), 1358–1363.
8. Baldi, I.; Cantagrel, A.; Lebailly, P.; Tison, F.; Dubroca, B.; Chrysostome, V.; Dartigues, J. F.; Brochard, P. Association Between Parkinson's Disease and Exposure to Pesticides in Southwestern France. *Neuroepidemiology* **2003,** *22*(5), 305–310.
9. Banerjee, R.; Starkov, A. A.; Beal, M. F.; Thomas, B. Mitochondrial Dysfunction in the Limelight of Parkinson's Disease Pathogenesis. *Biochim. Biophys. Acta Mol. Basis Dis.* **2009,** *1792*(7), 651–663.
10. Barranco Quintana, J. L.; Allam, M. F.; Del Castillo, A. S.; Navajas, R. F. n. -C. Parkinson's Disease and Tea: A Quantitative Review. *J. Am. Coll. Nutr.* **2009,** *28*(1), 1–6.
11. Bartus, R. T.; Dean, R. r.; Beer, B.; Lippa, A. S. The Cholinergic Hypothesis of Geriatric Memory Dysfunction. *Science* **1982,** *217*(4558), 408–414.
12. Bélanger, M.; Yang, J.; Petit, J. -M.; Laroche, T.; Magistretti, P. J.; Allaman, I. Role of the Glyoxalase System in Astrocyte-Mediated Neuroprotection. *J. Neurosci.* **2011,** *31*(50), 18338–18352.
13. Betarbet, R.; Sherer, T. B.; MacKenzie, G.; Garcia-Osuna, M.; Panov, A. V.; Greenamyre, J. T. Chronic Systemic Pesticide Exposure Reproduces Features of Parkinson's Disease. *Nat. Neurosci.* **2000,** *3*(12), 1301–1306.
14. Bieschke, J.; Russ, J.; Friedrich, R. P.; Ehrnhoefer, D. E.; Wobst, H.; Neugebauer, K.; Wanker, E. E. Egcg Remodels Mature A-Synuclein and Amyloid-B Fibrils and Reduces Cellular Toxicity. *Proc. Natl. Acad. Sci.* **2010,** *107*(17), 7710–7715.
15. Block, M. L.; Zecca, L.; Hong, J. -S. Microglia-Mediated Neurotoxicity: Uncovering the Molecular Mechanisms. *Nat. Rev. Neurosci.* **2007,** *8*(1), 57–69.
16. Boche, D.; Perry, V. H.; Nicoll, J. A. R. Review: Activation Patterns of Microglia and Their Identification in the Human Brain. *Neuropathol. Appl. Neurobiol.* **2013,** *39*(1), 3–18.
17. Boudet, A. -M. Evolution and Current Status of Research in Phenolic Compounds. *Phytochemistry* **2007,** *68* (22–24), 2722–2735.
18. Brondino, N.; Re, S.; Boldrini, A.; Cuccomarino, A.; Lanati, N.; Barale, F.; Politi, P. Curcumin as a Therapeutic Agent in Dementia: A Mini Systematic Review of Human Studies. *Sci. World J.* **2014,** *2014*, 23–29.
19. Bush, A. I. The Metallobiology of Alzheimer's Disease. *Trends Neurosci.* **2003**, *26*(4), 207–214.
20. Butterfield, D. A.; Boyd-Kimball, D. Oxidative Stress, Amyloid-B Peptide, and Altered Key Molecular Pathways in the Pathogenesis and Progression of Alzheimer's Disease. *J. Alzheimers Dis.* **2018,** *62*(3), 1345–1367.
21. Butterfield, P.; Valanis, B.; Spencer, P.; Lindeman, C.; Nutt, J. Environmental Antecedents of Young-Onset Parkinson's Disease. *Neurology* **1993,** *43*(6), 1150–1150.
22. Cassidy, L.; Fernandez, F.; Johnson, J. B.; Naiker, M.; Owoola, A. G.; Broszczak, D. A. Oxidative Stress in Alzheimer's Disease: A Review on Emergent Natural Polyphenolic Therapeutics. *Complement. Ther. Med.* **2020,** *49*, 102294.
23. Chen, W. W.; Zhang, X.; Huang, W. J. Role of Neuroinflammation in Neurodegenerative Diseases. *Mol. Med. Rep.* **2016,** *13*(4), 3391–3396.

24. Choi, J. -Y.; Park, C. -S.; Kim, D. -J.; Cho, M. -H.; Jin, B. -K.; Pie, J. -E.; Chung, W. -G. Prevention of Nitric Oxide-Mediated 1-Methyl-4-Phenyl-1,2,3,6-Tetrahydropyridine-Induced Parkinson's Disease in Mice by Tea Phenolic Epigallocatechin 3-Gallate. *Neurotoxicology* **2002,** *23*(3), 367–374.
25. Citron, M. Alzheimer's Disease: Strategies for Disease Modification. *Nat. Rev. Drug Discov.* **2010**, *9*(5), 387–398.
26. Clarke, C.E. Are Delayed-Start Design Trials to Show Neuroprotection in Parkinson's Disease Fundamentally Flawed? *Mov. Disord.* **2008,** *23*(6), 784–789.
27. Cohen, F. E.; Kelly, J. W. Therapeutic Approaches to Protein-Misfolding Diseases. *Nature* **2003,** *426*(6968), 905–909.
28. Costa, C.; Tsatsakis, A.; Mamoulakis, C.; Teodoro, M.; Briguglio, G.; Caruso, E.; Tsoukalas, D.; Margina, D.; Dardiotis, E.; Kouretas, D. Current Evidence on the Effect of Dietary Polyphenols Intake on Chronic Diseases. *Food Chem. Toxicol.* **2017,** *110*, 286–299.
29. Costello, S.; Cockburn, M.; Bronstein, J.; Zhang, X.; Ritz, B. Parkinson's Disease and Residential Exposure to Maneb and Paraquat from Agricultural Applications in the Central Valley of California. *Am. J Epidemiol.* **2009,** *169*(8), 919–926.
30. Cui, K.; Luo, X.; Xu, K.; Murthy, M. V. Role of Oxidative Stress in Neurodegeneration: Recent Developments in Assay Methods for Oxidative Stress and Nutraceutical Antioxidants. *Prog. Neuropsychopharmacol. Biol. Psychiatry* **2004,** *28*(5), 771–799.
31. Dai, Q.; Borenstein, A. R.; Wu, Y.; Jackson, J. C.; Larson, E. B. Fruit and Vegetable Juices and Alzheimer's Disease: The Kame Project. *Am. J. Med.* **2006,** *119*(9), 751–759.
32. Dajas, F.; Abin-Carriquiry, J. A.; Arredondo, F.; Blasina, F.; Echeverry, C.; Martínez, M.; Rivera, F.; Vaamonde, L. Quercetin in Brain Diseases: Potential and Limits. *Neurochem. Int.* **2015,** *89*, 140–148.
33. Danta, C. C.; Piplani, P. The Discovery and Development of New Potential Antioxidant Agents for the Treatment of Neurodegenerative Diseases. *Expert Opin. Drug Discov.* **2014,** *9*(10), 1205–1222.
34. de Andrade Teles, R. B.; Diniz, T. C.; Costa Pinto, T. C.; de Oliveira Júnior, R. G.; Gama e Silva, M.; de Lavor, É. M.; Fernandes, A. W. C.; de Oliveira, A. P.; de Almeida Ribeiro, F. P. R.; da Silva, A. A. M.; Cavalcante, T. C. F.; Quintans Júnior, L. J.; da Silva Almeida, J. R. G. Flavonoids as Therapeutic Agents in Alzheimer's and Parkinson's Diseases: A Systematic Review of Preclinical Evidences. *Oxid. Med. Cell. Longev.* **2018,** *2018*, 7043213.
35. de Queiroz Ferreira, R.; Greco, S.J.; Delarmelina, M.; Weber, K.C. Electrochemical Quantification of the Structure/Antioxidant Activity Relationship of Flavonoids. *Electrochim. Acta* **2015,** *163*, 161–166.
36. de Rus Jacquet, A.; Tambe, M. A.; Rochet, J. C. Dietary Phytochemicals in Neurodegenerative Disease. In *Nutrition in the Prevention and Treatment of Disease*; Coulston, A., Boushey, C., Ferruzzi, M., Delahanty, L., eds.; Academic Press: Cambridge, USA, 2017; pp 361–391.
37. Deng, J.; Cheng, W.; Yang, G. A Novel Antioxidant Activity Index (Aau) for Natural Products Using the Dpph Assay. *Food Chem.* **2011,** *125*(4), 1430–1435.
38. Elmann, A.; Beit-Yannai, E.; Telerman, A.; Ofir, R.; Mordechay, S.; Erlank, H.; Borochov-Neori, H. *Pulicaria incisa* Infusion Attenuates Inflammatory Responses of Brain Microglial Cells. *J. Funct. Foods* **2016,** *25*, 110–122.
39. Falcone Ferreyra, M. L.; Rius, S.; Casati, P. Flavonoids: Biosynthesis, Biological Functions, and Biotechnological Applications. *Front. Plant Sci.* **2012,** *3*, 222.

40. Farlow, M. R.; Miller, M. L.; Pejovic, V. Treatment Options in Alzheimer's Disease: Maximizing Benefit, Managing Expectations. *Dement. Geriatr. Cogn. Disord.* **2008,** *25*(5), 408–422.
41. Figueira, I.; Tavares, L.; Jardim, C.; Costa, I.; Terrasso, A. P.; Almeida, A. F.; Govers, C.; Mes, J. J.; Gardner, R.; Becker, J. D.; McDougall, G. J.; Stewart, D.; Filipe, A.; Kim, K. S.; Brites, D.; Brito, C.; Brito, M. A.; Santos, C. N. Blood–Brain Barrier Transport and Neuroprotective Potential of Blackberry-Digested Polyphenols: An in Vitro Study. *Eur. J. Nutr.* **2019,** *58*(1), 113–130.
42. Frandsen, J.; Narayanasamy, P. Flavonoid Enhances the Glyoxalase Pathway in Cerebellar Neurons to Retain Cellular Functions. *Sci. Rep.* **2017,** *7*(1), 1–10.
43. Frandsen, J. R.; Narayanasamy, P. Neuroprotection Through Flavonoid: Enhancement of the Glyoxalase Pathway. *Redox Biol.* **2018,** *14*, 465–473.
44. Funamoto, S.; Morishima-Kawashima, M.; Tanimura, Y.; Hirotani, N.; Saido, T. C.; Ihara, Y. Truncated Carboxyl-Terminal Fragments of B-Amyloid Precursor Protein Are Processed to Amyloid B-Proteins 40 and 42. *Biochemistry* **2004,** *43*(42), 13532–13540.
45. Ganguli, M.; Chandra, V.; Kamboh, M. I.; Johnston, J. M.; Dodge, H. H.; Thelma, B.; Juyal, R. C.; Pandav, R.; Belle, S. H.; DeKosky, S. T. Apolipoprotein E Polymorphism and Alzheimer Disease: The Indo-Us Cross-National Dementia Study. *Arch. Neurol.* **2000,** *57*(6), 824–830.
46. Gao, X.; Cassidy, A.; Schwarzschild, M.; Rimm, E.; Ascherio, A. Habitual Intake of Dietary Flavonoids and Risk of Parkinson Disease. *Neurology* **2012,** *78*(15), 1138–1145.
47. Goes, A. T.; Jesse, C. R.; Antunes, M. S.; Ladd, F. V. L.; Ladd, A. A. L.; Luchese, C.; Paroul, N.; Boeira, S. P. Protective Role of Chrysin on 6-Hydroxydopamine-Induced Neurodegeneration a Mouse Model of Parkinson's Disease: Involvement of Neuroinflammation and Neurotrophins. *Chem. Biol. Interact.* **2018,** *279*, 111–120.
48. Gorell, J. M.; Johnson, C.; Rybicki, B.; Peterson, E.; Richardson, R. The Risk of Parkinson's Disease with Exposure to Pesticides, Farming, Well Water, and Rural Living. *Neurology* **1998,** *50*(5), 1346–1350.
49. Greenamyre, J. T.; Maragos, W. F.; Albin, R. L.; Penney, J. B.; Young, A. B. Glutamate Transmission and Toxicity in Alzheimer's Disease. **1988,** *12*(4), 421–30.
50. Gu, F.; Chauhan, V.; Chauhan, A. Glutathione Redox Imbalance in Brain Disorders. *Curr. Opin. Clin. Nutr. Metab. Care* **2015,** *18*(1), 89–95.
51. Guerra-Araiza, C.; Álvarez-Mejía, A. L.; Sánchez-Torres, S.; Farfan-García, E.; Mondragón-Lozano, R.; Pinto-Almazán, R.; Salgado-Ceballos, H. Effect of Natural Exogenous Antioxidants on Aging and on Neurodegenerative Diseases. *Free Radic. Res.* **2013,** *47*(6–7), 451–462.
52. Haass, C.; Selkoe, D. J. Soluble Protein Oligomers in Neurodegeneration: Lessons From the Alzheimer's Amyloid B-Peptide. *Nat. Rev. Mol. Cell Biol.* **2007,** *8*(2), 101–112.
53. Halliwell, B.; Gutteridge, J. M. *Free Radicals in Biology and Medicine*. Oxford University Press: Oxford, USA, 2015; p 896.
54. Harborne, J. General Procedures and Measurement of Total Phenolics. *Methods Plant Biochem.* **1989,** *1*, 1–28.
55. Hardy, J. A.; Higgins, G. A. Alzheimer's Disease: The Amyloid Cascade Hypothesis. *Science* **1992,** *256*, 184–186.
56. Hernández-Rodríguez, P.; Baquero, L. P.; Larrota, H. R. Flavonoids: Potential Therapeutic Agents by Their Antioxidant Capacity. In *Bioactive Compounds*; Elsevier Inc.: Amsterdam, Netherlands, 2019; pp 265–288.

57. Huang, Y.; Mucke, L. Alzheimer Mechanisms and Therapeutic Strategies. *Cell* **2012,** *148*(6), 1204–1222.
58. Iriti, M.; Faoro, F., Health-Promoting Effects of Grape Bioactive Phytochemicals. In *Complementary and Alternative Therapies and the Aging Population*; Watson, R.; Ed.; Academic Press: Cambridge, USA, 2009; pp 445–474.
59. Jankovic, J.; Poewe, W. Therapies in Parkinson's Disease. *Curr. Opin. Neurol.* **2012,** *25*(4), 433–447.
60. Johnson, S. L.; Kirk, R. D.; DaSilva, N. A.; Ma, H.; Seeram, N. P.; Bertin, M. J. Polyphenol Microbial Metabolites Exhibit Gut and Blood–Brain Barrier Permeability and Protect Murine Microglia against Lps-Induced Inflammation. *Metabolites* **2019,** *9*(4), 78.
61. Kalt, W.; Blumberg, J. B.; McDonald, J. E.; Vinqvist-Tymchuk, M. R.; Fillmore, S. A.; Graf, B. A.; O'Leary, J. M.; Milbury, P. E. Identification of Anthocyanins in the Liver, Eye, and Brain of Blueberry-Fed Pigs. *J. Agric. Food Chem.* **2008,** *56*(3), 705–712.
62. Kasote, D. M.; Hegde, M. V.; Katyare, S. S. Mitochondrial Dysfunction in Psychiatric and Neurological Diseases: Cause (S), Consequence (S), and Implications of Antioxidant Therapy. *Biofactors* **2013,** *39*(4), 392–406.
63. Katiyar, S. K.; Agarwal, R.; Ekker, S.; Wood, G. S.; Mukhtar, H. Protection Against 12-O-Tetradecanoylphorbol-13-Acetate-Caused Inflammation in Sencar Mouse Ear Skin by Polyphenolic Fraction Isolated from Green Tea. *Carcinogenesis* **1993,** *14*(3), 361–365.
64. Khokhar, S.; Venema, D.; Hollman, P. C.; Dekker, M.; Jongen, W. A Rp-Hplc Method for the Determination of Tea Catechins. *Cancer Lett.* **1997,** *114*(1–2), 171–172.
65. Kim, G. H.; Kim, J. E.; Rhie, S. J.; Yoon, S. The Role of Oxidative Stress in Neurodegenerative Diseases. *Exp. Neurobiol.* **2015,** *24*(4), 325–340.
66. Kim, H.; Park, B. -S.; Lee, K. -G.; Choi, C. Y.; Jang, S. S.; Kim, Y. -H.; Lee, S. -E. Effects of Naturally Occurring Compounds on Fibril Formation and Oxidative Stress of Β-Amyloid. *J. Agric. Food Chem.* **2005,** *53*(22), 8537–8541.
67. Kim, K. H.; Lee, D.; Lee, H. L.; Kim, C. -E.; Jung, K.; Kang, K. S. Beneficial Effects of Panax Ginseng for the Treatment and Prevention of Neurodegenerative Diseases: Past Findings and Future Directions. *J. Ginseng Res.* **2018,** *42*(3), 239–247.
68. KoSIK, K. S.; Joachim, C. L.; Selkoe, D. J. Microtubule-Associated Protein Tau (Tau) Is a Major Antigenic Component of Paired Helical Filaments in Alzheimer Disease. *Proc. Natl. Acad. Sci.* **1986,** *83*(11), 4044–4048.
69. Kumar, A.; Naidu, P. S.; Seghal, N.; Padi, S. S. V. Neuroprotective Effects of Resveratrol Against Intracerebroventricular Colchicine-Induced Cognitive Impairment and Oxidative Stress in Rats. *Pharmacology* **2007,** *79*(1), 17–26.
70. Kumar, P.; Padi, S.; Naidu, P.; Kumar, A. Possible Neuroprotective Mechanisms of Curcumin in Attenuating 3-Nitropropionic Acid-Induced Neurotoxicity. *Methods Find. Exp. Clin. Pharmacol.* **2007,** *29*(1), 19–26.
71. Kurz, A.; Rabbani, N.; Walter, M.; Bonin, M.; Thornalley, P.; Auburger, G.; Gispert, S. Alpha-Synuclein Deficiency Leads to Increased Glyoxalase I Expression and Glycation Stress. *Cell. Mol. Life Sci.* **2011,** *68*(4), 721–733.
72. Lawrence, A. D.; Evans, A. H.; Lees, A. J. Compulsive Use of Dopamine Replacement Therapy in Parkinson's Disease: Reward Systems Gone Awry? *Lancet Neurol.* **2003,** *2*(10), 595–604.
73. Lee, K. W.; Lee, H. J. The Roles of Polyphenols in Cancer Chemoprevention. *Biofactors* **2006,** *26*(2), 105–121.

74. Lenta, B. N.; Vonthron-Sénécheau, C.; Weniger, B.; Devkota, K. P.; Ngoupayo, J.; Kaiser, M.; Naz, Q.; Choudhary, M. I.; Tsamo, E.; Sewald, N. Leishmanicidal and Cholinesterase Inhibiting Activities of Phenolic Compounds from Allanblackia Monticola and Symphonia Globulifera. *Molecules* **2007,** *12*(8), 1548–1557.
75. Letenneur, L.; Proust-Lima, C.; Le Gouge, A.; Dartigues, J. -F.; Barberger-Gateau, P. Flavonoid Intake and Cognitive Decline over a 10-Year Period. *Am. J Epidemiol.* **2007,** *165*(12), 1364–1371.
76. Levites, Y.; Weinreb, O.; Maor, G.; Youdim, M. B.; Mandel, S. Green Tea Polyphenol (–) Epigallocatechin-3-Gallate Prevents N-Methyl-4-Phenyl-1, 2, 3, 6-Tetrahydropyridine-Induced Dopaminergic Neurodegeneration. *J. Neurochem.* **2001,** *78*(5), 1073–1082.
77. Li, K.; Fan, H.; Yin, P.; Yang, L.; Xue, Q.; Li, X.; Sun, L.; Liu, Y. Structure-Activity Relationship of Eight High Content Flavonoids Analyzed with a Preliminary Assign-Score Method and Their Contribution to Antioxidant Ability of Flavonoids-Rich Extract from Scutellaria Baicalensis Shoots. *Arab. J Chem.* **2018,** *11*(2), 159–170.
78. Lindsay, J.; Laurin, D.; Verreault, R.; Hébert, R.; Helliwell, B.; Hill, G. B.; McDowell, I. Risk Factors for Alzheimer's Disease: A Prospective Analysis From the Canadian Study of Health and Aging. *Am. J Epidemiol.* **2002,** *156*(5), 445–453.
79. Luchsinger, J. A.; Mayeux, R. Dietary Factors and Alzheimer's Disease. *Lancet Neurol.* **2004,** *3*(10), 579–587.
80. Luchsinger, J. A.; Tang, M. -X.; Shea, S.; Mayeux, R. Antioxidant Vitamin Intake and Risk of Alzheimer Disease. *Arch. Neurol.* **2003,** *60*(2), 203–208.
81. Mandel, S.; Maor, G.; Youdim, M. B. Iron and A-Synuclein in the Substantia Nigra of Mptp-Treated Mice. *J. Mol. Neurosci.* **2004,** *24*(3), 401–416.
82. Markesbery, W. R. Oxidative Stress Hypothesis in Alzheimer's Disease. *Free Radic. Biol. Med.* **1997,** *23*(1), 134–147.
83. Massano, J.; Bhatia, K. P. Clinical Approach to Parkinson's Disease: Features, Diagnosis, and Principles of Management. *Cold Spring Harb. Perspect. Med.* **2012,** *2*(6), a008870.
84. Mazumder, M. K.; Choudhury, S.; Borah, A. An in Silico Investigation on the Inhibitory Potential of the Constituents of Pomegranate Juice on Antioxidant Defense Mechanism: Relevance to Neurodegenerative Diseases. *IBRO Rep.* **2019,** *6*, 153–159.
85. Menegon, A.; Board, P. G.; Blackburn, A. C.; Mellick, G. D.; Le Couteur, D. G. Parkinson's Disease, Pesticides, and Glutathione Transferase Polymorphisms. *Lancet* **1998,** *352*(9137), 1344–1346.
86. Mukhtar, H.; Ahmad, N. Tea Polyphenols: Prevention of Cancer and Optimizing Health. *Am. J. Clin. Nutr.* **2000,** *71*(6), 1698S–1702S.
87. Neagu, E.; Paun, G.; Albu, C.; Radu, G. -L. Assessment of Acetylcholinesterase and Tyrosinase Inhibitory and Antioxidant Activity of Alchemilla Vulgaris and Filipendula Ulmaria Extracts. *J. Taiwan Inst. Chem. Eng.* **2015,** *52*, 1–6.
88. Neves Carvalho, A.; Firuzi, O.; Joao Gama, M.; van Horssen, J.; Saso, L. Oxidative Stress and Antioxidants in Neurological Diseases: Is There Still Hope? *Curr. Drug Targets* **2017,** *18*(6), 705–718.
89. Nguyen, C. T.; Luong, T. T.; Kim, G. -L.; Pyo, S.; Rhee, D. -K. Korean Red Ginseng Inhibits Apoptosis in Neuroblastoma Cells Via Estrogen Receptor Β-Mediated Phosphatidylinositol-3 Kinase/Akt Signaling. *J. Ginseng Res.* **2015,** *39*(1), 69–75.
90. Omar, S. H. Biophenols: Impacts and Prospects in Anti-Alzheimer Drug Discovery. In *Discovery and Development of Neuroprotective Agents from Natural Products*; Elsevier Inc.: Amsterdam, Netherlands, 2018; pp 103-148.

91. Ono, K.; Hasegawa, K.; Naiki, H.; Yamada, M. Curcumin Has Potent Anti-Amyloidogenic Effects for Alzheimer's B-Amyloid Fibrils in Vitro. *J. Neurosci. Res.* **2004,** *75*(6), 742–750.
92. Ono, K.; Yoshiike, Y.; Takashima, A.; Hasegawa, K.; Naiki, H.; Yamada, M. Potent Anti-Amyloidogenic and Fibril-Destabilizing Effects of Polyphenols in Vitro: Implications for the Prevention and Therapeutics of Alzheimer's Disease. *J. Neurochem.* **2003,** *87*(1), 172–181.
93. Orgogozo, J.; Dartigues, J.; Lafont, S.; Letenneur, L.; Commenges, D.; Salamon, R.; Renaud, S.; Breteler, M. Wine Consumption and Dementia in the Elderly: A Prospective Community Study in the Bordeaux Area. *Rev. Neurol.* **1997,** *153*(3), 185–192.
94. Pandareesh, M.; Mythri, R.; Bharath, M. S. Bioavailability of Dietary Polyphenols: Factors Contributing to Their Clinical Application in Cns Diseases. *Neurochem. Int.* **2015,** *89*, 198–208.
95. Park, S. -Y.; Kim, H. -S.; Cho, E. -K.; Kwon, B. -Y.; Phark, S.; Hwang, K. -W.; Sul, D. Curcumin Protected Pc12 Cells against Beta-Amyloid-Induced Toxicity through the Inhibition of Oxidative Damage and Tau Hyperphosphorylation. *Food Chem. Toxicol.* **2008,** *46*(8), 2881–2887.
96. Pérez-H, J.; Carrillo-S, C.; García, E.; Ruiz-Mar, G.; Pérez-Tamayo, R.; Chavarría, A. Neuroprotective Effect of Silymarin in a Mptp Mouse Model of Parkinson's Disease. *Toxicology* **2014,** *319*, 38–43.
97. Perron, N. R.; Brumaghim, J. L. A Review of the Antioxidant Mechanisms of Polyphenol Compounds Related to Iron Binding. *Cell Biochem. Biophys.* **2009,** *53*(2), 75–100.
98. Petersen, R. C.; Smith, G. E.; Waring, S. C.; Ivnik, R. J.; Tangalos, E. G.; Kokmen, E. Mild Cognitive Impairment: Clinical Characterization and Outcome. *Arch. Neurol.* **1999,** *56*(3), 303–308.
99. Popat, R.; Van Den Eeden, S.; Tanner, C.; McGuire, V.; Bernstein, A.; Bloch, D.; Leimpeter, A.; Nelson, L. Effect of Reproductive Factors and Postmenopausal Hormone Use on the Risk of Parkinson Disease. *Neurology* **2005,** *65*(3), 383–390.
100. Porres-Martínez, M.; González-Burgos, E.; Accame, M. E. C.; Gómez-Serranillos, M. P. Phytochemical Composition, Antioxidant and Cytoprotective Activities of Essential Oil of Salvia Lavandulifolia Vahl. *Food Res. Int.* **2013,** *54*(1), 523–531.
101. Porres-Martínez, M.; González-Burgos, E.; Carretero, M. E.; Gómez-Serranillos, M. P. Protective Properties of Salvia Lavandulifolia Vahl. Essential Oil against Oxidative Stress-Induced Neuronal Injury. *Food Chem. Toxicol.* **2015,** *80*, 154–162.
102. Puspita, L.; Chung, S. Y.; Shim, J. -w. Oxidative Stress and Cellular Pathologies in Parkinson's Disease. *Mol. Brain* **2017,** *10*(1), 1–12.
103. Querfurth, H. W.; LaFerla, F. M. Mechanisms of Disease. *N. Engl. J. Med.* **2010,** *362*(4), 329–344.
104. Rabbani, N.; Thornalley, P. J. Dicarbonyl Stress in Cell and Tissue Dysfunction Contributing to Ageing and Disease. *Biochem. Biophys. Res. Commun.* **2015,** *458*(2), 221–226.
105. Ragonese, P.; D'amelio, M.; Salemi, G.; Aridon, P.; Gammino, M.; Epifanio, A.; Morgante, L.; Savettieri, G. Risk of Parkinson Disease in Women: Effect of Reproductive Characteristics. *Neurology* **2004,** *62*(11), 2010–2014.
106. Ramesh, B. N.; Rao, T.; Prakasam, A.; Sambamurti, K.; Rao, K. Neuronutrition and Alzheimer's Disease. *J. Alzheimers Dis.* **2010,** *19*(4), 1123–1139.
107. Ray, B.; Bisht, S.; Maitra, A.; Maitra, A.; Lahiri, D.K. Neuroprotective and Neurorescue Effects of a Novel Polymeric Nanoparticle Formulation of Curcumin (Nanocurc™) in the Neuronal Cell Culture and Animal Model: Implications for Alzheimer's Disease. *J. Alzheimers Dis.* **2011,** *23*, 61–77.

108. Rezai-Zadeh, K.; Shytle, D.; Sun, N.; Mori, T.; Hou, H.; Jeanniton, D.; Ehrhart, J.; Townsend, K.; Zeng, J.; Morgan, D. Green Tea Epigallocatechin-3-Gallate (Egcg) Modulates Amyloid Precursor Protein Cleavage and Reduces Cerebral Amyloidosis in Alzheimer Transgenic Mice. *J. Neurosci.* **2005,** *25*(38), 8807–8814.
109. Ringman, J. M.; Frautschy, S. A.; Cole, G. M.; Masterman, D. L.; Cummings, J. L. A Potential Role of the Curry Spice Curcumin in Alzheimer's Disease. *Curr. Alzheimer Res.* **2005,** *2*(2), 131–136.
110. Rodrigo, R.; Miranda, A.; Vergara, L. Modulation of Endogenous Antioxidant System by Wine Polyphenols in Human Disease. *Clin. Chim. Acta* **2011,** *412*(5–6), 410–424.
111. Rojas, J.; Buitrago, A. Antioxidant Activity of Phenolic Compounds Biosynthesized by Plants and its Relationship with Prevention of Neurodegenerative Diseases. In *Bioactive Compounds*; Elsevier Inc.: Amsterdam, Netherlands, 2019; pp 3–31.
112. Sanders, L. H.; Greenamyre, J. T. Oxidative Damage to Macromolecules in Human Parkinson Disease and the Rotenone Model. *Free Radic. Biol. Med.* **2013,** *62*, 111–120.
113. Sandoval-Yañez, C.; Mascayano, C.; Martínez-Araya, J. I. A Theoretical Assessment of Antioxidant Capacity of Flavonoids by Means of Local Hyper–Softness. *Arab. J Chem.* **2018,** *11*(4), 554–563.
114. Serrano-Pozo, A.; Frosch, M. P.; Masliah, E.; Hyman, B. T. Neuropathological Alterations in Alzheimer Disease. *Cold Spring Harb. Perspect. Med.* **2011,** *1*(1), a006189.
115. Sheeja Malar, D.; Pandima Devi, K. Dietary Polyphenols for Treatment of Alzheimer's Disease–Future Research and Development. *Curr. Pharm. Biotechnol.* **2014,** *15*(4), 330–342.
116. Shulman, J. M.; De Jager, P. L.; Feany, M. B. Parkinson's Disease: Genetics and Pathogenesis. *Ann. Rev. Pathol. Mech. Dis.* **2011,** *6*, 193–222.
117. Shulman, L.M.; Bhat, V. Gender Disparities in Parkinson's Disease. *Expert Rev. Neurother.* **2006,** *6*(3), 407–416.
118. Singh, B.; Parsaik, A. K.; Mielke, M. M.; Erwin, P. J.; Knopman, D. S.; Petersen, R. C.; Roberts, R. O. Association of Mediterranean Diet With Mild Cognitive Impairment and Alzheimer's Disease: A Systematic Review and Meta-Analysis. *J. Alzheimers Dis.* **2014,** *39*(2), 271–282.
119. Singh, B.; Singh, J. P.; Kaur, A.; Singh, N. Phenolic Composition and Antioxidant Potential of Grain Legume Seeds: A Review. *Food Res. Int.* **2017,** *101*, 1–16.
120. Singh, M.; Arseneault, M.; Sanderson, T.; Murthy, V.; Ramassamy, C. Challenges for Research on Polyphenols from Foods in Alzheimer's Disease: Bioavailability, Metabolism, and Cellular and Molecular Mechanisms. *J. Agric. Food Chem.* **2008,** *56*(13), 4855–4873.
121. Soto, C.; Sigurdsson, E. M.; Morelli, L.; Kumar, R. A.; Castaño, E. M.; Frangione, B. B-Sheet Breaker Peptides Inhibit Fibrillogenesis in a Rat Brain Model of Amyloidosis: Implications for Alzheimer's Therapy. *Nat. Med.* **1998,** *4*(7), 822–826.
122. Sousa Silva, M.; Gomes, R. A.; Ferreira, A. E.; Ponces Freire, A.; Cordeiro, C. The Glyoxalase Pathway: The First Hundred Years… and Beyond. *Biochem. J.* **2013,** *453*(1), 1–15.
123. Spillantini, M. G.; Schmidt, M. L.; Lee, V. M. -Y.; Trojanowski, J. Q.; Jakes, R.; Goedert, M. A-Synuclein in Lewy Bodies. *Nature* **1997,** *388*(6645), 839–840.
124. Stahl, S. M. *Essential Psychopharmacology: The Prescriber's Guide*. Cambridge University Press: Cambridge University, UK, 2005; p 562.
125. Talavéra, S.; Felgines, C.; Texier, O.; Besson, C.; Gil-Izquierdo, A.; Lamaison, J. -L.; Rémésy, C. Anthocyanin Metabolism in Rats and Their Distribution to Digestive Area, Kidney, and Brain. *J. Agric. Food Chem.* **2005,** *53*(10), 3902–3908.

126. Tan, X.; Gu, J.; Zhao, B.; Wang, S.; Yuan, J.; Wang, C.; Chen, J.; Liu, J.; Feng, L.; Jia, X. Ginseng Improves Cognitive Deficit Via the Rage/Nf-Kb Pathway in Advanced Glycation End Product-Induced Rats. *J. Ginseng Res.* **2015,** *39*(2), 116–124.
127. Tang, H.; Lu, D.; Pan, R.; Qin, X.; Xiong, H.; Dong, J. Curcumin Improves Spatial Memory Impairment Induced by Human Immunodeficiency Virus Type 1 Glycoprotein 120 V3 Loop Peptide in Rats. *Life Sci.* **2009,** *85*(1–2), 1–10.
128. Tangney, C. C. Dash and Mediterranean-Type Dietary Patterns to Maintain Cognitive Health. *Curr. Nutr. Rep.* **2014,** *3*(1), 51–61.
129. Tanzi, R. E. A Brief History of Alzheimer's Disease Gene Discovery. *J. Alzheimers Dis.* **2013,** *33*(s1), S5–S13.
130. Taylor, K. S. M.; Cook, J. A.; Counsell, C. E. Heterogeneity in Male to Female Risk for Parkinson's Disease. *J. Neurol. Neurosurg. Psychiatry* **2007,** *78*(8), 905–906.
131. Thanikachalam, P. V.; Ramamurthy, S.; Kumar, A.; Gupta, M.; Bansal, G., Protective Role of Epigallocatechin Gallate, a Dietary Antioxidant against Oxidative Stress in Various Diseases. In *Pathology*; Elsevier Inc.: Amsterdam, Netherlands, 2020; pp 213–223.
132. Van Sumere, C. Phenols and Phenolic Acids. In *Methods in Plant Biochemistry*; Dey, P. M., Ed.; Academic Press: Cambridge, USA, 1989; pp 29–73.
133. Walsh, D. M.; Selkoe, D. J. Aβ Oligomers–a Decade of Discovery. *J. Neurochem.* **2007,** *101*(5), 1172–1184.
134. Wang, J. Y.; Wen, L. L.; Huang, Y. N.; Chen, Y. T.; Ku, M. C. Dual Effects of Antioxidants in Neurodegeneration: Direct Neuroprotection against Oxidative Stress and Indirect Protection Via Suppression of Gliamediated Inflammation. *Curr. Pharm. Des.* **2006,** *12*(27), 3521–3533.
135. Wang, R.; Li, Y. -B.; Li, Y. H.; Xu, Y.; Wu, H. -l.; Li, X. -J. Curcumin Protects against Glutamate Excitotoxicity in Rat Cerebral Cortical Neurons by Increasing Brain-Derived Neurotrophic Factor Level and Activating Trkb. *Brain Res.* **2008,** *1210*, 84–91.
136. Wen, L.; Zhao, Y.; Jiang, Y.; Yu, L.; Zeng, X.; Yang, J.; Tian, M.; Liu, H.; Yang, B. Identification of a Flavonoid C-Glycoside as Potent Antioxidant. *Free Radic. Biol. Med.* **2017,** *110*, 92–101.
137. Wu, L.; Davies, G.; Roesler, W.; Juurlink, B. Regulation of the Glyoxalase Pathway in Human Brain Microvascular Endothelium: Effects of Troglitazone and Tertiary Butylhydroperoxide. *Endothelium* **2002,** *9*(4), 273–278.
138. Yang, F.; Lim, G. P.; Begum, A. N.; Ubeda, O. J.; Simmons, M. R.; Ambegaokar, S. S.; Chen, P. P.; Kayed, R.; Glabe, C. G.; Frautschy, S. A. Curcumin Inhibits Formation of Amyloid B Oligomers and Fibrils, Binds Plaques, and Reduces Amyloid in Vivo. *J. Biol. Chem.* **2005,** *280*(7), 5892–5901.
139. Youle, R. J.; Van Der Bliek, A. M. Mitochondrial Fission, Fusion, and Stress. *Science* **2012,** *337*(6098), 1062–1065.
140. Zhou, T.; Zu, G.; Zhang, X.; Wang, X.; Li, S.; Gong, X.; Liang, Z.; Zhao, J. Neuroprotective Effects of Ginsenoside Rg1 through the Wnt/B-Catenin Signaling Pathway in Both In vivo and In vitro Models of Parkinson's Disease. *Neuropharmacology* **2016,** *101*, 480–489.

CHAPTER 7

Neuroprotective Efficacy of Ethnomedicinal Plants in Mental Disorders with Animal Models

Adedamola Aminat Bayo-Olugbami, Joseph Enya, and Olorunfemi Samuel Tokunbo

ABSTRACT

The recent approach to traditional medicine and medicinal plants, especially with regards to the recommendation of the World Health Organization, is increasingly changing the world of ethnomedicinal research. Some drugs are synthesized from plants and there is a high possibility that more drugs will still be produced from plants. Medicinal plants are well documented to be helpful in the management of various disease conditions. Mental health is a vital component of human health. There are alarming reports on the rate at which people develop mental disorders (MDs) ranging from depression, anxiety, mood disorders, schizophrenia, and many others. There are also evidences that a good percentage of the world's population depend on herbal extract for treating various forms of mental diseases. Despite the aforementioned, there is paucity of preclinical evidence that support the effectiveness of herbal derivatives in the treatment of MDs. This review seeks to harvest ethnomedicinal plants with psychotherapeutic properties that can serve as suitable alternatives to the conventional antipsychotics, which are burdened with undesirable side effects, taking into consideration the several benefits of herbal plants. Our focus will be on plants (and its derivatives) that have animal-based experimental validations, whose behavioral and pharmacological actions have been characterized in the treatment of MDs, with emphasis on depression, anxiety, and schizophrenia.

Curative and Preventive Properties of Medicinal Plants: Research on Disease Management and Animal Model Studies. Megh R. Goyal, Junaid Ahmad Malik, & Ademola Olabode Ayeleso (Eds.)

7.1 INTRODUCTION

Mental disorders (MDs) also called mental illnesses are commonly used to describe a clinical condition known as psychological or psychiatric disorders.[46] MDs could be occasional or chronic (long-lasting), and usually characterized by abnormal thinking, conduct disorders (abnormal behaviors and social communication), anxiety, and depressive behavior, which place a wide range of emotional, health, and financial burden on victims, their families and the society at large[40] (Figure 7.1). Globally, increase in the occurrence of MDs is of major concern because of its impact on people's quality of life. According to WHO, MDs account for more than 11% of health problems in developed countries.[20] An epidemiological study conducted by ESEMeD/MHEDEA reported that major depressive disorder, alcohol abuse, dysthymia, and specific phobia were the most common MDs among European population, with estimated lifetime prevalence rates of 12.8%, 4.1%, and 4.1% and 7.7%, respectively.[16]

Different types of MDs including depression, anxiety, schizophrenia, attention deficit hyperactivity disorder, and obsessive-compulsive disorder,[71] have been shown to be associated with a wide range of structural and functional decline linked to neurodegenerative diseases such as Alzheimer's disease, Parkinson's diseases, and multiple sclerosis.[10] Thus, these disorders have been highlighted by WHO, as needing more research to unravel the burden.[72] Progress has been made in the search for novel pharmacotherapy from herbal extracts, of which some have been tested and proven effective using various experimental animal models including those correlating MDs.[79] Researchers have shown more interest in exploring the pharmacological actions of herbal extracts in order to confirm the veracity of claims made in various books of herbal medicine.[54] Herbal extracts play an important role in modern drug development, particularly as antibacterial, antioxidant, and anti-inflammatory agents. Interestingly, enormous number of compounds derived from herbal extract is in various stages of clinical development, suggesting the existing viability and the importance of using herbal extract as sources of novel drug production.[17] Up to 50% of the approved drugs during the last 30 years are either directly or indirectly from herbal extract.[68]

The use of synthetic materials has been the mainstay for drug development owing to their effectiveness and safety but with the major setback of undesirable side effects and some have failed to treat the fundamental causes of some of these diseases. As a result of this, the use of herbal extract in the therapeutic treatment and prevention of diseases has been encouraged.[12,68] Thus, this review seeks to harvest ethnomedicinal plants with psychotherapeutic properties that can serve as suitable alternatives to the conventional antipsychotics, which are burdened with undesirable side effects, taking into consideration the several benefits of herbal plants. Our focus will be on plants (and its derivatives) that have animal-based experimental validations, whose behavioral and pharmacological actions have been characterized in the treatment of MDs, with emphasis on depression, anxiety, and schizophrenia.

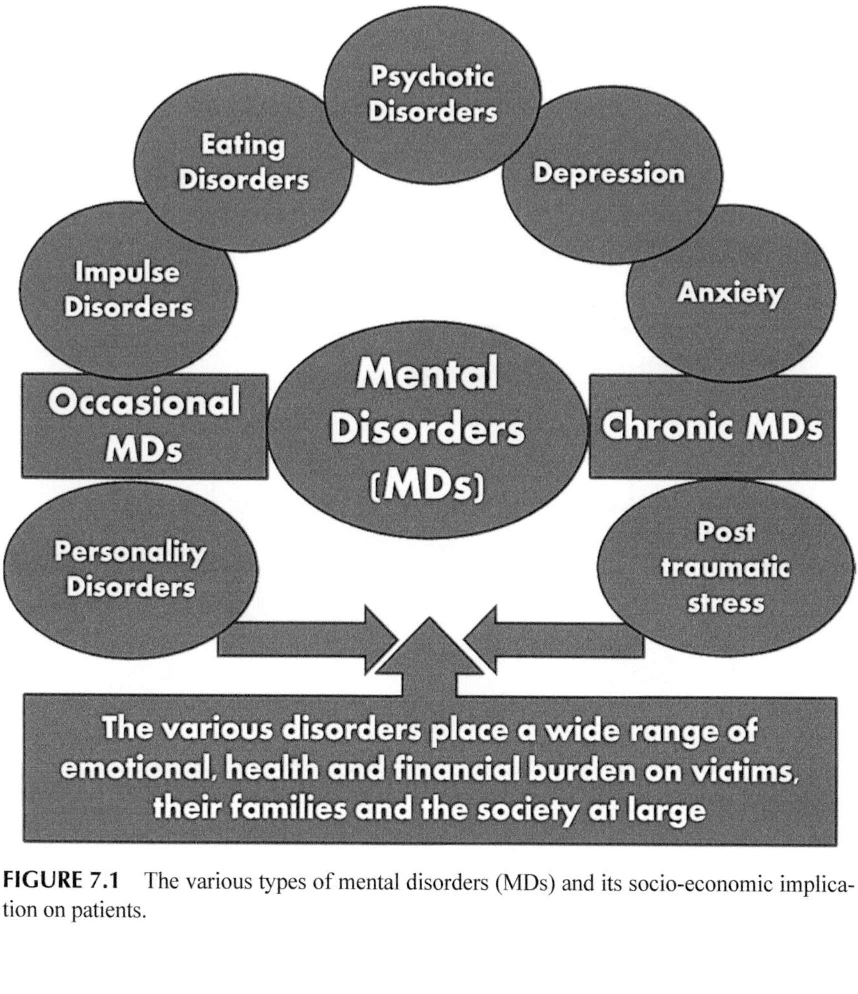

FIGURE 7.1 The various types of mental disorders (MDs) and its socio-economic implication on patients.

7.2 NEUROPROTECTIVE ACTIVITY OF HERBAL EXTRACTS CORRELATING ANIMAL MODELS

Medicinal herbs have played a crucial role in the treatment of various disease conditions especially in the rural areas due to their accessibility, availability, and cost.[68] The use of herbs continues to expand rapidly across the world with many people largely depending on herbal medicines as their source of primary health care.[70] According to the WHO report, the use of herbal remedies in primary health care exceeds that of conventional drugs.[52] A number of medicinal herbs with psychoprotective and psychotherapeutic properties have been identified including: *allium cepa* (onions), *abelmoschus esculentus* (okra), *curcuma longa* (turmeric), *zingiber officinalis* (ginger), *allium sativum* (garlic), *bacopa monnieri* (water hyssop), *lavandula angustifolia* (lavender), *garcinia kola* (bitter kola), *moringa oleifera* (drumstick tree), and *aloe barbadensis* Miller (aloe vera).[15] The psychoprotective and psychotherapeutic potentials of extracts from the leaves, stem, bark, roots, flowers, and fruits of the aforementioned medicinal herbs and their mechanism of action in the treatment of MDs[2,15] are shown in Figure 7.2.

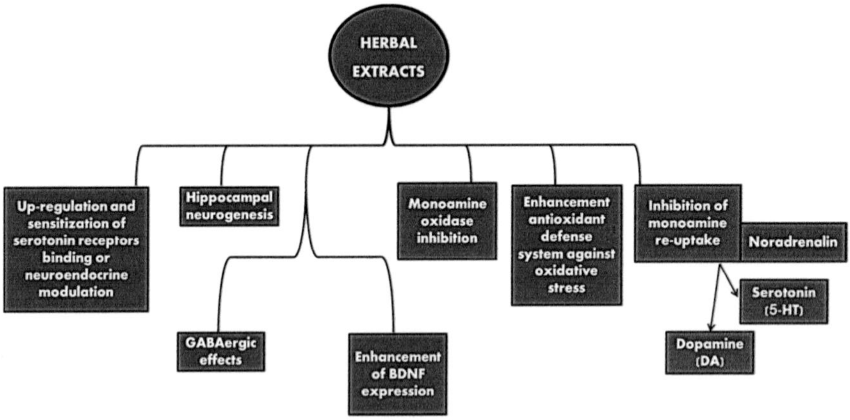

FIGURE 7.2 The mechanisms of action of medicinal plants in the management of mental disorders.

7.2.1 *ALLIUM CEPA*

A. cepa is a popular vegetable commonly known as onions or bulb onion. It is one of the most widely cultivated species of the genus *Allium* belonging to the family of *Alliaceae*.[13] The use of *A. cepa* as a dietary ingredient for food preparations in Africa and many parts of the world has been in existence for centuries. It contains essential flavonoids that include quercetin, isorhamnetin-4-glucoside, quercetin-3-glucoside, quercetin 4′-O-glucoside, organosulfur compounds, galactose, glucose, mannose, xylose, allyl sulfides, thiosulfinates, cyclically, selenium, flavonoids, and flavenols.[27]

A. cepa flavonoid has been reported for its numerous pharmacological activities such as antipsychotics,[65] antioxidant, anti-inflammatory,[38] neuroprotective,[37] cardioprotective,[32] anticonvulsant activities,[28] and antidiabetic effects.[66] Also, in African's traditional medicine, *A. cepa* extracts are being used as antioxidant/anticancer antimicrobial, asthma protective agent, for cardiovascular support and management of hypoglycemic conditions.[29] The antipsychotic effects of *A. cepa* have been supported by several in vitro and in vivo studies.[14,65] According to Kadian and Parle, *A. cepa* reportedly improved psyhotic behaviors as evidenced by climbing attempts, locomotor activity, and ketamine-induced stereotypic behavior. Treatment with *A. cepa* paste at 50 mg/kg significantly blocked dopaminergic neurotransmission and inhibited dopamine (D2) receptor, which inturn has been implicated in the pathophysiology of MDs indicating its effects in the management of MDs.[29] Singh and Shri identified quercetin 4′-O-glucoside extracted from *A. cepa* to be responsible for the antidepressant-like effects of *A. cepa* in mouse model of unpredictable chronic mild stress (UCMS) induced depressive behavior and to elucidated its mechanism of action. The authors used forced swim and open field behavioral test that are frequently used behavioral tests to evaluate the antidepressant-like activity of *A. cepa* on mouse model of UCMS. Mouse subjected to UCMS without intervention showed depressive state as depicted by an increase in the duration of immobility in the forced swim test. However, treatment with various doses of *A. cepa* showed a marked decrease in immobility time indicating a decline in depressive behavior.[65]

7.2.2 *ABELMOSCHUS ESCULENTUS*

A. esculentus also known as okra is a flowering plant and is popularly known as lady's fingers or gumbo in many English-speaking countries in the world.[33] It is also called *guino-gombo* in Spanish, *guibeiro* in Portuguese, *bhindi* in India, *Kenkase (Berta)* in Ethiopia.[7]

In Africa particularly Nigeria, it is one of the most essential flowering vegetables widely cultivated for its fruits (edible green seed pods) and fresh green leaves.[3] It has nutrients such as carbohydrates, vitamins, proteins, minerals, fats, enzymes, and large amount of mucilage that contains soluble dietary fibers.[60] Traditionally, *A. esculentus* extract has been used to treat many diseases like diabetes and MDs.[67] *A. esculentus* extract (particularly the green pods) has been proven to contained high content of polyphenols and flavonoids.[6] Likewise, it possesses antioxidant, anti-inflammatory, neuro-protective, antidiabetic, antidepressive, and antihyperlipidemic effects.[75]

Yan and his colleagues evaluated the anxiolytic and antidepressant effects of okra extract and its possible mechanism of action in Chronic Unpredictable Mild Stress (CUMS)-induced mouse model of depression.[75] Tail suspension, open field, and forced swim tests were used to determine anxiety and depressive behavior. The involvement of proinflammatory cytokines, TLR4/NF-κB, and MAPKs pathways in CUMS-induced mice were explored. The result showed that anxiety and depressive behaviors, as well as the activation of inflammatory cytokines in the colon, serum, and hippocampus of mice were alleviated following treatment with *A. esculentus.* Furthermore, *A. esculentus* treatment inhibited TLR4 and NF-κB expression and reduced the high level of proinflammatory cytokines, as well as enhanced MAPKs signaling activities. Thus, these findings suggest the potential anxiolytic and antidepressant effects Okra extract in mice.[75]

Similarly, the study carried out by Xia et al. determined the antidepressant activity of *A. esculentus* seed extract (AESE) using behavioral tests such as open field test, tail suspension test (TST), forced-swimming test (FST), and novelty suppressed feeding test, for subchronic treatment and chronic sleep-interruption (CSI) animal models.[73] The levels of neurotransmitters, total antioxidant capacity, superoxide dismutase (SOD), and malondialdehyde (MDA) were determined in the Prefrontal Cortex and hippocampus of experimental animals. In addition, the activities of Corticosterone (CORT) and adrenocorticotropic hormone (ACTH) were quantified in the serum. Based on the results, AESE showed antidepressant effect in both SCT and CSI animal models by reducing the hyper-activation of the hypothalamic-pituitary adrenal axis, alleviating oxidative stress and regulating the level of neurotransmitter.[73]

7.2.3 CURCUMA LONGA

C. longa well-known as turmeric is a perennial herb. This plant originated from India, Southeast Asia, and Indonesia, and it is extensively cultivated in tropical and subtropical regions of the world.[4] *C. longa* has been used traditionally as medicine for numerous disease conditions for many centuries.

Its pharmacological activity has been attributed to curcuminoids, which consists of curcumin (CUR) and two other related compounds known as bisdemethoxycurcumin and demethoxy curcumin.[30]

There are several clinical and animal-based experimental evidences to support the antipsychotic, antioxidant, anti-inflammatory, antitumor, anti-aging, and antimicrobial properties of CUR[74,78] (Fig. 7.3). CUR has been studied extensively studied and many clinical studies using human subject as well as preclinical trials using rat and mice models of depression have highlighted the antidepressant potential of CUR.[44] A meta-analysis done by Ng et al. summarized current evidence and generated hypotheses for further research on the clinical use of CUR as an antidepressant. Authors reviewed six clinical trials with a total of 377 patients, and compared their use of CUR with a placebo. The pooled standardized mean difference using Hamilton Rating Scale for Depression scores supports the significant clinical efficacy of CUR in ameliorating depressive symptoms in patients with depression. As such, CUR was concluded to be safe and effective in the management of depression.[44]

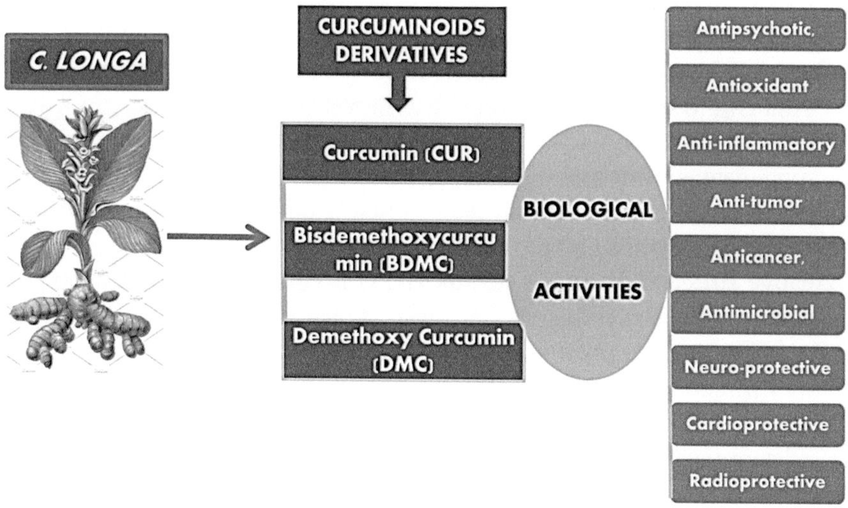

FIGURE 7.3 Schematic representation of Curcuminoid derivatives and their multiple biological activities.

Yu et al. investigated the effects of CUR extracted from *C. longa,* on the brain-gut axis in irritable bowel syndrome (IBS) rat models and the role of serotonin-cAMP response element-binding protein- Brain-derived neurotrophic factor (5-HT-CREB-BDNF) signaling pathway.[77] IBS is an intestinal disorder characterized by abdominal discomfort, and it is associated with features of disordered defecation such as diarrhea and constipation. It has been reported that 90% of patients with IBS can develop several types of depressive disorder.[76] Yu et al. established that IBS rats exhibited anxiety and depressive behavior which was associated with reduced expression of 5-HT, BDNF, and pCREB in the hippocampus following exposure to Chronic Acute Combined Stress.[77] However, administration of 40 mg/kg CUR reduced the immobility time in forces swim test and the number of buried marbles in the behavioral test. It also reduced the number of fecal output and abdominal withdrawal reflex scores. Furthermore, 5-HT, BDNF, and pCREB levels in the hippocampus were increased. Therefore, CUR exerted its effect by regulating the activities of 5-HT, BDNF, and CREB signaling both in the brain and peripheral intestinal system.[61]

In another study, the antidepressant effects of aqueous extract of *C. longa* was examined in a chronic mild stress mice model. Treatment using 140–560 mg/kg of the extract for 14 days produced a dose-dependent reduction of immobility time in forced swimming and tail suspension tests. In addition, the extract significantly inhibited the monoamine oxidase-A (MAO-A) activity in the mouse whole brain when compared to control. Thus, the antidepressant effect of *C. longa* extract was mediated via the inhibition of MAO-A.[73]

Animal model of Schizophrenia has shown the effectiveness of CUR as antipsychotic and antioxidant agent. A study investigated the neurotoxic effects of supramagnetic iron oxide nanoparticles (SPIONs) and the antagonizing potential of CurNPs (curcumin nanoparticles) on isolated mitochondria of schizophrenic rats. Exposure to SPION alters physiologic processes, causes oxidative stress, and induces neurotoxicity. It causes cellular mitochondria function, apoptosis, inflammation, generation of reactive oxygen species (ROS), increase in micronuclei, and chromosome condensation.[76] Using X-ray diffraction and dynamic laser light scattering methods, the authors characterized the physicochemical properties of Fe_3O_4 magnetic-CurNPs. They also determined the level of ROS generation, adenosine triphosphate (ATP), cytochrome c release, and histopathology of the cerebellum of schizophrenic rats. CUR-treated rats had reduced SPION-induced oxidative stress as well as mitochondrial dysfunction compared with the untreated group. Mitochondrial complex II activity, mitochondrial membrane potential, and ATP level were markedly reduced in the cerebellar mitochondria of CUR-treated group. In addition, there was a significant decline in ROS generation and cytochrome C. Therefore, the study concluded that CUR protected the cerebellum against neurotoxicity in rat model of Schizophrenia.[43]

In a clinical study, CUR extracted from *C. longa* was used as an adjuvant with regular antipsychotic medications in patients suffering from chronic schizophrenia. In a double blind, randomized, placebo-controlled study, 38 patients suffering from chronic schizophrenia were treated with 3000 mg curcumin for a duration of 24 weeks. There was a significant positive response to curcumin within 6 months. Also, patients did not complain of any form of adverse effect. The study concluded that CUR has promising potentials as an add-on to antipsychotics for the management and treatment of schizophrenia.[69]

7.2.4 *ZINGIBER OFFICINALIS*

Z. officinale is a dietary and herbaceous perennial plant, which is consumed for culinary and medicinal purposes. *Z. officinale* rhizomes are highly medicinal having *gingerols* (*6-gingerdiol, 8-gingerol, 10-gingerdione, and 6-shogaols*) as its active constituents. It is used for treating a wide range of ailments especially in South Asia and Africa. Iron (Fe), magnesium (Mg), calcium (Ca), Vitamin-C, flavonoids, sesquiterpenes, and paradols are other vital compounds present in *Z. officinale*.[64]

Olanrewaju et al.[48] reported the attenuating potentials of *Z. officinale* ethanolic extract against reserpine-induced anxiety and depressive condition and its associated hippocampal aberrations in rats. These findings validate an earlier report by Farzin et al., wherein antidepressant potential of *Z. officinale* methanolic extract in diabetic mice using FST was investigated. It was established that mice induced with diabetes (streptozotocin (STZ)-induced mice model) were significantly more immobile during the FST. While treatment with *Z. officinale* methanolic extract at 175 and 350 mg/kg did not change blood glucose, it rather altered immobility time by reducing it. However, both blood glucose levels and immobility were significantly reduced following treatment with *Z. officinale* methanolic extract at the dose of 700 mg/kg.[18]

7.2.5 *ALLIUM SATIVUM* (GARLIC)

A. sativum is a perennial bulbous herb also known as garlic in English. The biological activities of garlic including antipsychotic,[57] antioxidant, anticancer antimicrobial, antifungal, antihyperglycemic, antithrombotic, insulinotropic, cholesterol, and lipid-lowering properties have been attributed to *allicin*, which gives it a unique odor/scent. Other major organosulfur compounds derived from *A. sativum* are Diallyl Trisulfide (DATS). Rahmani et al. investigated the antipsychotic effects of *A. sativum* extract on anxiety and depressive behaviors as well as brain oxidative biomarkers in streptozotocin (STZ)-induced diabetes in rat model.[56] Oral administration of 0.5 g/kg *A. sativum* following STZ-induced diabetes was observed to deplete anxiety and depressive behavioral activities. It also significantly reduced MDA, increased SOD and GPx levels in rat brain. Therefore, its ability to attenuate cerebral oxidative stress indicates the possible mechanism of action.

Using adult rats, Marschollek et al. induced cortical spreading depression (CSD) by microinjection of KCl in to the somatosensory cortex in order to determine the effects of garlic oil on CSD rat model. *A. sativum* extract reportedly reduced the amplitude of CSD-induced model, in dose-dependent pattern, but had no effect on the duration and velocity in neocortical brain slices. It also lowered the amplitude of field excitatory postsynaptic potentials at concentration of 500 µl/L, and inhibited the induction of long-term potentiation. Additionally, the oil prevented CSD-induced reactive astrocytes in the neocortex. Its mechanism of action against CSD was attributed to inhibition of synaptic plasticity and prevention of harmful effects on astrocyte in the neocortex.[41]

In another study conducted by Huang et al. using forced swim test and UCMS to induce depression in rats, oral administration of 250 and 500 mg/kg garlic oil for 28 days significantly attenuated the duration of immobility in forced swim test. Furthermore, garlic oil and Duallyl disulfide, a major organosulphur component of garlic markedly increased sucrose preference index in the UCMS model. A total of 25 mg/kg dose of garlic essential oil effectively attenuated the turnover ratio of 5-HT and dopamine (DA) in the frontal cortex but not on the hippocampus. Chronic garlic essential oil treatment enhanced hippocampal BDNF, protein kinase B expression, and c-AMP response element binding protein (CREB). The garlic oil exhibited its antipsychotic effect in part by modulating monoamine neurotransmitter and BDNF-related signaling pathway.[25] In furtherance, another study documented that 100 mg/kg hydro-alcoholic extract of *A. sativum* reduced signs of depression in reserpine-induced rat model of depression.[1]

7.2.6 *BACOPA MONNIERI*

B. monnieri also known as water hyssop, is a creeping perennial plant commonly found in warm wetlands areas as a weed in rice fields. The leaves, stem, and flowers of this plant are often used either alone or in combination with other herbs in Ayurvedic medicine to manage and treat several mental health disorders like anxiety and depression.[59] Although, there are very few clinical reports on the impacts of *B. monnieri* on MD, yet there are some in vivo and in vitro studies that showed that *B. monnieri* enhances learning rate, retention ability and also attenuates amnesia and MD.[24]

The antidepressant property of *B. monnieri* extract as reported by Hazra et al. using stress-induced model of depression showed that treatment with 80 and 120 mg/kg of the extract significantly restored body weight, attenuated escape latency in shuttle box escape test (SBET) and the level of plasma corticosterone. These effects suggest the antidepressant property of *B. monnieri* extract and also propose the specific doses required for treating depression and other stress-related psychiatric disorders.[37] Similarly, the neuroprotective/antipsychotic, neurotrophic, and antioxidant effects of *B. monnieri* extract on chronic unpredictable stress-induced depression in Wistar rats has been documented.[35] Behavioral tests such as SBET, FST, and TST were conducted. To determine the potential mechanism of action, the plasma level of ACTH and corticosterone were measured to determine the impacts of *B. monnieri* extract on hypothalamic pituitary adrenal axis. The hippocampal expression of neurotrophin (BDNF) and neuronal marker double cortin (DCX) were measured. Its effect on hippocampal neurogenesis and oxidative stress was investigated. The results showed significant reduction in depressive behaviors, normalized ACTH and corticosterone, and enhanced BDNF, DCX, and BrdU/NeuN expression following treatment with *B. monnieri* extract. Also, the extract significantly enhanced the activity of antioxidant enzymes. Authors concluded that *B. monnieri* extract exerted its antipsychotic effects possibly by upregulating hippocampal neurogenesis, promoting BDNF expression and antioxidant defense system against oxidative stress.[35]

7.2.7 LAVANDULA ANGUSTIFOLIA

L. angustifolia is a famous medicinal herb that is native to East Africa, Southern and Northern part of Europe, middle East countries to Southwest Asia, and Southeast India.[9] The various categories of lavenders which include *L. angustifolia*, *Lavandula stoechas*, *Lavandula Latifolia*, and *Lavandula intermedia* contain the same ethnobotanical properties and essential chemical constituents. Lavender essential oil extracted by steam distillation from fresh *L. angustifolia,* and it is composed of linalyl acetate and linalool (mainly used in aromatherapy as a relaxant), lavandulyl acetate, 1,8-cineole, and camphor. Other medicinal effects of this essential oil include anxiolytic,[63] antidepressive,[39] and neuroprotective effect.[39]

A preliminary study done by Ogata and colleague evaluated the impact of *L. angustifolia* extract (lavender oil) inhalation on cortisol levels, pulse measurements, blood pressure, anxiety, and depressive disorders in human and animal models.[45] It was found that inhalation of *L. angustifolia* extract reduced the blood pressure and attenuated Self-Rating Depression Scale Score. From the *ex vivo* result, lavender oil promoted intracellular calcium ions concentration in the hypothalamic oxytocin neurons. Therefore, it has been suggested as a useful therapy for depression, stress relief, and anxiety. Its mechanism of action has been linked to activation of the central oxytocin neurons in the hypothalamus of rodents.[45] In another study conducted by Kianpour et al. wherein 140 postpartum women were grouped into aromatherapy and nonaromatherapy. The study explored the potential effect of *L. angustifolia* extract on postpartum depression, anxiety, and stress. The results showed that inhalation of *L. angustifolia* extract significantly attenuated stress, anxiety, and depression up to 12 weeks after delivery.[34]

In another study in which chronic administration of 40 mg/kg corticosterone was used to induce anxiety and depression. After 14 days of treatment with *L. angustifolia* extract, depressive behavior induced by chronic corticosterone administration was ameliorated. Also, the number of bromodeoxyuridine (BrdU) positive cells in the hippocampus and the subventricular zone were reduced. The level of BDNF was also upregulated, which demonstrate the neurogenic effect *L. angustifolia* extract. In furtherance, oxytocin concentration in serum was upregulated. These results provide evidence of the biological effect of *L. angustifolia* extract on neuroplasticity and neurogenesis, which are mechanisms that have been implicated in the pathophysiology of anxiety and depression.[62]

7.2.8 GARCINIA KOLA

G. kola also called bitter kola due to its bitter and resinous taste, is commonly consumed in Africa especially Nigeria for its health benefits and the belief that it promotes longevity.[49] *G. kola* seed extract *is rich in* flavonoids, phenolic acids, cardiac glycosides, vitamin C, saponins, and tannins. The predominant constituent in *G. kola* is Kolaviron (KV), which is responsible for most of its pharmacological actions.[47] KV contains biflavanones (GB1, GB2, and kola flavanone) and has been reported to elicit strong antidepressant, anti-inflammatory, antioxidant, antidiabetic, and neuro-protective effect[37,51] both in vivo and in vitro experimental models.

The 100 mg/kg of kolaviron reduced immobility time in the forced swimming and tail suspension tests suggesting its antidepressant effect. Furthermore, mice were pretreated with α1-adrenoceptor antagonist (prazosin), 5-HT2 receptor antagonist (metergoline), and muscarinic cholinergic antagonist (atropine) to determine the receptors involved in this pathway. α1-adrenoceptor antagonist (prazosin) abolished the antidepressant effect of KV. As such, the possible mechanism of action of KV could be by modulating the adrenergic pathway.[51]

7.2.9 MORINGA OLEIFERA

M. oleifera called by different names such as moringa, drumstick tree or horseradish tree is a versatile and exceptionally nutritious (minerals and vitamin rich) plant, which is medicinally useful to human, animals, and in various industrial applications.[50] The plant is characterized by its whitish-grey corky bark, yellow or white petioles, tripinnate leaves, and hanging 3-sided pods. Almost every part of *M. oleifera* tree is useful for food preparation, water purification, biodiesel production, and medicine. Quercetin, niazimicin, moringin, phenol, triterpenoids, cardiac glucose, flavonoids, alkaloids, anthraquinones, saponins, luteolin, and carotenoid are the phytochemicals present in *M. oleifera*.[5] Thus, various studies have ascertained that the extract possesses anxiolytic, antidepressant, antioxidant, anti-inflammatory, antirheumatic, bactericidal, and diuretic effects, likewise it serves as antidote, stimulant, emetic, tonic, purgative, and vesicant.[26]

Studies have been done on the anxiolytic activities of crude organic fractions of *M. oleifera* leaves,[8] ethanolic extract of *M. oleifera* leaves, and the effects of *M. oleifera* Pudding.[42] Kaur et al. evaluated the antidepressant effects of *M. oleifera* ethanolic extracts on the acute and chronic mouse models of depression. Authors reported a significant positive change in FST, TST, and locomotor activity of chronically dosed mice. The mechanism of action of KV was linked to the modulation of noradrenergic-serotonergic neurotransmission pathway, which is the hallmark of the treatment of depression with class of drugs.[31]

7.2.10 ALOE BARBADENSIS MILLER

This is commonly called Aloe vera (*A. barbadensis*) and it is characterized by its green color succulent, xerophytic, and shrubby or arborescent features.[58]

The plant latex or exudate has active compounds such as Barbaloin (also known as aloin A), aloesin (aloeresin B), isobarbaloin (aloin B), and aloeresin A, which are responsible for its various pharmacological actions including but not limited to immunomodulating, antimicrobial, and anti-inflammatory effects.[55]

Halder et al. explored the effect of *A. barbadensis* on mice models of depression, locomotion, learning, and memory. Result from this study showed significant reduction of transfer latency in the elevated plus-maze test following 400 mg/kg of *A. barbadensis,* while the duration of immobility in forced swimming and tail suppression tests significantly decreased at all doses (100, 200, and 400 mg/kg) administered. Also, 200 and 400 mg/kg of *A. barbadensis* administration significantly enhanced the acquisition and retention step-down latency.[22] Foadoddini and Alinejad reported the effects of pure *A. barbadensis* extract powder on depression in prediabetic patients. This study was conducted on 72 prediabetic patients who were randomly divided into 3 groups of *A. barbadensis*-300 mg, *A. barbadensis*-500 mg, and Placebo. Treatment was done twice a day for 8 weeks. The result showed that the mean depression scores was significantly ($P = 0.033$) decreased in *A. barbadensis*-500 mg treated group than *A. barbadensis*- 300 mg.[19]

7.3 SUMMARY

The use of herbal extracts offers promising outcomes in the management of MDs including depression, anxiety, and schizophrenia, via mechanisms such as neuroendocrine modulation, inhibition of monoamine oxidase enzyme and monoamine availability, and sensitization of serotoninergic receptors. Others are: modulation of GABAergic pathways, hippocampal neurogenesis, and enhancement of BDNF expression with antioxidant defense system against oxidative stress. Therefore, it is possible that these effects exhibited by herbal extract are related to their antioxidant and anti-inflammatory properties.

However, the methodological description and flow in some articles were not well grounded as some reports based their findings on neurobehavioral changes which may not be sufficient as evidence. As such, advanced methodological descriptions involving the use of molecular docking system and unraveling the mechanism of actions are needed to better understand neuro-protective efficacy of herbal extracts in the management of MDs. In addition, detailed methodological description will allow for reproducibility and possible consideration for clinical trials.

It was observed that most of the animal experimentation used stress in inducing MDs. Though stress is not classically a disorder, but the presence of stressors over a long time can stimulate depression and anxiety. Stress could induce accumulation of free radicals which is a culprit in the pathogenesis of many disease conditions including MDs. In stressed individuals, the antioxidant level has been shown to be reduced compared to those without anxiety or depression. Medicinal plants reportedly alleviate stress-induced diseases such as diabetes, cardiovascular and gastrointestinal disorders. As such, their anxiolytic and antidepressive effects could be attributable in part to their antioxidant activities. In conclusion, while the medicinal plants reviewed above appear promising in the management of depression and anxiety, more scientific evidences using robust methodology, biotechnologies, and good manufacturing practice are needed to promote more confidence in the use and acceptability of herbal products in the management of MDs.

KEYWORDS

- **experimental model**
- **herbs**
- **medicinal plants**
- **mental disorders**
- **neuroprotection**

REFERENCES

1. Aguiar, S.; Borowski, T. Neuropharmacological Review of the Nootropic Herb *Bacopa monnieri*. *Rejuvenation Res.* **2013,** *16* (4), 313–326.
2. Ahmed, M.; Azam, K.; Nur, M. Traditional Knowledge and Formulations of Medicinal Plants used by the Traditional Medical Practitioners of Bangladesh to Treat Schizophrenia Like Psychosis. *Schizophr. Res. Treatment*. **2014,** *2*, 14–19.
3. Alam, M. S.; Khan, G. A. Chemical Analysis of Okra Bast Fiber (*Abelmoschus Esculentus*) and its Physico-Chemical Properties. *J. Text. Appar. Technol. Manag.* **2007,** *5* (4).
4. Amalraj, A.; Pius, A.; Gopi, S.; Gopi, S. Biological Activities of Curcuminoids, Other Biomolecules from Turmeric and their Derivatives—A Review. *J. Tradit. Complement. Med.* **2017,** *7* (2), 205–233.
5. Aney, J. S.; Rashmi, T.; Maushumi, K.; Kiran, B. Pharmacological and Pharmaceutical Potential of Moringa Oleifera: A Review. *J. Pharm. Res.* **2019,** *2* (9), 1424–1426.

6. Arapitsas, P. Identification and Quantification of Polyphenolic Compounds from Okra Seeds and Skins. *Food Chem.* **2008,** *110* (4), 1041–1045.
7. Benjawan, C.; Chutichudet, P.; Kaewsit, S. Effects of Green Manures on Growth, Yield and Quality of Green Okra (*Abelmoschus esculentus* L.) Har Lium Cultivar. *Pak. J. Biol. Sci.* **2007,** *10* (7), 1028–1035.
8. Bhat, S. K.; Joy, A. E. Anti-Anxiety Effect of Ethanolic Extract of Leaves of Moringa Oleifera in Swiss Albino Mice. *Arch. Med. Health Sci*. **2014,** *2* (1), 5.
9. Bradley, B. F.; Starkey, N. J.; Brown, S. L.; Lea, R. W. Anxiolytic Effects of Lavandula Angustifolia Odour on the Mongolian Gerbil Elevated Plus Maze. *J. Ethnopharmacol.* **2007,** *111* (3), 517–525.
10. Caraci, F.; Copani, A.; Nicoletti, F.; Drago, F. Depression and Alzheimer's Disease: Neurobiological Links and Common Pharmacological Targets. *Eur. J. Pharmacol.* **2010,** *626* (1), 64–71.
11. Chen, T. H.; Tsai, M. J.; Fu, Y. S.; Weng, C. F. The Exploration of Natural Compounds for Anti-Diabetes from Distinctive Species Garcinia linii with Comprehensive Review of the Garcinia Family. *Biomolecules* **2019,** *9* (11), 641.
12. Chinedu, E.; Arome, D.; Ameh, F. S. A New Method for Determining Acute Toxicity in Animal Models. *Toxicol. Int.* **2013,** *20* (3), 224.
13. Chinnala, K. M.; Achanta, P.; Vangala, V. L.; Elsani, M. M. Evaluation for Nephroprotective Activity of Ethanolic Extract of Allium cepa linn. In Gentamicin-Induced Nephrotoxicity in Rats. *Asian J. Pharm. Clin. Res.* **2017,** *10* (3), 356–359
14. Chopra, R. N.; Nayar, S. L.; Chopra, I. C. *Glossary of Indian Medicinal Plants*; Council of Scientific & Industrial Research, Vol. 1, 1956, pp 138–139.
15. Donner, H.; Gao, L.; Mazza, G. Separation and Characterization of Simple and Malonylated Anthocyanins in Red Onions, *Allium cepa* L. *Food Res. Int*. **1997,** *30* (8), 637–643.
16. ESEMeD/MHEDEA 2000 Investigators, Alonso, J.; Angermeyer, M. C.; Bernert, S.; Bruffaerts, R.; Brugha, T. S.; Gasquet, I. Prevalence of Mental Disorders in Europe: Results from the European Study of the Epidemiology of Mental Disorders (ESEMeD) Project. *Acta Psychiatr. Scand*. **2004,** *109*, 21–27.
17. Farnsworth, N. R.; Morris, R. W. Higher Plants-The Sleeping Giant of Grug Development. *Am. J. Pharm. Educ*. **1976,** *148*, 46–52.
18. Farzin, D.; Fathiazad, F.; Fazellian, M. Antidepressant Effect of Methanolic Ginger Extract in Diabetic Mice Using Forced-Swim Test. *J. Maz. Univ. Med. Sci.* **2013,** *23* (98), 208–220.
19. Foadoddini, M.; Alinejad-Mofrad, S. Effect of Aloe Vera Extract on Depression in People with Prediabetes. *Mod. Care J.* **2020,** *17* (2), 49–58.
20. Gao, Q. H.; Fu, X.; Zhang, R.; Wang, Z.; Guo, M. Neuroprotective Effects of Plant Polysaccharides: A Review of the Mechanisms. *Int. J. Biol. Macromol.* **2018,** *106*, 749–754.
21. Gupta, S.; Jain, R.; Kachhwaha, S.; Kothari, S. L. Nutritional and Medicinal Applications of Moringa Oleifera Lam.—Review of Current Status and Future Possibilities. *J. Herb. Med.* **2018,** *11*, 1–11.
22. Halder, S.; Mehta, A. K.; Mediratta, P. K. Aloe Vera Improves Memory and Reduces Depression in Mice. *Nutr. Neurosci.* **2013,** *16* (6), 250–254.
23. Hashemi, H. J. The Effects of Hydro Alcoholic Extracts of Garlic on Depression Induced by Reserpine in Rats. *Indian J. Neurol.* **2009,** *7*, 527–534.

24. Hazra, S.; Banerjee, R.; Das, B. K.; Ghosh, A. K.; Banerjee, T. K.; Hazra, U. S.; Mondal, A. C. Evaluation of Antidepressant Activity of Bacopa Monnieri in Rat: A Study in Animal Model of Depression. *Drug Discov.* **2012,** *2*, 8–13.
25. Huang, Y. J.; Lu, K. H.; Lin, Y. E.; Panyod, S.; Wu, H. Y.; Chang, W. T.; Sheen, L. Y. Garlic Essential Oil Mediates Acute and Chronic Mild Stress-Induced Depression in Rats Via Modulation of Monoaminergic Neurotransmission and Brain-Derived Neurotrophic Factor Levels. *Food Funct.* **2019,** *10* (12), 8094–8105.
26. Islam, M. T.; Martins, N.; Imran, M.; Hameed, A.; Ali, S. W.; Salehi, B.; Sharifi-Rad, J. Anxiolytic-Like Effects of Moringa Oleifera in Swiss Mice. *Cell Mol. Biol.* **2020,** *66* (4), 59–79.
27. Jeje, S. O.; Adegbite, L. O.; Akindele, O. O.; Kunle-Alabi, O. T.; Raji, Y. Allium Cepa Linn Juice Protect Against Alterations in Reproductive Functions Induced by Maternal Dexamethsone Treatment During Lactation in Male Offspring of Wistar Rats. *Heliyon* **2020,** *6* (5), e03872.
28. Jevas, C. Anti-Diabetic Effects of Allium Cepa (Onions) Aqueous Extracts on Alloxan-Induced Siabetic Rattus Novergicus. *J. Med. Plants Res.* **2011,** *5* (7), 1134–1139.
29. Kadian, R.; Parle, M. Evaluation of Anti-Psychotic Effect of Allium Cepa. *World J. Pharm. Pharm. Sci.* **2014,** *3* (11), 1146–59.
30. Kalpravidh, R. W.; Siritanaratkul, N.; Insain, P.; Charoensakdi, R.; Panichkul, N.; Hatairaktham, S.; Fucharoen, S. Improvement in Oxidative Stress and Antioxidant Parameters in β-Thalassemia/Hb E Patients Treated with Curcuminoids. *Clin. Biochem.* **2010,** *43* (4–5), 424–429.
31. Kaur, G.; Invally, M.; Sanzagiri, R.; Buttar, H. S. Evaluation of the Antidepressant Activity of Moringa Oleifera Alone and in Combination with Fluoxetine. *J. Ayurveda Integr. Med.* **2015,** *6* (4), 273.
32. Kharadi, G. B.; Patel, K. J.; Purohit, B. M.; Baxi, S. N.; Tripathi, C. B. Evaluation of Cardioprotective Effect of Aqueous Extract of *Allium cepa* Linn. Bulb on Isoprenaline-Induced Myocardial Injury in Wistar Albino Rats. *Res. Pharm. Sci.* **2016,** *11* (5), 419.
33. Khomsug, P.; Thongjaroenbuangam, W.; Pakdeenarong, N.; Suttajit, M.; Chantiratikul, P. Antioxidative Activities and Phenolic Content of Extracts from Okra (*Abelmoschus Esculentus* L.). *Res. J. Biol. Sci.* **2010,** *5* (4), 310–313.
34. Kianpour, M.; Mansouri, A. Mehrabi, T.; Asghari, G. Effect of Lavender Scent Inhalation on Prevention of Stress, Anxiety and Depression in the Postpartum Period. *Iranian J. Nurs. Midwifery Res.* **2016,** *21* (2), 197–202.
35. Koulivand, P. H.; Khaleghi-Ghadiri, M.; Gorji, A. Lavender and the Nervous System. *Evid. Based Complement. Altern. Med.* **2013,** *2013*, 993–1023.
36. Kumar, A.; Bora, K. S.; Jaggi, A. S.; Shri, R. Comparative Evaluation of Neuroprotective Effect of Three Varieties of Allium Cepa in Chronic Constriction Injury Induced Neuropathic Pain. *Thai J.Pharm. Sci.* **2016,** *40*, 1–5.
37. Kumar, S.; Mondal, A. C. Neuroprotective, Neurotrophic and Anti-Oxidative Role of Bacopa Monnieri on CUS Induced Model of Depression in Rat. *Neurochem. Res.* **2016,** *41* (11), 3083–3094.
38. Lee, K. A.; Kim, K. T.; Kim, H. J.; Chung, M. S.; Chang, P. S.; Park, H.; Pai, H. D. Antioxidant Activities of Onion (*Allium Cepa* L.) Peel Extracts Produced by Ethanol, Hot Water, and Subcritical Water Extraction. *Food Sci. Biotechnol.* **2014,** *23* (2), 615–621.

39. López, V.; Nielsen, B.; Solas, M.; Ramírez, M. J.; Jäger, A. K. Exploring Pharmacological Mechanisms of Lavender (*Lavandula Angustifolia*) Essential Oil on Central Nervous System Targets. *Front. Pharmacol.* **2017,** *8*, 280.
40. Mäki-Marttunen, T.; Kaufmann, T.; Elvsåshagen, T.; Devor, A.; Djurovic, S.; Westlye, L. T.; Efrim-Budisteanu, M. Biophysical Psychiatry—How Computational Neuroscience Can Help to Understand the Complex Mechanisms of Mental Disorders. *Front. Psychiat.* **2019,** *10*, 534.
41. Marschollek, C.; Karimzadeh, F.; Jafarian, M.; Ahmadi, M.; Mohajeri, S. M. R.; Rahimi, S.; Gorji, A. Effects of Garlic Extract on Spreading Depression: In Vitro and In Vivo Investigations. *Nutr. Neurosci.* **2017,** *20* (2), 127–134.
42. Martatino, I. Y.; Habibie, R.; Sahrah, A.; Wardhana, A. A. The Innovative of Anxiety Disorder Healing: Nutri Moringa Pudding for HIV/AIDS-Infected Patients. *Int. J. Asian Soc. Sci.* **2014,** *4* (11), 1100–1109.
43. Miodownik, C.; Lerner, V.; Kudkaeva, N.; Lerner, P. P.; Pashinian, A.; Bersudsky, Y.; Bergman, J. Curcumin as Add-On to Antipsychotic Treatment in Patients with Chronic Schizophrenia: a Randomized, Double-Blind, Placebo-Controlled Study. *Clin. Neuropharmacol.* **2019,** *42* (4), 117–122.
44. Ng, Q. X.; Koh, S. S. H.; Chan, H. W.; Ho, C. Y. X. Clinical use of curcumin in depression: a meta-analysis. *J. Am. Med. Dir. Assoc.* **2017,** *18* (6), 503–508.
45. Ogata, K.; Ataka, K.; Suzuki, H.; Yagi, T.; Okawa, A.; Fukumoto, T.; Asakawa, A. Lavender Oil Reduces Depressive Mood in Healthy Individuals and Enhances the Activity of Single Oxytocin Neurons of the Hypothalamus Isolated from Mice: A Preliminary Study. *Evid. Based Complement. Alternat. Med.* ***2020,*** *2020*, 56–69.
46. Ogundele, M. O. Behavioral and Emotional Disorders in Childhood: A Brief Overview for Paediatricians. *World J. Clin. Pediatr.* **2018,** *7* (1), 9.
47. Ojo, O. B.; Amoo, Z. A.; Saliu, I. O.; Olaleye, M. T.; Farombi, E. O.; Akinmoladun, A. C. Neurotherapeutic Potential of Holaviron on Neurotransmitter Dysregulation, Excitotoxicity, Mitochondrial Electron Transport Chain Dysfunction and Redox Imbalance in 2-VO Brain Ischemia/Reperfusion Injury. *Biomed. Pharmacother.* **2019,** *111*, 859–872.
48. Olanrewaju, J. A.; Owolabi, J. O.; Awodein, I. P.; Enya, J. I.; Adelodun, S. T.; Olatunji, S. Y.; Fabiyi, S. O. Zingiber officinale Ethanolic Extract Attenuated Reserpine-Induced Depression-Like Condition and Associated Hippocampal Aberrations in Experimental Wistar Rats. *J. Exp. Pharmacol.* **2020,** *12*, 439.
49. Omotoso, G. O.; Olajide, O. J.; Gbadamosi, I. T.; Adebayo, J. O.; Enaibe, B. U.; Akinola, O. B.; Owoyele, B. V. Cuprizone Toxicity and Garcinia Kola Biflavonoid Complex Activity on Hippocampal morphology and Neurobehaviour. *Heliyon* **2019,** *5* (7), e02102.
50. Omotoso, G. O.; Gbadamosi, I. T.; Olajide, O. J.; Dada-Habeeb, S. O.; Arogundade, T. T.; Yawson, E. O. Moringa Oleifera Phytochemicals Protect the Brain against Experimental Nicotine-Induced Neurobehavioral Disturbances and Cerebellar Degeneration. *Pathophysiology* **2018,** *25* (1), 57–62.
51. Onasanwo, S. A.; Ilenre, K. O.; Faborode, S. O. The Impact of Kolaviron (a Bioflavonoid of Garcinia Kola Seed) on Depression Status in Laboratory Rodents: Roles of Monoaminergic Systems. *Ann. Depress. Anxiety*. **2015,** *2* (1), 1042.
52. Pal, S. K.; Shukla, Y. Herbal Medicine: Current Status and the Future. *Asian Pac. J. Cancer Prev.* **2003,** *4* (4), 281–288.

53. Pandit, S.; Kim, H. J.; Kim, J. E.; Jeon, J. G. Separation of an Effective Fraction from Turmeric Against Streptococcus Mutans Biofilms by the Comparison of Curcuminoid Content and Anti-Acidogenic Activity. *Food Chem.* **2011,** *126* (4), 1565–1570.
54. Pujari, R. R.; Vyawahare, N. S.; Thakurdesai, P. A. Neuroprotective and Antioxidant Role of Phoenix Dactylifera in Permanent Bilateral Common Carotid Occlusion in Rats. *J. Acute Dis.* **2014,** 104–114.
55. Radha, M. H.; Laxmipriya, N. P. Evaluation of Biological Properties and Clinical Effectiveness of Aloe Vera: A Systematic Review. *J. Tradit. Complement. Med.* **2015,** *5* (1), 21–26.
56. Rahmani, G.; Farajdokht, F.; Mohaddes, G.; Babri, S.; Ebrahimi, V.; Ebrahimi, H. Garlic (*Allium Sativum*) Improves Anxiety-and Depressive-Related Behaviors and Brain Oxidative Stress in Diabetic Rats. *Arch. Physiol. Biochem.* **2020,** *126* (2), 95–100.
57. Rehman, R.; Saif, S.; Hanif, M. A.; Riaz, M. Garlic. In *Medicinal plants of South Asia*; Hanif, M., Nawaz, H., Khan, M., Byrne, H., Eds.; Elsevier BV: Amsterdam, 2019; pp 301–315.
58. Reynolds, T.; Dweck, A. C. Aloe Vera Leaf Gel: A Review Update. *J. Ethnopharmacol* **1999,** *68* (1–3), 3–37.
59. Roodenrys, S.; Booth, D.; Bulzomi, S.; Phipps, A.; Micallef, C.; Smoker, J. Chronic Effects of Brahmi (Bacopa monnieri) on Human Memory. *Neuropsychopharmacology* **2002,** *27* (2), 279–281.
60. Sabitha, V.; Ramachandran, S.; Naveen, K. R.; Panneerselvam, K. Investigation of In Vivo Antioxidant Property of *Abelmoschus esculentus* (L) Moench. Fruit Seed and Peel Powders in Streptozotocin-Induced Diabetic Rats. *J. Ayurveda Integr. Med.* **2012,** *3* (4), 188.
61. Sadeghiani, N.; Barbosa, L. S.; Silva, L. P.; Azevedo, R. B.; Morais, P. C.; Lacava, Z. G. M. Genotoxicity and Inflammatory Investigation in Mice Treated with Magnetite Nanoparticles Surface Coated with Polyaspartic Acid. *J. Magn. Magn. Mater.* **2005,** *289*, 466–468.
62. Sánchez-Vidaña, D. I.; Po, K. K. T.; Fung, T. K. H.; Chow, J. K. W.; Lau, W. K. W.; So, P. K.; Tsang, H. W. H. Lavender Essential Oil Ameliorates Depression-Like Behavior and Increases Neurogenesis and Dendritic Complexity in Rats. *Neurosci. Lett.* **2019,** *701*, 180–192.
63. Seol, G. H.; Shim, H. S.; Kim, P. J.; Moon, H. K.; Lee, K. H.; Shim, I.; Min, S. S. Antidepressant-Like Effect of Salvia Sclarea is Explained by Modulation of Dopamine Activities in Rats. *J. Ethnopharmacol.* **2010,** *130* (1), 187–190.
64. Shahrajabian, M. H.; Sun, W.; Cheng, Q. Clinical Aspects and Health Benefits of Ginger (Zingiber Officinale) in Both Traditional Chinese Medicine and Modern Industry. *Acta Agric. Scand.* **2019,** *69* (6), 546–556.
65. Singh, V.; Chauhan, G.; Shri, R. Anti-Depressant Like Effects of Quercetin 4'-O-Glucoside from Allium Cepa Via Regulation of Brain Oxidative Stress and Monoamine Levels in Mice Subjected to Unpredictable Chronic Mild Stress. *Nutr. Neurosci.* 2021;24(1):35–44.
66. Stajner, D.; Varga, I. S. An Evaluation of the Antioxidant Abilities of Allium Species. *Acta Biol. Szeged.* **2003,** *47* (1–4), 103–106.
67. Tongjaroenbuangam, W.; Ruksee, N.; Chantiratikul, P.; Pakdeenarong, N.; Kongbuntad, W.; Govitrapong, P. Neuroprotective Effects of Quercetin, Rutin and Okra (Abelmoschus Esculentus Linn.) in Dexamethasone-Treated Mice. *Neurochem. Int.* **2011,** *59* (5), 677–685.

68. Veeresham, C. Natural Products Derived from Plants as a Source of Drugs. *J. Adv. Pharm. Technol. Res.* **2012,** *3* (4), 200.
69. White, B. Ginger: An Overview. *Am. Fam. Physician.* **2007,** *75* (11), 1689–1691.
70. WHO. *WHO Guidelines on Safety Monitoring of Herbal Medicines in Pharmacovigilance Systems*; World Health Organization: Geneva, 2004.
71. WHO. Mental Disorders [Online]. www.who.int/news-room/fact-sheets/detail/mental-disorders (accessed August 31, 2021).
72. WHO. Mental Health and Forced Displacement. https://www.who.int/news-room/fact-sheets/detail/mental-health-and-forced-displacement (accessed August 31 2021).
73. Xia, F.; Li, C.; Li, M.; Liao, Y.; Liu, X.; Si, J.; Pan, R. Antidepressant Activity of an Aqueous Extract from Okra Seeds. *RSC Adv.* **2018,** *8* (57), 32814–32822.
74. Xu, Y.; Ku, B. S.; Yao, H. Y.; Lin, Y. H.; Ma, X.; Zhang, Y. H.; Li, X. J. Antidepressant Effects of Curcumin in the Forced Swim Test and Olfactory Bulbectomy Models of Depression in Rats. *Pharmacol. Biochem. Behav.* **2005,** *82* (1), 200–206.
75. Yan, T.; Nian, T.; Liao, Z.; Xiao, F.; Wu, B.; Bi, K.; Jia, Y. Antidepressant Effects of a Polysaccharide from Okra (*Abelmoschus Esculentus* (L) Moench) by Anti-Inflammation and Rebalancing the Gut Microbiota. *Int. J. Biol. Macromol.* **2020,** *144*, 427–440.
76. Yu, Y.; Wu, S.; Li, J.; Wang, R.; Xie, X.; Yu, X.; Zheng, L. The Effect of Curcumin on the Brain-Gut Axis in Rat Model of Irritable Bowel Syndrome: Involvement of 5-HT-Dependent Signaling. *Metab. Brain Dis.* **2015,** *30* (1), 47–55.
77. Yu, Z. F.; Kong, L. D., Chen, Y. Antidepressant Activity of Aqueous Extracts of Curcuma Longa in Mice. *J. Ethnopharmacol.* **2002,** *83*(1–2), 161–165.
78. Zhan, P. Y.; Zeng, X. H.; Zhang, H. M.; Li, H. High-Efficient Column Chromatographic Extraction of Curcumin from Curcuma Longa. *Food Chem.* **2011,** *129* (2), 700–703.
79. Zhang, Z. J. Therapeutic Effects of Herbal Extracts and Constituents in Animal Models of Psychiatric Disorders. *Life Sci.* **2004,** *75*(14), 1659–1699.

CHAPTER 8

Translational Neuroprotective Activity of Curcumin in Neurodegenerative Diseases: An Overview with Animal Models

Adedamola Aminat Bayo-Olugbami, Kawkab Mujahidah Babalola and Aminat Omolola Imam-Fulani

ABSTRACT

The two commonest neurodegenerative diseases (NDDs), which affects the elderly people are Parkinson's (PD) and Alzheimer's diseases (AD). Protein aggregation, oxidative damage, apoptotic signaling, inflammation, and microgliosis are some of the mechanisms implicated in the pathophysiology of NDDs. The available conventional drugs provide symptomatic relief and not necessarily slow down the disease progression or prevent it. In addition, the attending side effects are a major concern. These two disorders are only best managed as effective cure is still lacking and as remained the focus of several researches.

There is a growing increase in exploration of plants of medicinal value for their effectiveness in the management of various diseases. The reason could be that herbal plants are cheaper, readily accessible, and culturally suitable alternatives to conventional therapies especially in the developing nations. Curcumin is a naturally-occurring polyphenol and the main curcuminoid found in Curcuma longa plant. Its pleiotropic effects on the nervous system have been documented. These include but not limited to its ability to mitigate against protein aggregation, oxidative stress, cell death, and inflammation.

Curative and Preventive Properties of Medicinal Plants: Research on Disease Management and Animal Model Studies. Megh R. Goyal, Junaid Ahmad Malik, & Ademola Olabode Ayeleso (Eds.)

8.1 INTRODUCTION

Neurodegenerative diseases (NDDs) result from the progressive deterioration of neurons, which over time leads to neurodegeneration and disabilities. NDDs diseases affect the quality of life and its prevalence in recent times is on the increase. These diseases have no known cure and are managed by treating the symptoms exhibited per time.[14] Clinically, there is loss of neurological functions (paralysis, akinesia, dementia) and gradual degeneration of the neuron. Example of these diseases are multiple sclerosis, Alzheimer's disease (AD), amyotrophic lateral sclerosis, Parkinson's disease (PD), and Huntington's disease. AD as well as PD are the common neurodegenerative disorders.[51] They have multiple etiologies and have no known cure yet.[55] Both AD and PD also share some pathogenic patterns, such as, age-related decline, diverse environmental risks, and multiple genetic defects distributed across the genome. The current drugs for managing AD and PD mostly reduce the disease symptoms and coadministration of several drugs may lead to severe side-effects and adverse reactions in patients.[8]

In the world, about 40–50 million individuals are affected by one form of dementia or the other in addition to AD.[37] In AD, there is impairment in cognition, language processing, and ability to solve problems. AD brain has hyperphosphorylated *tau* found in the amyloid beta and neurofibrillary tangles located within the amyloid plaques. Conversely, PD being the most common motor dysfunction and second most common NDDs may affect not less than 10 million people in the World. The most distinct pathologic feature in PD is the progressive destruction of dopamine-containing neurons along with the presence of collections known as Lewy bodies, consisting of proteinous alpha-synuclein.[51] NDDs are mostly found in elderly persons and have become one of the main causes of health challenges.[42]

AD, PD, and most neurodegenerative disorders have features of accumulated abnormal folded proteins, which are not only toxic but also interfere with neuronal metabolism. In these disorders, the existence of clusters of misfolded proteins indicates that protein control system is unable to maintain protein homeostasis, often called proteostasis.[21] Mechanisms like neuroinflammation, oxidative stress (OS), and the accumulation of certain aggregated proteins are implicated as the underlying causes of neurodegeneration in PD and AD.[12]

The constituents in medicinal plants have therapeutic effect or serves as precursor(s) for the synthesis of other drugs.[49] For a long time, medicinal plants have been harnessed traditionally to manage several ailments. The use of herbal plants is on the rise. This has been attributed to high cost and abnormal effects of orthodox medicines, affordability, accessibility, and availability of medicinal plants.[42] The phytochemical nature of medicinal plants enables them to work on multiple areas of associated physiologic dysfunction in a logical, multitactical approach to systems regulation.[20] For many years, many of these plants were used traditionally even without sufficient scientific data to ascertain their efficacy.[49]

The attention of researchers is gradually turning to herbal medicine as an alternative therapy in managing NDDs.[33] Phytochemicals have a wide range of biological impacts. They include but not limited to: flavonoids, steroids, tannins, phenols, saponins, glycosides, alkaloids, and terpenoids. Given that the causes of NDDs are multifactorial and complex, the multiple constituents of herbal plants may be more potent in treating these disorders since effective medication is still lacking.[42] Herbal medicines are reportedly effective in ameliorating various disease conditions. Also, some natural products have been discovered to have advantage over synthetic drugs.[9]

Curcumin is a naturally existing polyphenol extracted from rhizome of *Curcuma longa.*[7] It accounts for 2–5% of the spice. The powder form has a characteristic orange–yellow color, which is completely insoluble in water. It is harnessed in the treatment of various ailments, as coloring additives in foods and textiles, and spice for cooking. Traditionally, it is popularly used to treat inflammation, hepatic disorders, rheumatism, diabetes, cough, wounds, and sinusitis. Curcumin has a number of biological effects including its antimicrobial, anti-inflammatory, anticarcinogenic, hypercholesteremic, antimutagenic, antioxidant, antiprotozoal, antibacterial, antifibrotic, antiulcer, hypotensive, antiproliferative, and antidiabetic properties.[4,7] In spite of the fact that Curcumin has been confirmed to be generally safe for human consumption, its effectiveness is greatly challenged by its low bioavailability. Though curcumin is easily metabolized, yet it has low rate of absorption and short half-life.[7]

This chapter focuses on highlighting evidence-based preventive and neuroprotective properties of Curcumin, its mechanism of action and impact on neuronal communication, with focus on restoration of neural plasticity, integrity, cognitive and motor imbalances associated with AD and PD in various animal models

8.2 PATHOPHYSIOLOGY OF PD AND AD

In these NDDs, there is a characteristic continuous, steady and specific destruction of some neuronal cell types. Nitrosative and oxidative stresses, mitochondrial malfunction, inflammation, and apoptosis underlie these disorders. They can be described typically by neuronal vulnerability and accumulations of abnormally folded proteins. While the structures of protein clusters differ, these disorders have similar pattern of aggregation of proteins, along with formation of fibrils and amyloids.[13] According to Alois Alzheimer, the availability of amyloid plaques along with neurofibrillary tangles within neurons is the hallmark of AD.

Tau, which is a microtubule-associated protein, is not only a classic feature of AD, but is also observed in other similar NDDs. In transgenic mice showing tau, co-expression of Amyloid precursor protein or α-synuclein aggravates tau pathology, so that neurodegeneration caused by tau pathology may be activated downstream of disease pathways. The α-synuclein can be described as intracellular amyloid characterized in PD. Mutation of the gene coding for α-synuclein is the basis for familial form of PD. Increased cortical expression of synuclein has fundamental role in PD pathology, because α-synuclein gene mutation has been connected to autosomal-dominantly inherited Lewy body dementia and PD.[48] Similarly, transgenic mice conditioned to overexpress mutant α-synuclein showed α-synuclein pathology and neuronal degeneration.[32]

8.2.1 AGING

This is an essential predisposing factor in the development of NDDs. Aging impairs neuronal repair making them more susceptible to neurodegenerative assaults.[22] Many signaling pathways identified to be involved in aging and lifespan have also been identified in age-mediated cognitive impairments.[6,19] According to aging models using invertebrate organisms, there is a relationship between cognition and mitopathology because several genes involved in synaptic function, neural plasticity, and mitochondrial metabolism show structural changes during aging. Synaptic integrity is compromised in the axon defective in MEC-17. Such axons also show transport deficit and reduction in the quantity of mitochondria. In *Caenorhabditis elegans,* neuronal dysfunction caused by axonal degeneration has been attributed to the loss of MEC-17, an alpha-tubulin acetyltransferase gene.[19] With aging, a vicious circle of OS is generated whereby OS causes dyshomeostasis in various body systems, such as, the immune response which in turn leads to inflammatory response and further exacerbates OS..[15]

8.2.2 OXIDATIVE STRESS (OS)

Due to the different biological processes and metabolic activities required to sustain cellular homeostasis, the development of reactive oxygen stress (ROS) and nitrogen stress in cells is unavoidable. In biological systems, OS denotes the discrepancy between oxidants and antioxidants. This disparity is due to overwhelming ROS concentrations and redox-state dyshomeostasis. OS process includes overproduction of ROS, which is critically involved in the advancement of NDDs. The cortical tissue is susceptible to OS due to the elevated level of lipid in addition to increased oxygen breakdown. Self-oxidation of dopamine, iron and calcium dysmetabolism, glial activation, and mitochondrial dysregulation are due to ROS.[31]

The nervous tissue particularly the cortical structure is very susceptible to oxidative damage (attacks from free radicals) due to low quantity of enzymatic and nonenzymatic antioxidants), increased rate of metabolism, and elevated level of polyunsaturated fats. In addition, leaked oxygen metabolism electrons allow the removal of hydrogen ions and increase the occurrence of lipid peroxidation. Thus, OS damage is caused by few signaling pathways that affect enzymes that are essential to neuronal and glial functions.[26]

Research has shown that OS increases in the aging brain. AD, PD and most neurodegenerative disorders are age-driven; and in the pathogenesis of these diseases, ROS-induced oxidative damage is involved. Further, OS, increased lipid peroxidation, mitochondrial dysfunction, and energy failure resulting from plaque formation are caused by Aβ peptides and clustering of fibrillary tangles containing *aggregated* Tau and Aβ.[62] The AD brain has been shown to have an elevated level of Malondialdehyde, a pro-oxidant and a marker of lipid peroxidation.[1]

In another study, histological examination demonstrated dispersion in the arrangement of pyramidal neurons, and degenerating features while histochemical examination demonstrated deposition of Tau and Aβ in the hippocampus.[18] Cellular damage, mitochondrial dysfunction, and deficit in DNA repair are capable of playing pivotal role in the aging process, which in turn may influence the development of NDDs,[50] and it may be triggered by the release of ROS.

The relationship between the generation of ROS and pathological basis of NDDs has been identified in several research studies.[52] OS and ROS significantly affect AD because of the toxic impacts on protein and other molecules.[46] In AD brain, oxidative imbalance exacerbates neuronal damage as a result of oxidation of Tau and β-amyloid by ROS.[35] The buildup of Aβ aggregates causes mitochondrial malfunction, which is a major driver of AD.[38] Furthermore, calcium dysregulation, poor antioxidant status, Tau aggregation, excessive production of ROS, accumulation of amyloid plaque, mitochondrial dysfunction, and accumulation of iron have been implicated in OS, which in turn underlies the pathophysiology of PD and AD.[58] Excessive oxidative assaults to β-amyloid protein may cause misfolding and accumulation of abnormal proteins, which can worsen neurodegeneration and subsequently the loss of neuronal cells in AD. Likewise, deposition of alpha-synuclein protein, dopamine dysmetabolism, alteration in mitochondrial DNA, iron buildup, upsurge in the level of nigral lipid peroxidation, accumulation of OH^- radical, and reduced antioxidant status underlie the pathology of PD. All these promote dopaminergic neuronal death, which reduces dopaminergic transmission and consequently, loss of motor coordination and balance, which characterize PD.[31]

8.2.3 INFLAMMATION

Activation of the Immune system is a typical feature in ischemia, trauma, NDDs, infections, and immune-linked diseases. However, some immune responses in the CNS are protective as they promote regenerative processes and cellular repair. For instance, microglia are involved in the clearance of debris following myelin injury and any attempt to alter this process can hinder neuronal regeneration. Microglial activation can induce the release of both neuroprotective and neurotoxic molecules depending on the pathway activated. Upregulation of IL-1β and TNF-α, which are important markers of inflammation, has been demonstrated in the AD brain in various studies.[1] Activation of nuclear kappa factor-B controls the impact of TNF-α, which is one of the most important pro-inflammatory markers in these diseases.[2]

8.3 CURCUMIN AND NEURODEGENERATION

Curcumin is an anti-inflammatory and antioxidant biocompound, which has been documented as an effective in the treatment of cancers, inflammatory diseases and in laboratory models of PD and AD.[23] It is a pleiotropic compound, which does not only function to restore immune homeostasis, cleaves iron, promotes the response of antioxidant elements and removal of toxic collections, cleaves off free radicals, but also binds and reduces clustering of amyloid sheets.[23] Curcumin modulates Aβ by cleaving off the aggregated synuclein clusters thus, reducing its toxicity.[27] The diketone bridge in curcumin is responsible for binding amyloid and the consequent reduction in plaque formation.[56]

The fluorescence property of curcumin may be beneficial in reducing the disease progress by analysis of retina plaques, which can reflect plaque deposits in the brain;[23] and this property has been proposed as a noninvasive diagnostic tool.[29]

8.3.1 NEUROPROTECTIVE EFFECTS OF CURCUMIN

Curcumin attaches to the β-pleated sheet in the helical strands of Tau and also inhibits its collection. A pivotal mechanism via which curcumin acts is by binding to amyloid structure to reduce plaque deposition in most AD models.[11] It also reduces soluble phosphorylated Tau oligomers and rectifies synaptic deficits.[30] Curcumin reportedly attaches to neurofibrillary tangles in animal models and human AD brain. It also binds other aggregates β-sheets found in Prion and Huntington's disease.[39,40]

Curcumin may increase the life-cycle of *C. elegans* by inhibiting the release of ROS because of its antioxidative properties.[1] Cur2004-8, a man-made off-shoot of curcumin, was reportedly more potent in improving the lifespan and enhancing resistance against OS than the curcumin.[45] Likewise, the ability of curcumin to promote resistance to heat stress, irradiation, and OS, thereby improving the lifespan of Drosophila has been reported.[10] Kitani and others[28] revealed that curcumin derivative, tetrahydro-curcumin was able to improve the lifespan of mice when administered for 13 months.

8.3.2 ANTI-INFLAMMATORY EFFECTS OF CURCUMIN

The inflammatory cascade is a basic mediator in the pathophysiology of NDDs.[2] Curcumin was found to possess powerful anti-inflammatory response that was comparable with both steroidal and NSAIDs.[2] Its anti-inflammatory outcome is facilitated by preventing the initiation of transcription factors, such as NF-κB and AP-1, induction of iNOS, COX-2, and release of cytokines (e.g., interferon and TNF).[34] Curcumin reportedly modulates cytokines, redox status, transcription factors, and enzymes involved inflammatory processes.[2]

In another study, curcumin considerably attenuated amyloid level, IL-1β as well as oxidized proteins in APP/SW mice, thereby reducing cognitive impairment. The ability of Curcumin to interact with Cu and Fe in a metal-induced AD-like pathology showed its probable mechanism of action.[5]

Curcumin could also subdue inflammatory impairment by inhibiting metallic stimulation of NF-κB.[2]

Similarly, it reportedly decreased the Aβ formation. This effect was dependent on fibril conformation and not on the sequence of Aβ. When brain slices of Tg2576 and AD mice were incubated with curcumin, these preferentially labeled the amyloid plaques. In addition, it permeated the blood-brain barrier (BBB) to bind senile deposits when introduced peripherally in aged Tg2576 mice. Administration of curcumin to Tg2576 mice having advanced amyloid deposit improved the amyloid pathology by labeling the plaques and reduction in amyloid levels. This implies that curcumin binds directly to beta-amyloid sheets to prevent formation and accumulation of fibril in in vivo and in vitro models.[2]

Furthermore in AD brain, Aβ phagocytosis by macrophages and its clearance are defective.[36] However, curcumin enhanced the Aβ uptake by macrophages.[61] Similarly in an animal model of AD, curcumin significantly improved the structure of distorted neurites and disrupted amyloid plaques by crossing the BBB to cleave off senile plaques in APPswe/PS1dE9 mice.[16]

8.3.3 ANTIOXIDANT PROPERTIES OF CURCUMIN

Curcumin possesses antioxidant properties for reducing and scavenging ROS radicals.[3] Curcumin protected against OS in the rotenone, 6-OHDA, and MPTP models of PD.[43,47] Curcumin reduces OS, inflammatory cytokines, and upregulates antioxidant defense mechanisms. In primary astrocytes, it inhibited the formation of MPP^+, suppressed the activities of IL-6, TNF, and ROS as well as augmented GSH level and IL-10.[59]

Curcumin prevents Aβ fibril formation by binding to it and inhibiting Aβ clusters.[27] In rat's cortical neurons, maturation of the precursor for Aβ was inhibited following intake of curcumin.[60] Huang and others[24] reported that curcumin offers a more potent protective effect against oxidative assaults when administered prior to injection of Aβ toxin as evidenced by decline in ROS generation. As shown by Xiao et al., curcumin and its variant Cur1 promoted the expression of hTERT, which guarded the cells from Aβ-mediated toxicity.[54] Similarly in rat primary neurons, curcumin inhibited death of Aβ-induced cells.[57] Another curcumin derivative, di-O-demethylcurcumin reportedly preserved neuroblastoma cells from Aβ-mediated oxidative toxicity. Not only does curcumin inhibited peroxide-mediated neuronal death, it also caused decline in lipid peroxidation, intracellular caspase-3 and -9 in H_2O_2-driven OS prototype of SH-SY5Y neurons.[53] Pogue et al.[41]

showed that curcumin could permeate the BBB via in vivo multiphoton microscopy. Curcumin was protective against glutamate-mediated oxidative toxicity. In transgenic models of AD and streptozotocin model of dementia, curcumin showed neuroprotective impacts. In the Streptozotocin-induced dementia model of AD, curcumin reversed OS [44] and amplified glutathione antioxidant enzyme.[25] In APP/PS1 mice genetically modified for AD, curcumin modulated degenerative changes in synapse and increased the expression of proteins associated with mitochondrial functions.[17]

8.4 SUMMARY

Scientific data on the role of curcumin in NDDs using animal models depicted that this extract has potent antioxidant and anti-inflammatory properties. It could regulate apoptotic signaling and cell cycle, modifies gene expression, and alters several molecular targets and signaling pathways. Curcumin has been subjected to preclinical investigation in various neurodegenerative models but its low bioavailability is a limitation for oral administration. Curcumin bioavailability has been enhanced by several nano-formulations. Data from several animal studies recommended that curcumin might be beneficial in treating NDDs. While applicable molecular targets are unclear, nevertheless, curcumin is safe, inexpensive natural extract and a promising candidate for further investigation and development in the quest for PD and AD therapeutics. As such, from the above potential mechanism of action of curcumin extract, it may be effective as cognitive enhancer as well as delaying the progression of AD and PD. However, further molecular investigations will provide more mechanistic evidences that will scientifically support its use in the management of NDDs.

KEYWORDS

- **Alzheimer's disease**
- **animal models**
- **curcumin**
- **neurodegenerative disease**
- **neuroprotection**
- **Parkinson's disease**

REFERENCES

1. Aboelwafa, H. R.; El-Kott, A. F.; Abd-Ella, E. M.; Yousef, H. N. The Possible Neuroprotective Effect of Silymarin against Aluminum Chloride-Prompted Alzheimer's-Like Disease in Rats. *Brain Sc.* **2020,** *10* (9), 628–634.
2. Aggarwal, B. B.; Harikumar, K. B. Potential Therapeutic Effects of Curcumin, the Anti-Inflammatory Agent, against Neurodegenerative, Cardiovascular, Pulmonary, Metabolic, Autoimmune and Neoplastic Diseases. *Int.J. Biochem. Cell Biol.* **2019,** *41* (1), 40–59.
3. Babu, A.; Mohammed, S.; Harikumar, K. B. Antioxidant Properties of Curcumin: Impact on Neurological Disorders. In *Curcumin for Neurological and Psychiatric Disorders: Neurochemical and Pharmacological Properties*; Farooqui, T.; Farooqui, A., Eds.; Academic Press: Cambridge, 2019; pp 155–167.
4. Bagchi, A. Extraction of Curcumin. *Toxicol. Food Technol.* **2012,** *1* (3), 25–36.
5. Baum, L.; Ng, A. Curcumin Interaction with Copper and Iron Suggests One Possible Mechanism of Action in Alzheimer's Disease Animal Models. *J. Alzheimer's Dis.* **2004,** *6* (4), 367–377.
6. Berkowitz, B. A.; Podolsky, R. H.; Childers, K. L.; Gow, A.; Schneider, B. L.; Lloyd, S. C.; Bosse, K. E.; Conti, A. C.; Roberts, R.; Berri, A. M.; Graffice, E.; Sinan, K.; Eliwat, W.; Shen, Y. Age-Related Murine Hippocampal CA1 Laminae Oxidative Stress Measured In Vivo by QUEnch-assiSTed (QUEST) MRI: Impact of Isoflurane Anesthesia. *GeroScience* **2020,** *42* (2), 563–574.
7. Bhat, A.; Mahalakshmi, A. M.; Ray, B.; Tuladhar, S.; Hediyal, T. A.; Manthiannem, E.; Sakharkar, M. K. Benefits of Curcumin in Brain Disorders. *BioFactors* **2019,** *45* (5), 666–689.
8. Cacabelos, R. Pharmacogenomics of Alzheimer's and Parkinson's Diseases. *Neurosci. Lett.* **2018,** *2018*, E-article 133807.
9. Calixto, J. B. The Role of Natural Products in Modern Drug Discovery. *An. Acad. Bras. Ciênc.* **2019,** *91*, 75–83.
10. Chen, Y.; Liu, X.; Jiang, C.; Liu, L.; Ordovas, J. M.; Lai, C. Q.; Shen, L. Curcumin Supplementation Increases Survival and Lifespan in Drosophila under Heat Stress Conditions. *BioFactors* **2018,** *44* (6), 577–587.
11. Cheng, K. K.; Yeung, C. F.; Ho, S. W.; Chow, S. F.; Chow, A. H. L.; Baum, L. Highly Stabilized Curcumin Nanoparticles Tested in an In Vitro Blood-Brain Barrier Model and in Alzheimer's Disease Tg2576 Mice. *AAPS J.* **2013,** *15* (2), 324–336.
12. Concetta-Scuto, M.; Mancuso, C.; Tomasello, B.; Laura Ontario, M.; Cavallaro, A.; Frasca, F.; Calabrese, V. Curcumin, Hormesis and the Nervous System. *Nutrients* **2019,** *11* (10), 2417–2429.
13. Dugger, B. N.; Dickson, D. W. Pathology of Neurodegenerative Diseases. *Cold Spring Harb. Perspect. Biol.* **2017,** *9* (7), a028035.
14. Durães, F.; Pinto, M.; Sousa, E. Old Drugs as New Treatments for Neurodegenerative Diseases. *MDPI* **2018,** *2*, 44–49.
15. Fuente, M.; Miquel, J. An Update of the Oxidation-Inflammation Theory of Aging: The Involvement of the Immune System in Oxi-Inflamm-Aging. *Curr. Pharm. Des.* **2009,** *15* (26), 3003–3026.
16. Garcia-Alloza, M.; Borrelli, L. A.; Rozkalne, A.; Hyman, B. T.; Bacskai, B. J. Curcumin Labels Amyloid Pathology In Vivo, Disrupts Existing Plaques, and Partially Restores Distorted Neurites in an Alzheimer Mouse Model. *J. Neurochem.* **2007,** *102* (4), 1095–1104.

17. Gerenu, G.; Liu, K.; Chojnacki, J. E.; Saathoff, J. M.; Martínez-Martín, P.; Perry, G.; Zhu, X.; Lee, H. G.; Zhang, S. Curcumin/Melatonin Hybrid 5-(4-Hydroxy-phenyl)-3-Oxo-Pentanoic Acid [2-(5-Methoxy-1H-indol-3-yl)-Ethyl]-Amide Ameliorates AD-Like Pathology in the APP/PS1 Mouse Model. *ACS Chem. Neurosci.* **2015,** *6* (8), 1393–1399.
18. Gibbons, G. S.; Banks, R. A.; Kim, B.; Changolkar, L.; Riddle, D. M.; Leight, S. N.; Lee, V. M. Detection of Alzheimer Disease (AD)-Specific Tau Pathology in AD and Non-AD Tauopathies by Immunohistochemistry with Novel Conformation-Selective Tau Antibodies. *J. Neuropathol. Exp. Neurol.* **2018,** *77* (3), 216–228.
19. Gkikas, I.; Petratou, D.; Tavernarakis, N. Longevity Pathways and Memory Aging. *Front. Genet.* **2015,** *5*, 155–160.
20. Hedayat, K. M.; Lapraz, J.C.; Schuff, B. Medicinal Plants in Clinical Practice. In *The Theory of Endobiogeny*; Hedayat, K. M., Lapraz, J. C., Schuff, B. Eds.; Academic Press: Cambridge, 2020; pp 57–60.
21. Hoozemans, J. J.; Van Haastert, E. S.; Nijholt, D. A.; Rozemuller, A. J.; Scheper, W. Activation of the Unfolded Protein Response is an Early Event in Alzheimer's and Parkinson's Disease. *Neurodegener. Dis.* **2012,** *10* (1–4), 212–215.
22. Hrelia, P.; Sita, G.; Ziche, M.; Ristori, E.; Marino, A.; Cordaro, M.; Molteni, R.; Spero, V.; Malaguti, M.; Morroni, F.; Hrelia, S. S. S. Common Protective Strategies in Neurodegenerative Disease: Focusing on Risk Factors to Target the Cellular Redox System. *Oxid. Med. Cell. Longev.* **2020,** *2020*, 24–36.
23. Hu, S.; Maiti, P.; Ma, Q.; Zuo, X.; Jones, M. R.; Cole, G. M.; Frautschy, S. A. Clinical Development of Curcumin in Neurodegenerative Disease. *Expert Rev. Neurother.* **2015,** *15* (6), 629–637.
24. Huang, H. C.; Chang, P.; Dai, X. L.; Jiang, Z. F. Protective Effects of Curcumin on Amyloid-b-Induced Neuronal Oxidative Damage. *Neurochem. Res.* **2012,** *37* (7), 1584–1597.
25. Huang, H. C.; Zheng, B. W.; Guo, Y.; Zhao, J.; Zhao, J. Y.; Ma, X. W.; Jiang, Z. F. Antioxidative and Neuroprotective Effects of Curcumin in an Alzheimer's Disease Rat Model Co-Treated with Intracerebroventricular Streptozotocin and Subcutaneous D-Galactose. *J. Alzheimer's Dis.* **2016,** *52* (3), 899–911.
26. Huang, W. J.; Zhang, X. I. A.; Chen, W. W. Role of Oxidative Stress in Alzheimer's Disease. *Biomed. Rep.* **2016,** *4* (5), 519–522.
27. Kim, H.; Park, B. S.; Lee, K. G.; Cheol, Y. C.; Sung, S. J.; Kim, Y. H.; Lee, S. E. Effects of Naturally Occurring Compounds on Fibril Formation and Oxidative Stress of β-Amyloid. *J. Agric. Food Chem.* **2005,** *53* (22), 8537–8541.
28. Kitani, K.; Osawa, T.; Yokozawa, T. The Effects of Tetrahydrocurcumin and Green Tea Polyphenol on the Survival of Male C57BL/6 Mice. *Biogerontology* **2007,** *8* (5), 567–573.
29. Koronyo, Y.; Salumbides, B. C.; Black, K. L.; Koronyo-Hamaoui, M. Alzheimer's Disease in the Retina: Imaging Retinal Aβ Plaques for Early Diagnosis and Therapy Assessment. *Neurodegener. Dis.* **2012,** *10* (1–4), 285–293.
30. Ma, Q. L.; Zuo, X.; Yang, F.; Ubeda, O. J.; Gant, D. J.; Alaverdyan, M.; Teng, E.; Hu, S.; Chen, P. P.; Maiti, P.; Teter, B.; Cole, G. M.; Frautschy, S. A. Curcumin Suppresses Soluble Tau Dinners and Corrects Molecular Chaperone, Synaptic, and Behavioral Deficits in Aged Human Tau Transgenic Mice. *J. Biol. Chem.* **2013,** *288* (6), 4056–4065.
31. Madireddy, S.; Madireddy, S. Protection from the Pathogenesis of Neurodegenerative Disorders, including Alzheimer's Disease, Amyotrophic Lateral Sclerosis, Huntington's Disease, and Parkinson's Diseases, through the Mitigation of Reactive Oxygen Species. *J. Neurosci. Neurol. Disord.* **2019,** *3* (2), 148–161.

32. Maries, E.; Dass, B.; Collier, T. J.; Kordower, J. H.; Steece-Collier, K. The Role of α-Synuclein in Parkinson's Disease: Insights from Animal Models. *Nat. Rev. Neurosci.* **2003,** *4* (9), 727–738.
33. Mehta, D.; Jackson, R.; Paul, G.; Shi, J.; Sabbagh, M. Why do Trials for Alzheimer's Disease Drugs Keep Failing? A Discontinued Drug Perspective for 2010-2015. *Expert Opin. Investig. Drugs* **2017,** *26* (6), 735–739.
34. Menon, V. P.; Sudheer, A. R. Antioxidant and anti-inflammatory properties of curcumin. *Adv. Exp. Med. Biol.* **2007,** *595*, 105–125.
35. Morzelle, M. C.; Salgado, J. M.; Massarioli, A. P.; Bachiega, P.; Rios, A. de O.; Alencar, S. M.; Schwember, A. R.; Camargo, A. C. Potential Benefits of Phenolics from Pomegranate Pulp and Peel in Alzheimer's Disease: Antioxidant Activity and Inhibition of Acetylcholinesterase. *J. Food Bioact.* **2019,** *5*, 466–483.
36. Neumann, H.; Kotter, M. R.; Franklin, R. J. M. Debris Clearance by Microglia: An Essential Link between Degeneration and Regeneration. *Brain* **2009,** *132* (2), 288–295.
37. Nichols, E.; Szoeke, C. E. I.; Vollset, S. E.; Abbasi, N.; Abd-Allah, F.; Abdela, J.; Aichour, M. T. E.; Akinyemi, R. O.; Alahdab, F.; Asghu-Edom, S. W.; Awasthi, A.; Barker-Collo, S. L.; Baune, B. T.; Béjot, Y.; Belachew, A. B.; Bennett, D. A.; Biadgo, B.; Bijani, A.; Bin Sayeed, M. S.; Murray, C. J. L. Global, Regional, and National Burden of Alzheimer's Disease and Other Dementias, 1990–2016: A Systematic Analysis for the Global Burden of Disease Study 2016. *Lancet Neurol.* **2019,** *18* (1), 88–106.
38. Oliver, D. M. A.; Reddy, P. H. Small Molecules as Therapeutic Drugs for Alzheimer's Disease. *Mol. Cell. Neurosci.* **2019,** *96*, 47–62.
39. Ono, K.; Yamada, M. Antioxidant Compounds have Potent Anti-Fibrillogenic and Fibril-Destabilizing Effects for α-Synuclein Fibrils In Vitro. *J. Neurochem.* **2006,** *97* (1), 105–115.
40. Pavan, R.; Jain, S.; Kumar, S.A. Properties and Therapeutic Application of Bromelain: A Review. *Biotechnol. Res. Int.* **2012,** *2012*, 1–6.
41. Pogue, A. I.; Percy, M. E.; Cui, J. G.; Li, Y. Y.; Bhattacharjee, S.; Hill, J. M.; Kruck, T. P. A.; Zhao, Y.; Lukiw, W. J. Up-Regulation of NF-kB-Sensitive miRNA-125b and miRNA-146a in Metal Sulfate-Stressed Human Astroglial (HAG) Primary Cell Cultures. *J. Inorg. Biochem.* **2011,** *105* (11), 1434–1437.
42. Prasansuklab, A.; Brimson, J. M.; Tencomnao, T. Potential Thai Medicinal Plants for Neurodegenerative Diseases: A Review Focusing on the Anti-Glutamate Toxicity Effect. *J. Tradit. Complement. Med.* **2020,** *1* (3), 301–308.
43. Ramkumar, M.; Rajasankar, S.; Gobi, V. V.; Janakiraman, U.; Manivasagam, T.; Thenmozhi, A. J.; Essa, M. M.; Chidambaram, R.; Chidambaram, S. B.; Guillemin, G. J. Demethoxycurcumin, a Natural Derivative of Curcumin Abrogates Rotenone-Induced Dopamine Depletion and Motor Deficits by its Antioxidative and Anti-Inflammatory Properties in Parkinsonian Rats. *Pharmacogn. Mag.* **2018,** *14* (53), 9–16.
44. Samy, D. M.; Ismail, C. A.; Nassra, R. A.; Zeitoun, T. M.; Nomair, A. M. Downstream Modulation of Extrinsic Apoptotic Pathway in Streptozotocin-Induced Alzheimer's Dementia in Rats: Erythropoietin Versus Curcumin. *Eur. J. Pharmacol.* **2016,** *770*, 52–60.
45. Shen, L. R.; Xiao, F.; Yuan, P.; Chen, Y.; Gao, Q. K.; Parnell, L. D.; Meydani, M.; Ordovas, J. M.; Li, D.; Lai, C. Q. Curcumin-Supplemented Diets Increase Superoxide Dismutase Activity and Mean Lifespan in Drosophila. *Age* **2013,** *35* (4), 1133–1142.
46. Singh, A.; Kukreti, R.; Saso, L.; Kukreti, S. Oxidative Stress: A Key Modulator in Neurodegenerative Diseases. *Molecules* **2019,** *24* (8), 236–243.

47. Singh, S.; Kumar, P. Neuroprotective Potential of Curcumin in Combination with Piperine against 6-Hydroxy Dopamine Induced Motor Deficit and Neurochemical Alterations in Rats. *Inflammopharmacology* **2017,** *25* (1), 69–79.
48. Singleton, A. B.; Farrer, M.; Johnson, J.; Singleton, A.; Hague, S.; Kachergus, J.; Hulihan, M.; Peuralinna, T.; Dutra, A.; Nussbaum, R.; Lincoln, S.; Crawley, A.; Hanson, M.; Maraganore, D.; Adler, C.; Cookson, M. R.; Muenter, M.; Baptista, M.; Miller, D.; Gwinn-Hardy, K. α-Synuclein Locus Triplication Causes Parkinson's Disease. *Science* **2003,** *302* (5646), 841–846.
49. Sofowora, A.; Ogunbodede, E.; Onayade, A. The Role and Place of Medicinal Plants in the Strategies for Disease Prevention. *Afr. J. Tradit. Complement. Altern. Med.* **2013,** *10* (5), 210–229.
50. Song, P.; Zou, M. H. Roles of Reactive Oxygen Species in Physiology and Pathology. In *Atherosclerosis: Risks, Mechanisms, and Therapies*; Wang, H.; Patterson, C., Eds.; **2015**; pp 379–392.
51. Soo, S. K.; Rudich, P. D.; Traa, A.; Harris-Gauthier, N.; Shields, H. J.; Van Raamsdonk, J. M. Compounds that Extend Longevity are Protective in Neurodegenerative Diseases and Provide a Novel Treatment Strategy for these Devastating Disorders. *Mech. Ageing Dev.* **2020,** *190*, 111297.
52. Tan, S. H.; Karri, V.; Tay, N. W. R.; Chang, K. H.; Ah, H. Y.; Ng, P. Q.; Ho, H. S.; Keh, H. W.; Candasamy, M. Emerging Pathways to Neurodegeneration: Dissecting the Critical Molecular Mechanisms in Alzheimer's Disease, Parkinson's Disease. *Biomed. Pharmacother.* **2019,** *111*, 765–777.
53. Uğuz, A. C.; Öz, A.; Naziroğlu, M. Curcumin Inhibits Apoptosis by Regulating Intracellular Calcium Release, Reactive Oxygen Species and Mitochondrial Depolarization Levels in SH-SY5Y Neuronal Cells. *J. Recept. Signal Transduct.* **2016,** *36* (4), 395–401.
54. Xiao, Z.; Zhang, A.; Lin, J.; Zheng, Z.; Shi, X.; Di, W.; Qi, W.; Zhu, Y.; Zhou, G.; Fang, Y. Telomerase: A Target for Therapeutic Effects of Curcumin and a Curcumin Derivative in aβ1-42 Insult In Vitro. *PLoS ONE* **2014,** *9* (7), 101251.
55. Xie, A.; Gao, J.; Xu, L.; Meng, D. Shared Mechanisms of Neurodegeneration in Alzheimer's Disease and Parkinson's Disease. *Biomed. Res. Int.* **2014,** *2014*, 239–246.
56. Yanagisawa, D.; Ibrahim, N. F.; Taguchi, H.; Morikawa, S.; Hirao, K.; Shirai, N.; Sogabe, T.; Tooyama, I. Curcumin Derivative with the Substitution at C-4 Position, but not Curcumin, is Effective against Amyloid Pathology in APP/PS1 Mice. *Neurobiol. Aging* **2015,** *36* (1), 201–210.
57. Ye, J.; Zhang, Y. Curcumin Protects against Intracellular Amyloid Toxicity in Rat Primary Neurons. *Int. J. Clin. Exp. Med.* **2012,** *5* (1), 44–49.
58. Youssef, S. B.; Doucet-Beaupré, H.; Castonguay, A. M.; Gora, C.; Amri, M.; Lévesque, M. Neuroprotective Benefits of Grape Seed and Skin Extract in a Mouse Model of Parkinson's Disease. *Nutr. Neurosci.* **2021,** *24* (3), 197–211.
59. Yu, S.; Wang, X.; He, X.; Wang, Y.; Gao, S.; Ren, L.; Shi, Y. Curcumin Exerts Anti-Inflammatory and Antioxidative Properties in 1-Methyl-4-Phenylpyridinium ion (MPP^+)-Stimulated Mesencephalic Astrocytes by Interference with TLR4 and Downstream Signaling Pathway. *Cell Stress Chaperones*. **2016,** *21* (4), 697–705.
60. Zhang, C.; Browne, A.; Child, D.; Tanzi, R. E. Curcumin Decreases Amyloid-β Peptide Levels by Attenuating the Maturation of Amyloid-β Precursor Protein. *J. Biol. Chem.* **2010,** *285* (37), 28472–28480.

61. Zhang, L.; Fiala, M.; Cashman, J.; Sayre, J.; Espinosa, A.; Mahanian, M.; Zaghi, J.; Badmaev, V.; Graves, M. C.; Bernard, G.; Rosenthal, M. Curcuminoids Enhance Amyloid-β Uptake by Macrophages of Alzheimer's Disease Patients. *J. Alzheimer's Dis.* **2006,** *10* (1), 1–7.
62. Zhang, X.; Fu, Z.; Meng, L.; He, M.; Zhang, Z. The Early Events that Initiate β-Amyloid Aggregation in Alzheimer's Disease. *Front. Aging Neurosci.* **2018,** *10*, 359–363.

PART III

POTENTIAL OF MEDICINAL PLANTS IN ORGAN DYSFUNCTION

CHAPTER 9

Multiple Organ Dysfunction Syndrome: Phytotherapeutic Evidences from Intestinal Ischemia-Reperfusion Models

Akinleye Stephen Akinrinde and Joseph Fadeyemi Akinrinmade

ABSTRACT

The gastrointestinal mucosa serves important barrier and defensive function against pathogenic bacteria, aggressive factors, and other inflammatory stimuli present within the lumen of the intestines. A variety of pathological and/or surgical conditions, known to produce intestinal ischemia-reperfusion injury (IIRI) (e.g., hemorrhagic shock, strangulated bowel, intussusception, and volvulus) has been associated with breaches in gut barrier integrity, increased barrier permeability, and translocation of viable enteric bacteria and their toxins into the portal and systemic circulation. The result of this trans-epithelial movement of enteric microbes and/or toxins is the generalized activation of various inflammatory mediators (e.g., cytokines, reactive oxygen species, leucocytes, and the complement system) in what is often described as the systemic inflammatory response syndrome (SIRS). One devastating consequence of IIRI and SIRS is widespread injury to remote organs, a condition often referred to as multiple organ dysfunction syndrome (MODS), which is the leading cause of mortality in critically ill patients, especially those in intensive care units. From the foregoing, it has been suggested that compounds that combine multiple activities such as antibacterial, anti-inflammatory, antioxidant, and immune-modulatory roles, may provide new approaches to the prevention and/or treatment of MODS. In this regard, medicinal herbs with their composite nature of bioactivities

Curative and Preventive Properties of Medicinal Plants: Research on Disease Management and Animal Model Studies. Megh R. Goyal, Junaid Ahmad Malik, & Ademola Olabode Ayeleso (Eds.)

due to several inherent phytochemicals are being investigated at different levels for their benefits in modulating the pathophysiological mechanisms that give rise to MODS. With particular reference to animal models of IIRI, this chapter explores available experimental evidence of the potential prophylactic and/or therapeutic benefits of medicinal herbs in limiting remote organ dysfunction arising from gastrointestinal injury.

9.1 INTRODUCTION

Multiorgan dysfunction syndrome (MODS) is the term used to describe the progressive, simultaneous impairment of two or more separate organ systems, usually requiring the need for external support (e.g., mechanical ventilation, respiratory and extracorporeal supports, nutritional and fluid supplementation) to maintain homeostasis.[46] It is the leading cause of mortality in critically ill patients, especially those monitored in intensive care units, with mortality rates as high as 76%,[38] and rising to as much as 98% within a week when failure of more than three organs is experienced. Organ dysfunction may manifest as acute respiratory distress, kidney injury, cardiomyopathy, gastrointestinal dysfunction, encephalopathy, hepatic dysfunction, and coagulopathy.

Classical MODS is usually underlined by the development of a systemic inflammatory response usually to infection[65] or sequelae of traumatic, ischemic or hypoxic injuries, giving rise to the so-called systemic inflammatory response syndrome (SIRS).[5] This is often complicated by inadvertent iatrogenic events as the sustenance of the life of the critically ill patient is highly dependent on the provision of external support that necessitates intensive care in the first place.[39] Consequently, the ultimate outcome of managing a patient with MODS will depend on the prevention and/or treatment of organ injury related to the primary inflammatory insult as well as the effectiveness or shortcomings of the exogenous support provided to the critically ill patient.

The pathophysiology of MODS is still poorly understood but is believed to involve complex interactions that may include tissue hypoxia, bacterial translocation from the gut, dysregulation of immune and hormonal systems, induction of apoptosis, and alterations in neural signaling.[40] Several animal models have provided useful insights into potential mechanisms underlying the development of MODS, while a number of promising therapeutic strategies have also been investigated.[26] Conditions that create blockade of blood supply (ischemia) to the intestines followed by restoration of blood flow to such ischemic tissues (reperfusion) produce both local and remote organ injuries and have been linked to the development and progression of MODS. Consequently, animal models of intestinal ischemia-reperfusion injury (IIRI) hold promise for proving both pathophysiological and therapeutic insights into what is currently known about MODS. The important goal of management of MODS must be the early recognition and control of the underlying cause (usually inflammatory foci resulting in tissue injury), as well as comorbidities and complications arising from the management of the patient.

The potential application of several therapeutic modalities such as antioxidant, antileucocyte, anticomplement, and amino acid therapies as well as ischemic preconditioning against IIRI has been comprehensively reviewed.[37] Many of these approaches are, often directed at specific targets in the cascade of inflammation and MODS. However, what appears to be clear from studies on the pathophysiology of sepsis, inflammatory syndrome, and multiorgan dysfunction is the involvement of numerous/multiple targets.[46] It is, therefore, likely that the achievement of clinical success in the management of the inflammatory syndrome and MODS will depend on agents that can target multiple inflammatory and immune mediators. In this regard, alternative medicine through the use of medicinal herbs has been proposed as a useful strategy with plants as viable sources of multiple bioactivities—antibacterial, antioxidant, anti-inflammatory, immuno-modulatory; etc.—which could potentially protect the host from sepsis and the inflammatory syndrome, while increasing the chances of survival of the patient.[42]

For example, recent clinical trials of Septimeb™, a combination of extracts from herbs and phytochemicals (including *Rosa canina*, *Urtica dioica*, *Tanacetum vulgare*, selenium, flavonoids, and carotenes) showed positive effects in reducing the severity of sepsis and mortality rates in ICU patients.[49] Excellent reviews also exist on the potentials of several medicinal plants in the treatment of polymicrobial sepsis induced by cecal ligation and puncture.[42,57] The values of traditional Chinese and Western medicine against MODS have also been highlighted in reports by Zhang et al..[69]

The chapter aims to highlight the promise of medicinal plants as prophylactic and/or therapeutic alternatives against tissue damage in experimental IIRI as novel animal models of MODS.

9.2 ANIMAL MODELS FOR MULTIORGAN DYSFUNCTION SYNDROME

Several advances have been made in human and veterinary medicine toward understanding the complex pathophysiology of MODS as well as developing treatment strategies, although several aspects are yet unknown. Studies in animal models have contributed greatly to what is currently known about the pathophysiology, progression, and potential leads for novel therapies against MODS. Among other considerations, animal studies offer the benefit of being largely reproducible. Because experimental animals are often selected to reflect fairly homogenous populations with minimal genetic diversity, it is often possible to generate reasonably consistent outcomes in animal studies.[38] In contrast, human trials are often complicated by high genetic diversity of human populations, the presence of co-morbidities, and the great difficulty in obtaining biological samples such as biopsies and tissues for histological examinations.[58] These factors make replication of human diseases in animals very complicated, and are even more so in complex disorders such as MODS. Commonly utilized animal models for eliciting the exaggerated inflammatory response that characterizes MODS include: (i) the zymosan-induced inflammation (peritonitis) model,[13,41] and (ii) the cecal ligation and puncture model.[6] The former requires the injection of Zymosan, a polysaccharide derivative of the cell wall of yeast cells (*Saccharomyces cerevisiae*) with potential to activate innate immunity and recruit neutrophils and monocytes into the peritoneal cavity, as well as production of several inflammatory mediators, including pro-inflammatory cytokines.[17] Cecal ligation and puncture, on the other hand, involves surgery on anesthetized animals, where the caecum is punctured to allow fecal extrusion into the peritoneal cavity, and then subsequently sutured.[66] The marked differences in the modes of induction between the two models imply the likelihood of variations in the pathological responses that accompany them. It, therefore, means that our understanding of the pathophysiology of MODS is likely to be improved as new animal models are developed. In the last few decades, the search for novel therapeutically effective strategies against MODS has continued to drive the discovery of new animal models for better understanding of cellular and molecular mechanisms that give rise to remote organ dysfunction. One of the most frequently encountered clinical causes of MODS is ischemia-reperfusion (IR) injury arising from occlusion of blood supply and subsequent re-establishment of blood flow to critical organs such as the intestines, liver, or skeletal muscles.[18] The next sections describe IIRI and its link to development of multiple organ dysfunction, as well as animal models used to replicate clinical injury in experimental situations.

9.3 INTESTINAL ISCHEMIA-REPERFUSION INJURY (IIRI)

Clinically, IIRI occurs in a wide variety of unrelated conditions that compromise intestinal blood flow occurring either by mechanical vascular obstruction (e.g., strangulated hernias, volvulus, and intussusceptions) or as a result of diseases that reduce systemic blood flow to various organs (e.g., hemorrhagic shock and cardiopulmonary disease).[22,55] Severe IIRI may stimulate local and systemic inflammatory events, SIRS and consequently the widespread dysfunction of remote organs (Figure 9.1). In fact, certain hypotheses suggest that IIRI, with associated intestinal dysfunction is a major contributor to the development of MODS.[16] The pathophysiological mechanisms of IIRI and its relationship to the development of MODS are discussed in subsequent sections of this chapter.

Studies by Haward et al.[28] in human patients had suggested that re-establishment of blood flow to "chronically ischemic bowel" resulted in loss of hepatic, pulmonary, and coagulation function, and these were attributed to injury induced by intestinal ischemia and reperfusion. IIRI is particularly relevant for the development of septicaemia due to the great tendency for the compromise of the gut barrier and consequent trans-epithelial translocation of enteric bacteria across the compromised gut barrier.[64]

The extent to which IIRI induction (utilizing a surgical approach involving occlusion of major mesenteric arterial supply) in animals may model MODS in humans is still a subject for further investigation.[22] However, as described earlier, there seem to be no perfect animal model for human diseases, and further research may help us with better understanding on additional lessons that can be learnt from IIRI models of MODS, as compared with the earlier described models.

9.3.1 EXPERIMENTAL ANIMAL MODELS OF IIRI

Animal models have been extensively utilized in the understanding of the pathogenesis of organ damage following IIRI. A recent review by Gonzalez et al.[22] provides an excellent summary of the available techniques for producing IIRI. A few are summarized in this chapter. Models available for the induction of mesenteric ischemia and reperfusion can generally be classified into two types: Occlusive and Nonocclusive techniques. While occlusive approaches offer the specific reduction of blood flow to specific areas of the intestines, nonocclusive approaches tend to produce more generalized limitation to blood flow affecting various organs, including the intestines. Generally, there is no standardization of the times allowed for ischemia or reperfusion. However, it is generally believed that the severity of injury increases with longer ischemia or reperfusion times.

9.3.1.1 OCCLUSIVE TECHNIQUES

- *Occlusion of the Superior Mesenteric artery (SMA)/Cranial Mesenteric Artery (CMA):* The SMA arises as a direct branch arising from the abdominal aorta and supplying oxygenated blood to portions of the mid-gut, from the duodenum to the proximal colon. Occlusion and subsequent release of the Superior Messenteric Artery (SMA) or CMA is the most commonly employed technique for the induction of IIRI in rodent models. It involves the approach of the SMA via a ventral midline incision and occlusion of the vessel by means of atraumatic microvascular clips for varying durations of time, before their removal to establish reperfusion, which is also allowed for varying lengths of time.[26] This technique suffers the drawback of not producing consistent injury across all intestinal segments. A variation of this technique employs the occlusion of the SMA, along with occlusion of collateral vessels, helping to achieve more consistent, reliable, and reproducible injury outcomes across the different segments of the small and large intestine.[9]
- *Aortic clamping models:* Techniques involving occlusion of the abdominal aorta are intended to achieve a more global ischemia affecting various organs, rather than selective clamping of vessels directed to achieve ischemia in specific organs.[53] Dissecting through a ventral midline abdominal incision, the aorta could be clamped in a region between the diaphragm the aortic bifurcation (supraceliac occlusion) or just above the emergence of the renal artery (suprarenal occlusion).[19]

9.3.1.2 NONOCCLUSIVE TECHNIQUES

- *Hemorrhagic shock models:* Hypovolemic shock resulting from dramatic blood losses often stimulates inflammatory responses, which if untreated, could lead to multiorgan failure and death. Hemorrhagic shock models involve withdrawal of blood from an artery until mean arterial blood pressure falls to about 40 mmHg, followed by resuscitation and re-infusion of the same blood after a specific duration of ischemia.[29] This model is often described under the general approaches of low-flow ischemia in which blood flow falls to about 20% of baseline level.[22] It should be noted, however, that the hemorrhagic shock model may not adequately reproduce clinical situations as the damage due to ischemia alone may be very severe, making it difficult to evaluate the significance of reperfusion injury. Expectedly, injuries induced by hemorrhagic shock are not limited to the intestines, although the intestinal mucosa is often more highly susceptible compared with other organs.

- *Hypoxia and Re-oxygenation models:* Hypoxia may be induced by placing animals in an air-tight chamber with 100% CO_2 for about 5 min followed by re-oxygenation for another 5 min by administration of 100% oxygen.[47] Induction of hypoxia in the animals is evident by cyanosis and gasping for breath.

9.4 PATHOPHYSIOLOGICAL LINKS BETWEEN IIRI AND MULTIPLE ORGAN DYSFUNCTION

The intestinal epithelium has a particularly large requirement for oxygen and energy required for the maintenance of its absorptive and secretory functions. As such, interruptions to blood supply (ischemia), and in effect, oxygen supply to this critical organ predisposes enterocytes to a high tendency for cellular injury and death.[30] Susceptibility to ischemic and hypoxic injury is believed to be highest in cells on the villi tips, closest to the lumen, progressing to those in the crypts with increasing duration of ischemia, although this degree of susceptibility is now thought to depend on the state of differentiation of the enterocyte.[22]

To prevent irreversible tissue injury to ischemic intestinal tissues, it is required that interventions are instituted to re-establish blood flow to the organ (reperfusion). However, reperfusion often results in further exacerbation of tissue injury with the primary involvement of oxygen-derived free radicals, along with activation of pro-inflammatory cytokines, neutrophils, nitric oxide, the complement system, and a host of other molecular mediators.[37]

The potential mechanisms involved in the exacerbation of tissue injury after reperfusion of ischemic tissues have been described elsewhere.[12,37] When tissues are deprived of blood supply, they continue to depend on available ATP as energy source, like normal cells. The cellular utilization of ATP in normal or ischemic tissues results in the formation of a by-product—hypoxanthine—which, under normal conditions, is subjected to degradation by the enzyme, xanthine dehydrogenase.[33] As ATP continues to get depleted in the ischemic tissue, there is continual loss of regulation of the activity of ATP-dependent ion channels (especially the Na^+-K^+-ATPase) as well as Ca^{2+} channels that normally maintain the flux of ions across cellular membranes. The shifts in specific ion gradients caused by ATP depletion causes influx of Na^+ and Ca^{2+} as well as efflux of K^+ and Mg^{2+}, with resultant inflow of water and cell swelling.[60] Under this hypoxic condition, xanthine dehydrogenase is converted by certain Ca^{2+}-dependent proteases to xanthine oxidase, an oxygen-radical producing enzyme, causing the accumulation of hypoxanthine.[27,43]

However, in the presence of excess oxygen following reperfusion of the tissue, xanthine oxidase metabolizes the already accumulated hypoxanthine to produce a burst of superoxide radicals with subsequent generation of other reactive oxygen species (e.g., hydrogen peroxide) which are involved in direct injury to tissues. The severity of tissue injury is further complicated by the activation and mobilization of neutrophils to the sites of injury via production of leucocyte adhesion molecules including ICAM-1 and P-selectin.[56] Neutrophils are equipped with granules that contain a number of cytotoxic metabolites (e.g., hydrogen peroxide and hypochlorous acid) and enzymes (e.g., proteases and myeloperoxidase). Extensive connective tissue damage induced by these factors along with mechanical injury associated with neutrophil migration have all been associated with the increased severity of tissue injury following reperfusion of ischemic tissues.[71] In IIRI, these events ultimately lead to increased para-cellular permeability of the monolayer of intestinal epithelial cells.[21]

The foregoing suggests that IIRI may be intricately linked to the development of MODS as the effects of intestinal tissue hypoxia and subsequent re-oxygenation predisposes to alterations in normal gastrointestinal flora, disruption of the mucosal barrier, and increased intestinal permeability, which promotes the translocation of viable bacteria and/or their metabolites from the gastrointestinal tract to mesenteric lymph nodes and then into blood circulation with widespread dissemination to several organs leading to sepsis and multiorgan dysfunction.[11,20,23,67] Experimental evidence linking IIRI with MODS has been obtained in animal models where remote organ injury (e.g., in the lungs) was observed after ischemia and reperfusion of the intestines.[54] It should be pointed out, however, that in more realistic terms, MODS arises from a complex combination of several other factors such as induction of cellular apoptosis, dysregulation of the immune system (due to an imbalance between pro-inflammatory and anti-inflammatory mechanisms and a cascading activation of cytokines) and mitochondrial dysfunction, in addition to bacterial translocation.[40,46]

A "two-hit" hypothesis has been proposed for the pathogenesis of MODS. The first hit comprises the establishment of septicemia and sepsis leading to the activation of neutrophils, monocytes, and macrophages and the production of cytokines (cytokine storm) that leads to organ damage.[51] In effect, interventions that inhibit the migration and adherence of leucocytes could prevent initial organ damage. The second hit comprises the continued promotion of SIRS via sustained activation of macrophages and release of pro-inflammatory mediators leading to incremental derangements of physiologic functions in various organs (MODS) and eventually death.[52] Other mechanisms thought to contribute to the development of MODS include stimulation of the coagulation cascade (triggered by cytokines, bacterial endotoxins and antigens), neuroendocrine factors (such as adrenal hormones involved in mediating stress responses) and mitochondrial dysfunction (causing reduced ATP production and increased apoptosis).[22]

9.5 APPLICATIONS OF MEDICINAL PLANTS FOR PREVENTION OF IIRI AND MULTIORGAN DYSFUNCTION

Severe intestinal ischemia and reperfusion injury often results in compromise of barrier function with translocation of luminal bacteria and their toxins to the underlying vasculature, the sequelae being septicemia and multiple organ dysfunction,[35,50] as shown in Figure 9.1. Most clinical conditions that involve interruptions to intestinal blood flow are often difficult to prevent, hence it appears more beneficial that research attention be directed toward the advancement of novel interventions and modalities to combat injury resulting from ischemia and reperfusion of the tissues.[37] However, many of the proposed therapies for the management of reperfusion injury have demonstrated limited clinical effectiveness because they only tend to target specific mechanisms involved in the multifaceted pathogenesis of the condition.[36] While some interventions are mainly targeted at inhibition of oxidant-mediated injury or leucocyte activation, others have specific activity toward modulation of inflammatory pathways (inhibition of endotoxin or cytokine production) or cell death (apoptotic) pathways involved in ischemia-reperfusion injury.[24]

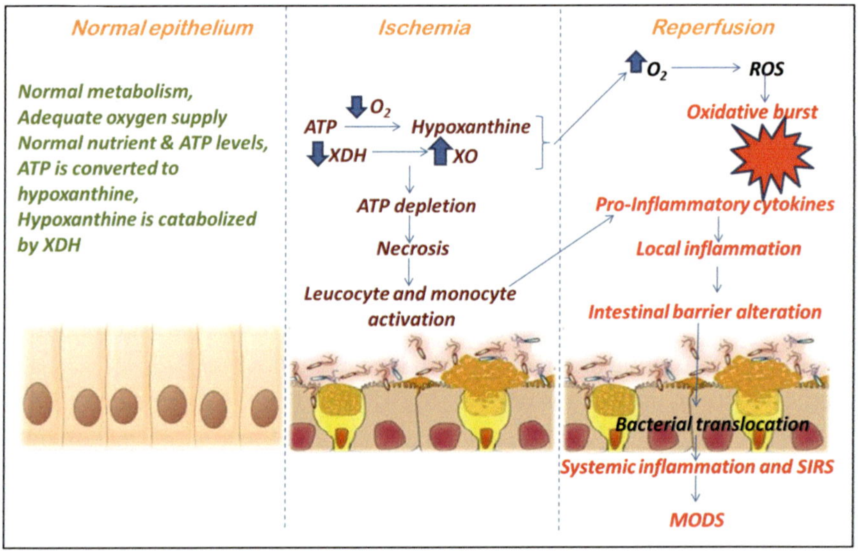

FIGURE 9.1 Pathophysiological links between IIRI and Multiple Organ Dysfunction Syndrome (MODS): ATP, adenosine triphosphate; XDH, xanthine dehydrogenase; XO, xanthine oxidase; ROS, reactive oxygen species; SIRS, systemic inflammatory response syndrome; MODS, multiple organ dysfunction syndrome.

Medicinal plants, on the other hand, are composites of compounds with a vast array of bioactivities that offer promise for better therapeutic efficacy against ischemia-reperfusion injury, as well as prevention of complications of multiorgan dysfunction.[10] To date, majority of evidence in support of potential application of medicinal plants for the management of IIRI, sepsis, and MODS have been obtained from animal models, while only a few studies report clinical trials in humans.[7]

Table 9.1 summarizes a list of medicinal plants that have been employed in IIRI models and the major pharmacological activities associated with their beneficial actions. The intricate pathophysiological links between IIRI and MODS suggests that interventions that show promise of effectiveness against IIRI may also prove beneficial in preventing MODS. In effect models of IIRI may be useful in finding appropriate therapeutic interventions against MODS. With particular reference to injuries related to

TABLE 9.1 Medicinal Plants for Prevention of Intestinal Ischemia-Reperfusion Injury.

Medicinal plant distribution	Part of plant used and route of administration	Method of IIR induction	Major outcome of intervention	Major Pharmacological activities	Reference
Artemisia vulgaris *(Asia, Europe, North America)*	Hexane partition of the water extract of *Artemisia vulgaris* leaves was administered via bolus injection prior to occlusion of the SMA.	Occlusion of the SMA for 10 min was used to produce ischemia and reperfusion was observed for 30 min	Anti-inflammatory activities of *Artemisia vulgaris* extract was demonstrated by reduction in leucocyte adherence on venules and this was thought to possibly play a role in preventing tissue damage following IR induction.	Antiinflammatory Antileucocyte	62
Astragalus membranaceus (Native to china)	Root preparation obtained commercially was diluted in normal saline and administered intravenously 3 minutes prior to reperfusion	Hemorrhagic shock via bleeding of the femoral artery for 60 min. followed by re-infusion with blood for 90 minutes	*Astragalus membranaceus* protected against small intestinal mucosa injury via antioxidative effects, involving reduction in MDA concentration and stimulation of GPx and SOD.	Antioxidant	8, 29
Camellia sinensis *(Green tea)*	Green tea infusion given as pretreatment for two weeks before induction of IR.	Ischemia induced for 30 min followed by reperfusion for 1 h.	Antioxidant effects including increased expressions of SOD, GPx and CAT protected intestinal mucosal integrity	Antioxidant	1
Chelidonium majus (Europe and western Asia)	Ukrain (NSC 631570), a semisynthetic compound from the plant extract	Ischemia induced by clamping of the SMA for 30 min and reperfusion allowed for 120 min.	Inhibition of cell apoptosis by up-regulation of Bcl-xL mRNA expressions and decreased mRNA expressions of Caspase-3 and iNOS.	Antiapoptotic	2
Chromolena odorata (North and Central America, Tropical Asia and West Africa)	Methanol extract from the leaves was administered orally for 7 days before IR induction.	Ischemia was produced by clamping of the SMA for 30 min followed by reperfusion for 45 min.	Attenuation of tissue intestinal tissue injury; reduced MDA and elevated GSH levels.	Antioxidant	3

TABLE 9.1 *(Continued)*

Medicinal plant distribution	Part of plant used and route of administration	Method of IIR induction	Major outcome of intervention	Major Pharma-cological activities	Reference
Gingko biloba (Found in Asia, Europe, North America, New Zealand and Argentina)	Standardized extract (EGb 761) from the leaves of *Gingko biloba.* Administration has been attempted intra-gastrically[34] and subcutaneously[47]	1. Clamping of SMA for 1 hour, reperfusion for 3 hours 2. Hypoxia induced by placing rats in air-tight chamber with 100% CO_2 for 5 min and re-oxygenation for 5 min with 100% oxygen.	EGb 761 produced significant reduction in MDA levels with enhancement of SOD activity pointing to an antioxidant effect which was believed to have mediated an antiapoptotic activity in intestinal epithelial cells.[34] Ozdemir et al.[47] reported reduction in MDA, nitric oxide and histo-pathologic injury scores by EGb 761 in intestinal samples.	Antioxidant Antiapoptotic	34,45,47,48
Hypericum perforatum (Europe, North Africa, North America and China)	*1. Hypericum perforatum* extract was given by bolus i.v. injection 15 min before reperfusion[14] 2. *Hypericum perforatum* extract was administered intra-peritoneally 10 min before reperfusion.[44]	Splanchnic arterial occlusion (SAO) shock was induced by clamping of the SMA for 45 min, while reperfusion was allowed for 1 h[14] Ischemia was induced by clamping of the SMA for 60 min, while reperfusion was allowed for 24 h[44]	*Hypericum perforatum* extract caused substantial reduction in rat intestinal IR injury by mechanisms involving reduction of the nitration of proteins, poly-ADP ribosylated proteins (PARP) activation, pro-inflammatory cytokine levels, and reduced expression of the adhesion molecule ICAM-1 and P-selectin. *Hypericum* extract also produced reduction in MDA and MPO levels, while also increasing levels of GSH and cardiotrophin-1 in protection of the intestine against in a model of IR-induced injury in hamsters.	Antiinflammatory Antileucocyte Antioxidant	14,44

TABLE 9.1 *(Continued)*

Medicinal plant distribution	Part of plant used and route of administration	Method of IIR induction	Major outcome of intervention	Major Pharmacological activities	Reference
Lycium barbarum (South eastern Europe and Asia)	*Lycium barbarum* Polysaccharides obtained commercially was administered by gastric gavage 2 days before IR induction	Ischemia induced by clamping of the SMA for 30 min. and reperfusion allowed for 6 h.	Suppression of oxidative stress (reduction of MDA levels and increased activities of SOD, CAT and GPx) and inflammation (reduced small intestinal PMN accumulation, reduced ICAM expression, reduced NF-KB activation and decreased TNFα levels	Antioxidant Antileucocyte Antiinflammatory	68
Nigella sativa (Blackseed) (Middle east and North East Asia)	Volatile oils extracted from the seeds were administered intraperitoneally prior to induction of IR.	Ischemia induced by clamping of Superior mesenteric artery for 30 min and reperfusion for 60 min.	*Nigella sativa* produced reduction in MPO and total oxidant status, with increase in total antioxidant capacity	Antiinflammatory Antioxidant	61
Panax notoginseng	*Panax notoginseng* saponins (PNS) obtained commercially was administered as an i.v. bolus injection 15 min before reperfusion	Ischemia was achieved by clamping the SMA for 1 h and reperfusion was allowed for 3 h.	Administration of PNS prevented acute lung injury induced by IR. PNS caused reduction in MDA production, xanthine oxidase and myeloperoxidase, while preventing IR-induced decreases in GPx and SOD activities; decreased leucocyte adhesion and infiltration as well as inhibition of IR-induced increases in plasma levels of TNF-α and IL-8	Antioxidant Antileucocyte Antiinflammatory	54

TABLE 9.1 *(Continued)*

Medicinal plant distribution	Part of plant used and route of administration	Method of IIR induction	Major outcome of intervention	Major Pharma-cological activities	Reference
Parquetina nigrescens (West Africa)	Methanol extract from the leaves was administered orally for 7 days before IR induction	Ischemia was produced by clamping of the SMA for 30 min followed by reperfusion for 45 min.	Prevention of intestinal tissue injury with reduction in MDA content and increased GSH and GST.	Antioxidant	4
Vaccinium myrtillus (Billberry)	Bilberry powder was incorporated in standard chow and fed orally to mice in combination with lactic acid bacteria for 10 days prior to induction of IR	Ischemia was achieved by clamping of the SMA for 30 or 40 min followed by reperfusion for 120 or 240 min.	Supplementation of feed with bilberry caused reduction in MDA in the ileum and colon; preservation of mucosal integrity and morphology.	Antioxidant	31

intestinal ischemia and reperfusion, this section highlights the roles of some promising medicinal plants that may be evaluated further for their potential clinical applications in the treatment of critically ill patients with multiple organ dysfunction.

9.5.1 BIOACTIVITIES OF MEDICINAL PLANTS RELEVANT FOR PREVENTION OF IIRI AND MODS

Plant extracts with their constituent phytochemicals have the ability to target multiple pathways involved in the complex pathogenesis of MODS, rather than single pathways and this could make them more attractive for the development of novel therapies. In experimental studies, the best results have often been obtained when plant extracts are administered through various routes prior to ischemia or reperfusion. A variety of pharmacological activities (e.g., antioxidant, anti-inflammatory, antiapoptotic, antibacterial, and immune-modulatory activities) exhibited by phytochemicals suggest their applicability as potential therapeutic candidates against IIRI and MODS.

9.5.1.1 ANTIOXIDANT ACTIVITIES

Relevant antioxidant activities demonstrated by phytochemicals for protection against IIRI could include free radical scavenging activities, inhibition of lipid peroxidation, metal-chelating, as well as hydrogen-donating activities. Many experimental animal models have documented the involvement of antioxidative effects as a major component of the protective actions of plant extracts and their bioactive components against IIRI.

Liu et al.[34] and Ozdemir et al.[47] reported that extracts from *Gingko biloba* caused reduction in lipid peroxidation as indicated by the lowering of malondialdehyde (MDA) content of intestinal tissues following induction of ischemia and reperfusion by clamping of the SMA or via induction of hypoxia and re-oxygenation. Other studies reported antioxidative effects (reduction of MDA content and stimulation of SOD and GPx activities) of *Astragalus membranaceus* extracts in the protection of intestinal injury arising from hemorrhagic shock induced by bleeding of the femoral artery and subsequent re-infusion with blood.[8,29] Among other activities, the associated antioxidant effect of plant extracts is believed to play a major role in prevention of intestinal tissue and mucosal damage during histopathological examination of tissues.[3,4]

9.5.1.2 ANTI-INFLAMMATORY AND ANTILEUCOCYTE ACTIVITIES

Inflammatory events brought about by the activation and migration of leucocytes and macrophages to the site of tissue injury have been identified as major contributors to the promotion of organ damage following ischemia and reperfusion, as well as progression of MODS. Antileucocyte activities may involve the inhibition of inflammatory mediators that stimulate leucocyte activation (e.g., Platelet Activating Factor and Tumor Necrosis Factor (TNF-α)), inhibition of synthesis of leucocyte adhesion molecules via modulation of transcription factors such as NF-κB) and limiting adhesions between leucocytes and the endothelium.[37] Several plant extracts have been associated with anti-inflammatory and antileucocyte activities in models of IIRI, further justifying their potential application in clinical situations.

Tigno and Gumila[62] reported the protection of intestinal tissues against IIRI by *Artemisia vulgaris* extracts via anti-inflammatory activities involving the reduction of leucocyte adherence in venules. The hexane partition of an aqueous extract of *A. vulgaris* was administered as a bolus injection prior to induction of ischemia by clamping of the SMA. In the same vein, anti-inflammatory mechanisms such as reduction of pro-inflammatory cytokines and reduced expression of ICAM-1, P-selectin, and MPO by extracts from *Hypericum perforatum* have been reported in other studies.[14,44]

An interesting IIRI model with implications for development of MODS was conducted by Rong et al.,[54] where the effect of SMA occlusion on remote lung injury was evaluated. This study observed that saponins extracted from *Panax-noto ginseng* exhibited anti-inflammatory activities including reduction in leucocyte adhesion and infiltration as well as inhibition of IR-induced increases in plasma levels of TNF-α and IL-8, while reducing lung injury in the process. *Lycium barbarum* Polysaccharides were found to be able to reduce intestinal inflammation in a model of IIRI induced by SMA clamping via reduction of polymorpho-neutrophil accumulation, reduced ICAM expression, reduced NF-KB activation and decreased TNFα levels.[68]

9.5.1.3 ANTIAPOPTOTIC ACTIVITY

An increase in proapoptotic gene expression (e.g., caspase 3) has been reported in intestinal tissues following ischemia and reperfusion and studies suggest that this apoptosis plays a key role during initiation and progression of IIRI.[70] Ukrain (NSC 631570), a semisynthetic anticancer compound from

Chelidonium majus was found to offer protective actions against IIRI by inhibition of cell apoptosis via up-regulation of Bcl-xL mRNA expressions and decreased mRNA expressions of Caspase-3 and iNOS.[2]

Similarly the extract from *Gingko biloba* (EGb 761) was also shown to inhibit intestinal epithelial apoptosis via mechanisms related to an inhibition of ceramide generation.[34] More studies are required to identify the effects of medicinal plants on various apoptotic markers in the development of organ injury following IIRI.

9.5.1.4 IMMUNO-MODULATORY ACTIVITY

The syndrome of multiple organ dysfunction is often an immuno-compromised state, in which compensatory anti-inflammatory responses to the primary insult often result in immunosuppression. There is growing interest in exploiting the immuno-modulatory potentials of herbal medicines, especially in the critically-ill patient as most available immune modulators (immunostimulants or immune-depressants) are associated with various side effects. Plant bioactive agents are thought to be potent and safe alternatives for the modulation of immune system over- or under-activation during response to injury.[32] Studies on the application of plant extracts in immune-modulation during IIRI and MODS are, however, still lacking and this could be an area of interest for research either for prophylactic or therapeutic applications.

9.6 FUTURE HORIZONS AND RECOMMENDATIONS

Several studies have shown that intestinal ischemia and reperfusion result in disruption of intestinal barrier, bacterial translocation, and associated septicemia resulting in the development of widespread inflammatory responses associated with multiple organ dysfunction. Current knowledge on the involvement of gut-derived bacteria in the injury associated with gut ischemia and reperfusion is still very scanty and results are somewhat controversial.[59] Some results have suggested that broad-spectrum antibiotics and antibacterials may be useful to combat bacterial translocation resulting from impaired intestinal permeability.[63] The role of plant-derived antibacterial agents in these situations should be the focus of future research in the development of novel therapies against MODS.

Furthermore, recent reports have indicated that modulation of gut microbiota may be an important event during intestinal ischemia and reperfusion. The influence of microbiota shifts on gut barrier integrity has been extensively reviewed elsewhere.[15,25] Nevertheless, knowledge on the influence of ischemia and reperfusion on gut microbiota, including the role of phytochemicals in the modulation of gut microbiome characteristics during intestinal ischemia and reperfusion, is still at the very early stages of development and represents an important direction for future research. Next-generation sequencing approaches may be useful for the identification of bacteria and their involvement in the induction and progression of IIRI and MODS.

9.7 SUMMARY

Withdrawal of blood flow to and subsequent reperfusion of the intestine produces local intestinal injuries with ramifications for damage to remote organs. The critical pathophysiological links between intestinal ischemia and reperfusion injury and multiple organ dysfunction suggests that animal models of IIRI could prove useful in delineating critical aspects of the MODS, such as the role of the gut microbiome, that are as yet unclear. Crucially, the application of IIRI models may also promote the development of novel treatment strategies, especially from herbal medicines. A complex pathogenetic process as MODS is highly unlikely to be effectively managed by a single treatment strategy. Hence, plant remedies with constituent phytochemicals which show activities against multiple steps in the cascade of ischemia and reperfusion injury should be investigated and developed for the identification of new drugs as complementary or alternative medicines against multiple organ dysfunctions. Experimental studies reviewed in this chapter have highlighted multiple pharmacological activities exhibited by plant extracts, which makes them important candidates for further exploration. Specifically, bioactive agents from plants with potential for limiting tissue damage during IIRI should be isolated and investigated using contemporary molecular biology approaches in order to identify their specific modulation of various molecular pathways involved in MODS. Overall, it should be noted that plants themselves are not absolute repositories of all agents that could offer benefit against either IIRI or MODS. Their effectiveness would also be boosted when combined with other agents such as probiotics in order to provide more rounded solution to multiple organ dysfunction resulting from gastrointestinal injury.

KEYWORDS

- **inflammation**
- **intensive care**
- **ischemia-reperfusion**
- **medicinal herbs**
- **multiorgan dysfunction**
- **sepsis**

REFERENCES

1. Abdeen, S. M.; Mathew, T. C.; Dashti, H. M.; Asfar, S. Protective Effects of Green Tea on Intestinal Ischemia-Reperfusion Injury. *Nutrition* **2011,** *27* (5), 598–603.
2. Akcılar, R.; Akcılar, A.; Koçak, C.; Koçak, F. E.; Bayat, Z.; Şimşek, H.; Şahin, S.; Savran, B. Effects of Ukrain on Intestinal Apoptosis Caused by Ischemia-Reperfusion Injury in Rats. *Int. J. Clin. Exp. Med.* **2015,** *8* (12), 22158–22166.
3. Akinrinmade, J. F.; Akinrinde, A. S.; Odejobi, A.; Oyagbemi, A. A. Evidence of Attenuation of Intestinal Ischemia-Reperfusion Injury Following Pre-Treatment with Methanolic Extracts from *Chromolena Odorata* in Rats. *J. Complement. Integr. Med.* **2015,** *12* (1), 23–32.
4. Akinrinmade, J. F.; Akinrinde, A. S.; Soyemi, O. O.; Oyagbemi, A. A. Antioxidant Potential of the Methanol Extract of *Parquetina Nigrescens* Mediates Protection Against Intestinal Ischemia-Reperfusion Injury in Rats. *J. Diet. Suppl.* **2016,** *13* (4), 420–432.
5. Bitto, A.; Polito, F.; Altavilla, D.; Irrera, N.; Giuliani, D.; Ottani, A.; Minutoli, L.; Spaccapelo, L.; Galantucci, M.; Lodi, R.; Guzzo, G.; Guarini, S.; Squadrito, F. Melanocortins Protect Against Multiple Organ Dysfunction Syndrome in Mice. *Br. J. Pharmacol.* **2011,** *162* (4), 917–928.
6. Bura, J. A.; Holzmann, B.; Sitkovsky, M. Animal Models of Sepsis: Setting the Stage. *Nat. Rev. Drug Discov.* **2005,** *4* (10), 854–865.
7. Cerqueira, N. F.; Hussni, C. A.; Yoshida, W. B. Pathophysiology of Mesenteric Ischemia/Reperfusion: A Review. *Acta Cir. Bras.* **2005,** *20* (4), 336–343.
8. Chen, R.; Shao, H.; Lin, S.; Zhang, J.-J.; Xu, K.-Q. Treatment with *Astragalus Membranaceus* Produces Antioxidative Effects and Attenuates Intestinal Mucosa Injury Induced by Intestinal Ischemia Reperfusion in Rats. *Am. J. Chin. Med.* **2011,** *39* (5), 879–887.
9. Chen, S. H.; Tang, Y. B.; Chen, H. C. Survival of Transferred Ileum after Ischemia Time Longer Than 1 Hour: A Clinical Result Different from Animal Studies. *J. Am. Coll. Surg.* **2013,** *217* (2), 300–305.
10. Chikezie, P. C.; Ibegbulem, C. O.; Mbagwu, F. N. Bioactive Principles from Medicinal Plants. *Res. J. Phytochem.* **2015,** *9* (3), 88–115.

11. Cicalese, L.; Sileri, P.; Green, M.; Abu-Elmagd, K.; Kocoshis, S.; Reyes, J. Bacterial Translocation in Clinical Intestinal Transplantation. *Transplantation* **2001,** *71* (10), 1414–1417.
12. Collard, C. D.; Gelman, S. Pathophysiology, Clinical Manifestations and Prevention of Ischemia-Reperfusion Injury. *Anesthesiology* **2001,** *84,* 1133–1138.
13. Cuzzocrea, S.; McDonald, M. C.; Mazzon, E.; Filipe, H. M.; Centorrino, T.; Lepore, V.; Terranova, M. L.; Ciccolo, A.; Caputi, A. P.; Thiemermann, C. Beneficial Effects of Tempol, a Membrane-Permeable Radical Scavenger, on the Multiple Organ Failure Induced by Zymosan in the Rat. *Crit. Care Med.* **2001,** *29* (1), 102–11
14. De Paola, R.; Muia, C.; Mazzon, E.; Genovese, T.; Crisafulli, C.; Menegazzi, M.; Caputi, A. P.; Suzuki, H.; Cuzzocrea. S. Effects of *Hypericum Perforatum* Extract in a Rat Model of Ischemia and Reperfusion Injury. *Shock* **2005,** *24* (3), 255–263.
15. Deitch, E. A. Gut-Origin Sepsis: Evolution of a Concept. *Surgeon* **2012,** *10,* 350–356.
16. Deitch, E. A.; Sambol, J. T. The Gut-Origin Hypothesis of MODS. In *Sepsis and Multiple Organ Dysfunction: A Multidisciplinary Approach;* Deitch, E. A., Vincent, J.-L., Windsor, A., Eds.; W. B. Saunders: London, 2002; pp 105–116.
17. Dillon, S.; Agrawal, S.; Banerjee, K.; Letterio, J.; Denning, T. L.; Oswald-Richter, K.; Kasprowicz, D. J.; Kellar, K.; Pare, J.; van Dyke, T.; Ziegler, S.; Unutmaz, D.; Pulendran, B. Yeast Zymosan, a Stimulus for TLR2 and Dectin-1, Induces Regulatory Antigen-Presenting Cells and Immunological Tolerance. *J. Clin. Investig.* **2006,** *116* (4), 916–928.
18. Eltzschig, H. K.; Collard, C. D. Vascular Ischemia and Reperfusion Injury, *Br. Med. Bull.* **2004,** *70* (1), 71–86.
19. Erling Junior, N.; Montero, E. F.; Sannomiya, P.; Poli-de-Figueiredo, L. F. Local and Remote Ischemic Preconditioning Protect Against Intestinal Ischemic/Reperfusion Injury After Supraceliac Aortic Clamping. *Clinics (Sao Paulo).* **2013,** *68* (12), 1548–1554.
20. Farrell, C. P.; Barr, M.; Mullin, J. M.; et al. Epithelial Barrier Leak in Gastrointestinal Disease and Multiorgan Failure. *J. Epithelial Biol. Pharmacol.* **2012,** *5,* 13–18.
21. Gayle, J.; Jones, S. L.; Argenzio, R. A.; Blikslager, A. T. Neutrophils Increase Paracellular Permeability of Restituted Ischemic-Injured Porcine Ileum. *Surgery* **2002,** *132,* 461–470.
22. Gonzalez, L. M.; Moeser, A. J.; Blikslager, A. T. Animal Models of Ischemia-Reperfusion-Induced Intestinal Injury: Progress and Promise for Translational Research. *Am. J. Physiol. Gastrointest. Liver Physiol.* **2015,** *308* (2), G63–G75.
23. Grootjans, J.; Lenaerts, K.; Deritx, J. P.; Matthijsen, R. A.; de Bruine, A. P.; van Bijnen, A. A.; van Dam, R. M.; Dejong, C. H.; Buurman, W. A. Human Intestinal Ischemia-Reperfusion-Induced Inflammation Characterized: Experiences from a New Translational Model. *Am. J. Pathol.* **2010,** *176* (5), 2283–2291.
24. Gu, C.; Qiao, W.; Wang, L.; Li, M.; Song, K. Identification of Genes and Pathways Associated with Multiple Organ Dysfunction Syndrome by Microarray Analysis. *Mol. Med. Rep.* **2018,** *18* (1), 31–40.
25. Guarner, F.; Malagelada, J. R. Gut Flora in Health and Disease. *Lancet* **2003,** *361,* 512–519.
26. Gubernatorova, E. O.; Perez-Chanona, E.; Koroleva, E. P.; Jobin, C.; Tumanov, A. V. Murine Model of Intestinal Ischemia-reperfusion Injury. *J. Vis. Exp.* **2016,** *111,* 53881.
27. Harrison, R. Structure and Function of Xanthine Oxidoreductase: Where are We Now? *Free Radic. Biol. Med.* **2002,** *33* (6), 774–797.

28. Haward, T. R. S.; Brooks, D. L.; Flynn, T. C.; Seeger, J. M. Multiple Organ Dysfunction after Mesenteric Artery Revascularization. *J. Vasc. Surg.* **1993,** *18* (3), 459–469.
29. Hei, Z.-Q.; Huang, H.-Q.; Zhang, J.-J.; Chen, B.-X.; Li, X.-Y. Protective Effect of *Astragalus Membranaceus* on Intestinal Mucosa Reperfusion Injury after Hemorrhagic Shock in Rats. *World J. Gastroenterol.* **2005,** *11* (32), 4986–4991.
30. Hinnebusch, B. F.; Ma, Q.; Henderson, J. W.; Siddique, A.; Archer, S. Y.; Hodin, R. A. Enterocyte Response to Ischemia is Dependent on Differentiation State. *J. Gastrointest. Surg.* **2002,** *6* (3), 403–409.
31. Jakesevic, M.; Xu, J.; Aaby, K.; Jeppsson, B.; Ahrné, S.; Molin, G. Effects of Bilberry (*Vaccinium myrtillus*) in Combination with Lactic Acid Bacteria on Intestinal Oxidative Stress Induced by Ischemia-Reperfusion in Mouse. *J. Agric. Food Chem.* **2013,** *61* (14), 3468–3478.
32. Jantan, I.; Ahmad, W.; Bukhari, S. N. Plant-Derived Immunomodulators: An Insight on their Preclinical Evaluation and Clinical Trials. *Front. Plant Sci.* **2015,** *6*, 655.
33. Lee, M.-C.; Velayutham, M.; Komatsu, T.; Hille, R.; Zweier, J. L. Measurement and Characterization of Superoxide Generation from Xanthine Dehydrogenase: A Redox-Regulated Pathway of Radical Generation in Ischemic Tissues. *Biochemistry* **2014,** *53*, 6615–6623.
34. Liu, K. X.; He, W.; Rinne, T.; Liu, Y.; Zhao, M. Q.; Wu, W. K. The Effect of *Gingko Biloba* Extract (EGb 761) Pre-Treatment on Intestinal Apoptosis Induced by Intestinal Ischemia/Reperfusion in Rats: Role of Ceramide. *Am. J. Chin. Med.* **2007,** *35* (5), 805–819.
35. Louis, K.; Netea, M. G.; Carrer, D. P.; Kotsaki, A.; Mylona, V.; Pistiki, A.; Savva, A.; Roditis, K.; Alexis, A.; Van der Meer, J. W.; Giamarellos-Bourboulis, E. J. Bacterial Translocation in an Experimental Model of Multiple Organ Dysfunctions. *J. Surg. Res.* **2013,** *183* (2), 686–694.
36. Maier, R. V. Pathogenesis of Multiple Organ Dysfunction Syndrome-Endotoxin, Inflammatory Cells, and their Mediators: Cytokines and Reactive Oxygen Species. *Surg. Infect. (Larchmt).* **2000,** *1* (3), 197–204.
37. Mallick, H.; Yang, W.; Winslet, M. C.; Seifalian, A. M. Ischemia–Reperfusion Injury of the Intestine and Protective Strategies against Injury. *Dig. Dis. Sci.* **2004**, *49* (9), 1359–1377.
38. Marshall, J. C. Modeling MODS: What can be Learned from Animal Models of the Multiple-Organ Dysfunction Syndrome? *Intensive Care Med.* **2005,** *31*, 605–608
39. Marshall, J. C. Critical Illness is an Iatrogenic Disorder. *Crit. Care Med.* **2011,** *38*, S582–S589.
40. Mizock, B. A. The Multiple Organ Dysfunction Syndrome. *Disease-a-Month* **2009,** *55*, 476–526.
41. Monroe, L. L.; Armstrong, M. G.; Zhang, X.; Hall, J. V.; Ozment, T. R.; Li, C.; Williams, D. L.; Hoover, D. B. Zymosan-Induced Peritonitis: Effects on Cardiac Function, Temperature Regulation, Translocation of Bacteria, and Role of Dectin-1. *Shock* **2016,** *46* (6), 723–730.
42. Mousavi, S.; Ghannadi, A.; Meidani, M. New Horizon in the Treatment of Sepsis: A Systematic Review of Alternative Medicine. *J. Complement. Integr. Med.* **2016,** *13* (4), 317–332.
43. Nilsson, U. A.; Schoenberg, M. H.; Aneman, A.; Poch, B.; Magadum, S.; Beger, H. G.; Lundgren, O. Free Radicals and Pathogenesis during Ischemia and Reperfusion of the Cat Small Intestine. *Gastroenterology* **1994,** *106*, 629–636.

44. Ocak, T.; Duran, A.; Ozyalvacli, G.; Ocak, Z.; Terzi, E. H.; Tosun, M.; Erdem, K. Protective Effects of Motelukast and *Hypericum Perforatum* Against Intestinal Ischemia-Reperfusion Injury in Hamsters. *Turk. J. Med. Sci.* **2014,** *44*, 381–386.
45. Onen, A.; Deveci, E.; Inaloz, S. S.; Isik, B.; Kilin, C. Histopathological Assessment of the Prophylactic Effect of *Gingko Biloba* Extract on Intestinal Ischemia-Reperfusion Injury. *Acta Gastroenterol. Belg.* **1999,** *62* (4), 386–389.
46. Osterbur, K.; Mann, F. A.; Kuroki, K.; De Clue, A. Multiple Organ Dysfunction Syndrome in Humans and Animals. *J. Vet. Intern. Med.* **2014,** *28* (4), 1141–1151.
47. Ozdemir, O. M.; Ergin, H.; Yenisey, C.; Turk, N. S. Protective Effects of *Gingko Biloba* Extract in Rats with Hypoxia/Reoxygenation-Induced Intestinal Injury. *J. Pediatr. Surg.* **2011,** *46* (4):685–690.
48. Pehliran, M.; Dalbeler, Y.; Hazinedaroglu, S.; Arikan, Y.; Erkek, A. B.; Gurnal, O.; Turkcapar, N.; Turkcapar, A. G. (2002) An Assessment of the Effect of *Gingko Biloba* EGb 761 on Ischemia-Reperfusion Injury of Intestine. *Hepatogastroenterology* **2002,** *49* (43), 201–204.
49. Pourdast, A.; Sanaei, M.; Jafari, S.; Mohammadi, M.; Khalili, H.; Shafiee, G.; Ahadi, Z.; Rostami, M.; Alizad, S.; Heshmat, R.; Mohraz, M. Effect of Septimeb™ as a New Natural Extract on Severe Sepsis: A Randomized Clinical Trial. *Caspian J. Intern. Med.* **2017,** *8* (1), 35–43
50. Puleo, F.; Arvanitakis, M.; Van Gossum, A.; Preiser, J. C. Gut Failure in the ICU. *Semin. Respir. Crit. Care Med.* **2011,** *32* (5), 626–38.
51. Riaz, A. A.; Wan, M. X.; Schaefer, T.; Schramm, R.; Ekberg, H.; Menger, M. D.; Jeppsson, B.; Thorlacius, H. Fundamental and Distinct Roles of P-Selectin and LFA-1 in Ischemia/Reperfusion-Induced Leukocyte-Endothelium Interactions in the Mouse Colon. *Ann. Surg.* **2002a,** *236* (6), 777–784.
52. Riaz, A. A.; Wan, M. X.; Schafer, T.; Dawson, P.; Menger, M. D.; Jeppsson, B.; Thorlacius, H. Allopurinol and Superoxide Dismutase Protect Against Leucocyte-Endothelium Interactions in a Novel Model of Colonic Ischaemia-Reperfusion. *Br. J. Surg.* **2002b,** *89* (12), 1572–1580.
53. Rocha, B. D.; Mendes, R. R. D; Lima, G. V.; Albuquerque, G. D.; Araujo, L. L.; De Jesus, M. N. D.; Dos Santos, W. L. C.; Carreiro, M. C. Experimental Model of Mesenteric Ischemia/Reperfusion by Abdominal Aorta Clamping in Wistar Rats. *Rev. Col. Bras. Cir.* **2012,** *39* (3), 207–210.
54. Rong, L.; Chen, Y.; He, M.; Zhou, X. *Panax Notoginseng* Saponins Attenuate Acute Lung Injury Induced by Intestinal Ischemia/Reperfusion in Rats. *Respirology* **2009,** *14* (6), 890–898.
55. Sastry, P.; Hardman, G.; Page, A.; Parker, R.; Goddard, M.; Large, S.; Jenkins, D. P. Mesenteric Ischemia Following Cardiac Surgery: The Influence of Intraoperative Perfusion Parameters. *Int. Cardiovasc. Thoracic. Surg.* **2014,** *19* (3):419–424.
56. Shafik, A. N. Febuxostat Improves the Local and Remote Organ Changes Induced by Intestinal Ischemia/Reperfusion in Rats. *Dig. Dis. Sci.* **2013,** *58* (3), 650–659.
57. Silva, J. B. N. F.; De Almeida, K. N. A.; Barbosa, B. B. Medicinal Plants and Polymicrobial Sepsis: A Review. *J. Med. Plants Stud.* **2017,** *5* (3), 42–46.
58. Solinas, P.; Isola, M.; Lilliu, M. A.; et al. Animal Models are Reliably Mimicking Human Diseases? A Morphological Study that Compares Animal with Human NAFLD. *Microsc. Res. Tech.* **2014,** *77* (10), 790–796.

59. Souza, D. G.; Vieira, A. T.; Soares, A. C.; Pinho, V.; Nicoli, J. R.; Vieira, L. Q.; Teixeira, M. M. The Essential Role of the Intestinal Microbiota in Facilitating Acute Inflammatory Responses. *J. Immunol.* **2004,** *173* (6), 4137–4146.
60. Taylor, C. T.; Colgan, S. P. Hypoxia and Gastrointestinal Disease. *J. Mol. Med. (Berl.)* **2007,** *85* (12), 1295–1300.
61. Terzi, A.; Coban, S.; Yildiz, F.; Ates, M.; Bitiren, M.; Taskin, A.; Askoy, N. Protective effects of Nigella sativa on intestinal ischemia-reperfusion injury in rats. *J. Invest. Surg.* **2010,** *23* (1), 21–27.
62. Tigno, X. T.; Gumila, E. In vivo microvascular actions of Artemisia vulgaris L. in a model of ischemia-reperfusion injury in the rat intestinal mesentery. *Clin. Hemorheol. Microcirc.* **2000,** *23* (2–4), 159–165.
63. Tilsed, J. V.; Casamassima, A.; Kurihara, H.; Mariani, D.; Martinez, I.; Pereira, J.; Ponchietti, L.; Shamiyeh, A.; Al-Ayoubi, F.; Barco, L.A.; Ceolin, M.; D'Almeida, A. J.; Hilario, S.; Olavarria, A. L.; Ozmen, M. M.; Pinheiro, L. F.; Poeze, M.; Triantos, G.; Fuentes, F. T.; Sierra, S. U.; Soreide, K.; Yanar, H. ESTES Guidelines: Acute Mesenteric Ischemia. *Eur. J. Trauma Emerg. Surg.* **2016,** *42* (2), 253–270.
64. Trompeter, M.; Brazda, T.; Remy, C. T.; Vestring, T.; Reimer P Non-Occlusive Mesenteric Ischemia: Etiology, Diagnosis, and Interventional Therapy. *Eur. Radiol.* **2002,** *12* (5), 1179–87.
65. Vincent, J. L.; Sakr, Y.; Sprung, C. L.; et al. Sepsis in European Intensive Care Units: Results of the SOAP Study. *Crit. Care Med.* **2006,** *34*, 344–353.
66. Wen, H. Sepsis Induced by Cecal Ligation and Puncture. *Methods Mol. Biol.* **2013,** *1031*, 117–124.
67. Wiest, R.; Rath, H. C. Gastrointestinal Disorders of the Critically-Ill. Bacterial Translocation in the Gut. *Best Pract. Res. Clin. Gastroenterol.* **2003,** *17*, 397–425.
68. Yang, X.; Bai, H.; Cai, W.; Li, J.; Zhu, Q.; Wang, Y.; Han, J.; Zhu, X.; Dong, M.; Hu, D. *Lycium Barbarum* Polysaccharides Reduce Intestinal Ischemia-Reperfusion Injuries in Rats. *Chem. Biol. Interact.* **2013,** *204*,166–172.
69. Zhang, S.-W.; Ren, A.-M.; Zhang, L. X.; et al. Study on Diagnosis and Treatment of Infectious Multiple Organ Dysfunction Syndrome by Integrated Traditional Chinese and Western Medicine. *Chin. J. Integr. Med.* **2001,** *7* (2), 120–120.
70. Zhang, F.; Tong, L.; Qiao, H.; Dong, X.; Qiao, G.; Jiang, H.; Sun, X. Taurine Attenuates Multiple Organ Injury Induced by Intestinal Ischemia Reperfusion in Rats. *J. Surg. Res.* **2008,** *149* (1), 101–109.
71. Zimmerman, B. J.; Granger, D. N. Reperfusion-Induced Leukocyte Infiltration: Role of Elastase. *Am. J. Physiol. Heart Circ. Physiol.* **1990,** *259*, H390–H394.

CHAPTER 10

Biochemical Constituents and Toxicological Effects of *Syzygium aromaticum* on Selected Organs in Male Wistar Rats

Mosebolatan Victoria Adegbola, Ademola Olabode Ayeleso, and Temitope Adenike Oyedepo

ABSTRACT

Syzygium aromaticum (Cloves) possesses various phytochemicals with several health-related benefits. Due to the increased use of herbal preparations for health benefits, cloves are being consumed without adequate information on the possible toxic effects from consuming large doses. This study was aimed at evaluating the potential for toxicity of *S. aromaticum* at varied doses in selected organs of Albino Wistar rats. Twenty-four (24) male Wistar rats, weighing between 150 and 200 g, were grouped into four groups of six (6) animals each. The animals were orally given varied concentration of the extract (500, 1000, and 1500 mg/kg bodyweight of the extract, and the control). The blood, kidney, and liver rats were obtained for hematological analysis and histological examinations. Hematological parameters, including the red blood cell count and platelets, showed no significant difference at 500 mg per kg bodyweight, as compared to control. However, on histological examination of kidney and liver cells, there was considerable histological distortion, such as, congestion of the portal vein, cytoplasmic vacuolation, and lipid/fat infiltration of cytoplasm of the hepatocytes. Inflammation was also observed in the hepatocytes and kidney cells on administration of *S.*

Curative and Preventive Properties of Medicinal Plants: Research on Disease Management and Animal Model Studies. Megh R. Goyal, Junaid Ahmad Malik, & Ademola Olabode Ayeleso (Eds.)

aromaticum extract at concentrations of 1000 mg per kg bodyweight and 1500 mg per kg bodyweight after 14 days of administration. It is recommended that the consumption of cloves be limited in duration and dose to reduce the possibility of toxicity.

10.1 INTRODUCTION

Spices have from time immemorial, played an important in food and the culinary industry due to their ability to add flavor, aroma color, and even enhance taste. They have also been recognized for their nutritional and medicinal properties.[26,37,38] The chemopreventive and bioprotective properties of spices such as garlic, turmeric, and cinnamon against cancer cells have been reported.[6,30] A large population of humans still depends on herbal medicine for majority of their health problems.[9] Hence, the investigation of these readily available plant materials is constantly being done to verify and ascertain their efficacy and possible toxic effects.

Cloves (*Syzygium aromaticum*) are aromatic dried buds of a tree used as a spice. It is a common kitchen spice used in places like Russia, Scandinavia, Greece, India, China, and in Africa. It is believed to have antidiabetic,[33] anti-inflammatory, antimicrobial,[5] and aphrodisiac activities.[39] The antioxidant activities of cloves are comparable to the activities of the synthetic antioxidant, butylated hydroxyl anisole, and Pyrogallol.[8]

FIGURE 10.1 Dried *Syzygium aromaticum* buds.

Phytochemicals obtained from the methanolic extraction of *S. aromaticum* have revealed that it consisted of alkaloids, aldehydes, glycoside, and carbohydrate. Other phytochemicals include phenolic compounds, flavonoids, ketones, and also steroids.[14] Clove buds contain essential oils, such as, eugenol, eugenyl acetate, and β-caryophyllene.[24] These phytochemicals are responsible for conferring the medicinal properties inherent in this spice. Cloves are highly priced for the essential oils, which may be obtained from the buds by simple distillation. The major component of these oils is Eugenol, which may make up to 85% of the total oils.[20] Other oils may include eugenyl acetate and β-caryophyllene.[30] Eugenol is used in traditional medicine to treat minor wounds. In dental care, it is used as a mild and local anesthesia for aches and minor tooth infections.[4,30]

Despite these advantages obtainable from the consumption of spices, studies have shown that there is also a risk of toxicity.[16] Lin et al.[19] reported the possibility of lead poisoning in children, who regularly ingested Indian spices. In humans, the safe daily dose of clove consumption is about 2.5 mg/kg body weight.[27] However, due to lack of information on the possible toxic effects of cloves, they are used indiscriminately in many herbal preparations.

This study was conducted to determine the phytochemical components, the proximate and mineral content of Cloves. It was also aimed at evaluating the safety profile of the ethanolic extract of *S. aromaticum* buds in selected organs of albino Wistar rats.

10.2 MATERIALS AND METHODS

10.2.1 PLANT MATERIAL COLLECTION

Dried buds of *S. aromaticum* were purchased at a local market in Osun state, Nigeria. The buds were authenticated at the Plant Biology Laboratory, Department of Basic Sciences, Adeleke University, Osun State, Nigeria.

10.2.2 EXTRACT PREPARATION

All extraneous materials in the buds were removed. The *S. aromaticum* buds were blended to powdery form using a blender. The sample, weighing 100 g was soaked in 400 mL of 50% ethanol for 72 h. The resultant mixture was filtered with a cheesecloth. This was then completely evaporated using a rotary evaporator. The resulting residue was stored at 4 °C.

10.2.3 PROXIMATE ANALYSIS OF *S. AROMATICUM* EXTRACTS

The proximate analysis was done according to the method prescribed by AOAC (2003). Moisture content was determined by weighing 2 g of extract in a crucible (W_1). This was then dried in an oven at about 105 °C and weighed intermittently. The crucible was cooled in a desiccator until a constant weight was obtained (W_2). The percent moisture is given as the difference between both weights, W_1 and W_2, divided by the weight of the sample and then multiplied by 100.

10.2.3.1 ASH DETERMINATION

Determination of ash was done by determining the weight of an empty crucible (W_1). A gram of the sample was put in the crucible and the total weight was noted as W_2. The sample was burned until it was charred. The gray-white ash was seen from the muffle furnace (550 °C), it indicated that a total oxidation of all organic matter in the sample had occurred. The crucible was cooled and the weight determined (W_3).

$$\% Ash = \frac{\text{difference in weight of sample}}{\text{Weight of sample}} \times 100$$

where, difference in wt. of Ash = $W_3 - W_1$.

10.2.3.2 PROTEIN DETERMINATION

The Kjeldahl method was employed in analyzing protein in the sample. Digestion was done by heating. This involved the use of concentrated H_2SO_4accompanied with of the mixture of (K_2SO_4 and $CuSO_4$ (8:1). It was ensured that the mixture was alkaline in nature. Ammonium sulfate formed liberated ammonia. This was collected in 2% boric acid solution and titrated against standard HCl. Total protein was determined by the product of the value of nitrogen with a given constant (6.25).

10.2.3.3 FAT DETERMINATION

Crude fat was determined by ether extract method using Soxhlet apparatus. Moisture free sample was wrapped in filter paper. It was put in thimble and then in the extraction tube. A receiving beaker was filled with petroleum ether and fitted into the apparatus. The extract was poured into a clean glass dish and evaporated on water bath. The percent crude fat was calculated by

$$\% Crude\, Fat = \frac{\text{Weight of ether extract}}{\text{Weight of sample}} \times 100$$

10.2.3.4 FIBER DETERMINATION

To determine crude fiber, digestion was done with dilute $H_2SO_{4,}$ followed by dilute KOH solution. After the digestion, the residue was collected and ignited. The difference in weight was and taken as the crude fiber.

10.2.3.5 CARBOHYDRATE DETERMINATION

Carbohydrate was determined by taking the difference. The formula given below was used. Carbohydrate = (100% moisture + % crude protein + % crude fiber ++ % crude fat % ash)

10.2.4 PHYTOCHEMICAL SCREENING AND QUANTIFICATION

The phytochemical screening was done using the methods described by Sofowora.[36] The spectrophotometric method was used to quantify the phytochemicals.[25]

10.2.5 MINERAL CONTENTS DETERMINATION

Mineral contents of *S. aromaticum* (Cloves) were analyzed using Atomic Absorption Spectrometry and Flame Photometry and Spectrophotometry following the methods described by AOAC.[2] This was done after the samples had undergone wet digestion.

10.2.6 ANIMALS

Male Wistar albino rats of 5–7 weeks of age weighing between 150 and 200 g were procured from the Department of Anatomy, Adeleke University,

Ede, Nigeria. The rats were acclimatized for a period of 2 weeks in a well-ventilated animal house and fed with standard rat chow and clean water *ad libitum*. The animals were subjected to natural photoperiod of 12 h daylight and 12 h night throughout the study period. All the animals received the humane care whenever needed.

10.2.6.1 ANIMAL GROUPING FOR TOXICOLOGICAL EFFECT

Twenty-four (24) male Wistar rats, weighing between 150 and 200 g were purchased and grouped into four groups of six (6) animals each. The animals were orally given a varied concentration of the extract (0, 500, 1000, and 1500 mg/kg bodyweight of the extract). The experiment was conducted for 14 days.

Group 1: Control (DMSO) with zero extract of clove buds.
Group 2: 500 mg per kg bodyweight of *S. aromaticum* extract only.
Group 3: 1000 mg per kg bodyweight *S. aromaticum* extract only.
Group 4: 1500 mg per kg bodyweight of *S. aromaticum* extract only.

10.2.7 PLASMA AND ORGAN COLLECTION

Animals were sacrificed 24 h after the last treatment under 0.3 mL sodium thiopentone/kg body weight anesthesia. The animals were dissected (cut open from the abdominal cavity to the thoracic cavity) and blood samples were received from the apex of the heart into heparinized sample bottles. The blood sample bottles were spun at the rate of 3000 revolutions per minute for 15 min. Plasma was collected and stored at low temperature (−20°C). The right kidney and a lobe from the liver from each group were removed immediately and were fixed in 10% formalin fluid and were stained with hematoxylin and eosin for histopathological examinations.

10.2.8 HEMATOLOGICAL PARAMETERS TEST

Dilution of the blood sample (50 μL) with cell pack was done in the White Blood Cell counting chamber. A defined volume of Stromatolyzer-WII solution (1 volume of stromatolyzer WH to 2 volumes of cell pack) was added to result in a final dilution of 1:500. The addition of stromatolyzer-WH caused the lysing of the red blood cell. The cell stroma left is not detectable by the instrument. At the same time, the White Blood Cell stabilized at a level that the instrument can detect. The Direct Current Method was used to count the cells. When the red blood cells lysed, they released hemoglobin which was converted to the methemoglobin and read photo metrically at 555 nm. A part of the diluted sample was conveyed automatically to the hemoglobin detector where the absorbance of the red pigment was measured to give blood hemoglobin level. Blood cell count is evaluated by counting the pulses.

10.2.9 HISTOLOGICAL ANALYSIS

The right kidney and a lobe from the liver were grossed and cut into small pieces of 3 mm thick in a labeled tissue cassette. They were processed using automatic tissue processor (LEICA TP 1020) and were passed through various reagents including alcohol to various concentration ranging from 0 to 100% of molten paraffin wax at 65 °C[3] and the stain used was hematoxylin and eosin for histopathological examinations. The photomicrographs of the stained kidney and liver sections were taken with the aid of a light microscope.

10.2.10 STATISTICAL ANALYSIS

The data were stated as average and standard deviation (mean ± SD). The one-way analysis of variance was used to determine if the values between the groups were statistically significance. Differences in means were determined to be significant when $p < 0.05$. These analyses were performed using GraphPad Prism 8.

10.3 RESULTS

10.3.1 PHYTOCHEMICAL CONSTITUENTS OF S. AROMATICUM EXTRACT

The presence of alkaloids, tannins, cardiac glycosides, and anthraquinones was established in the extract of *S. aromaticum.* Other phytochemical found in the ethanolic extract include flavonoids and saponins.

TABLE 10.1 Phytochemicals Present in 50% Ethanolic Extract of *Syzygium aromaticum*

Phytochemicals	Ethanolic extract
Alkaloid	+
Reducing sugar	−
Steroid	−
Saponin	+
Flavonoid	+
Tannin	+
Cardiac glycoside	+
Anthraquinone	+

+ *means present;* − *means absent.*

10.3.2 ESTIMATION OF PHYTOCHEMICAL CONSTITUENTS OF S. AROMATICUM EXTRACT

The phytochemicals that are present in the ethanolic extract of *S. aromaticum* were quantified as provided in Table 10.2.

TABLE 10.2 Estimation of Phytochemicals Present in Ethanolic Extract of *Syzygium aromaticum.*

Alkaloids (mg/g)	Tannins (mg/g)	Saponins (mg/100 g)	Cardiac glycosides (mg/100 g)	Anthraqunone (mg/100 g)
17.33 ± 2.31	1.93 ± 0.03	113.67 ± 1.53	10.30 ± 03	11.6 ± 0.1

10.3.3 PROXIMATE ANALYSIS OF S. AROMATICUM

The proximate analysis of *S. aromaticum* is graphically illustrated as shown in Figure 10.2. Carbohydrate had the highest percentage value (71.85%) while crude fat was lowest with a value of 0.77%.

10.3.4 MINERAL COMPOSITION OF S. AROMATICUM BUDS

The mineral content of the ethanolic extract of *S. aromaticum* is presented in Table 10.3. The values were presented in mg/100 g of the sample. Zinc (Zn) was observed to have the highest concentration (25.33 mg/100 g) while Lead (Pb) concentration was negligible at 0.001 mg/100 g of the sample.

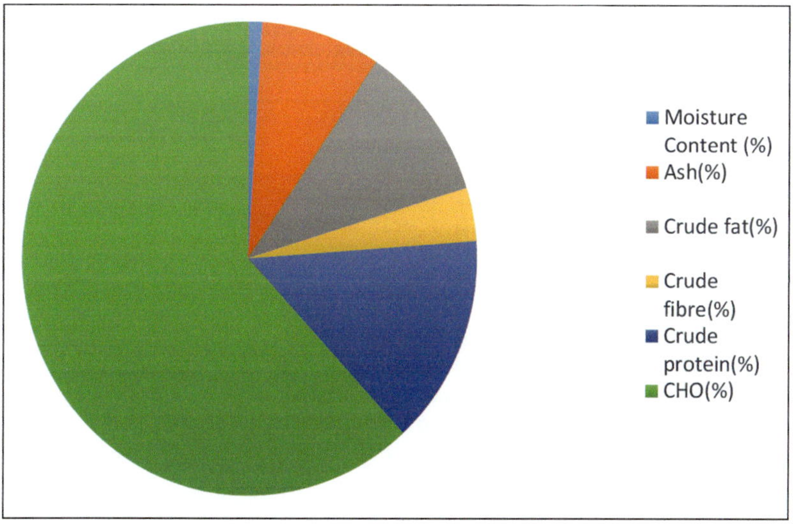

FIGURE 10.2 Pie chart showing values of proximate components present in cloves.

TABLE 10.3 Estimation of Minerals Present in *Syzygium aromaticum*.

Minerals	Concentration (mg/100 g)
Zinc	25.33 ± 0.15
Copper	0.012 ± 0.001
Iron	1.11 ± 0.017
Manganese	0.031 ± 0.0015
Phosphorus	15.04 ± 0.12
Selenium	1.33 ± 0.058
Nickel	5.01 ± 0.015
Molybdenum	3.0 ± 0.1
Magnesium	22.04 ± 0.15
Lead	0.001 ± 0[a]
Chromium	0.1 ± 0

Each value is a mean of three values ± SD.

[a]*negligible.*

10.3.5 RESULTS FOR S. AROMATICUM ON HEMATOLOGICAL PARAMETERS AT DIFFERENT CONCENTRATIONS

Hematological parameters after the administration of 500, 1000, and 1500 mg/kg bodyweight of *S. aromaticum* to experimental animals show that there were no significant variances in all the parameters across the groups. The results are displayed in Figure 10.3 and Table 10.4.

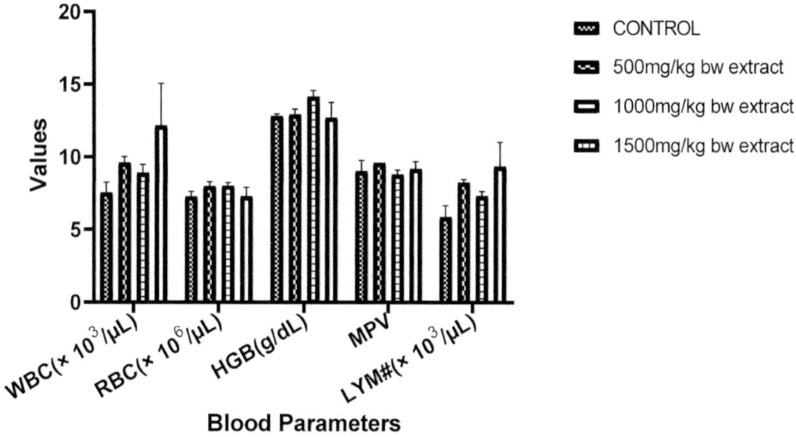

FIGURE 10.3 Effect of varying concentration of *Syzygium aromaticum* extract on selected hematological parameters.

TABLE 10.4 Effect of *Syzygium aromaticum* on Hematological Parameters at Varying Concentrations.

Group/parameter	Control	500 mg/kg bw extract	1000 mg/kg bw extract	1500 mg/kg bw extract
HCT(%)	51.07 ± 0.79	54.4 ± 0.65	51.7 ± 3.47	51 ± 5.72
MCV(fL)	70.77 ± 3.63	64.77 ± 2.17	64.67 ± 3.03	69.95 ± 1.84
MCH(pg)	17.73 ± 0.77	16.2 ± 0.25	17.6 ± 0.32	17.45 ± 0.04
MCHC(g/dL)	25.07 ± 0.30	24.37 ± 1.07	27.53 ± 1.18	25 ± 0.73
PLT(× 10³/µL)	710 ± 112.2	810 ± 97.31	864 ± 90.64	915.5 ± 162.9
LYM%	77.6 ± 2.93	80.8 ± 2.43	88.87 ± 5.15	78.75 ± 4.86
RDW_SD (fL)	54.3 ± 7.99	51.10 ± 13.19	37.5 ± 3.16	40.6 ± 20.98
rdW_CV	21.8 ± 2.35	22.85 ± 5.49	16.17 ± 1.03	22.3 ± 3.67
PDW	12.37 ± 1.04	10.9 ± 0.0	12.67 ± 1.374	11.05 ± 0.28
P_LCR	19.33 ± 3.47	17.6 ± 0.0	18.17 ± 2.21	13.6 ± 4.08

Data is represented as mean ± SEM.

10.3.6 HISTOPATHOLOGICAL STUDIES ON THE LIVER OF EXPERIMENTAL RATS

The effects of varying concentrations of *S. aromaticum* extract liver of experimental rats were investigated (Figure 10.4). The control (Figure 10.4A) and

500 mg/kg bodyweight of extract (Figure 10.4B) showed normal architecture, no infiltration, and no lesions in the liver.

The experimental rats treated with 1000 mg per kg bodyweight of the *S. aromaticum* extract (Figure 10.4C) revealed mildly congested portal vein and the hepatocytes displayed cytoplasmic vacuolation and lipid/fat infiltration of cytoplasm. A mild hemorrhage was also observed. The rats treated with 1500 mg/kg bodyweight of the *S. aromaticum* extract showed congestion of the portal vein, and the parenchyma show focal area of inflammatory cells aggregate as seen in Figure 10.4D. The sinusoids were also infiltrated by inflammatory cells.

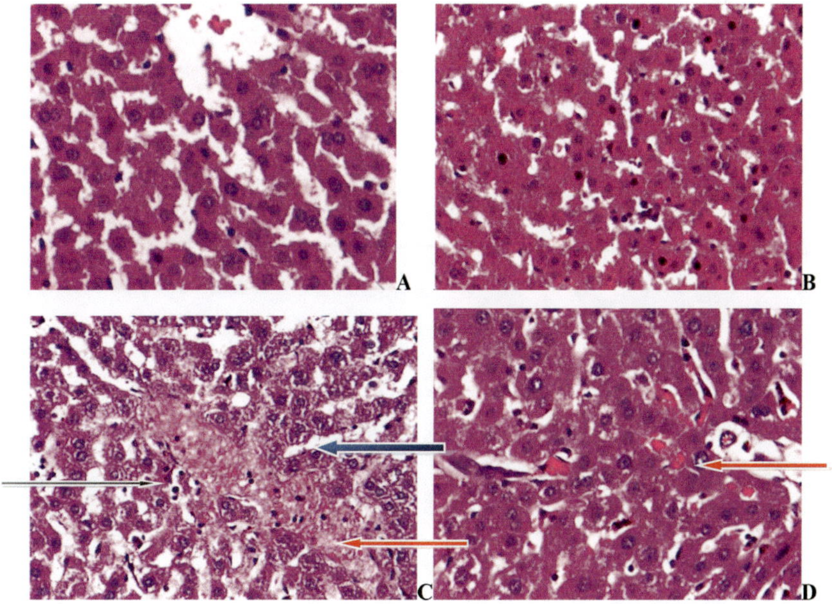

FIGURE 10.4 Photomicrograph of the liver section of Male Wistar rats stained by Hematoxylin and Eosin. The control group is represented as (A); (B) 500 mg/kg bodyweight of the *S. aromaticum* extract; (C) 1000 mg/kg bodyweight of the *S. aromaticum* extract; (D) 1500 mg/kg bodyweight of the *S. aromaticum* extract; Magnification ×400)

Note: Black arrow, inflammatory cells aggregate; Red arrow, mild hemorrhage; Blue arrow, fat infiltration of cytoplasm.

10.3.7 HISTOPATHOLOGICAL STUDIES ON THE KIDNEY OF EXPERIMENTAL RATS

The effects of varied concentrations of *S. aromaticum* on kidney of experimental rats were investigated (Figure 10.5). The control (Figure 10.5A) and the 500 mg/kg bodyweight of extract (Figure 10.5B), normal glomeruli with normal capsular spaces were observed in the renal cortex. Normal tubules are also seen and the interstitial spaces appear normal.

Figure 10.5C and 10.5D shows the groups treated with 1000 mg/kg bodyweight and 1500 mg/kg bodyweight of *S. aromaticum* extract, respectively. These groups had renal tubules that showed mild tubular necrosis and some collapsed tubules with lack of luminar space. Also, the interstitial spaces had mild vascular congestion. The group treated with 1500 mg/kg bodyweight of *S. aromaticum* (Figure 10.5D) also showed the interstitial spaces with areas of mild infiltration of inflammatory cells.

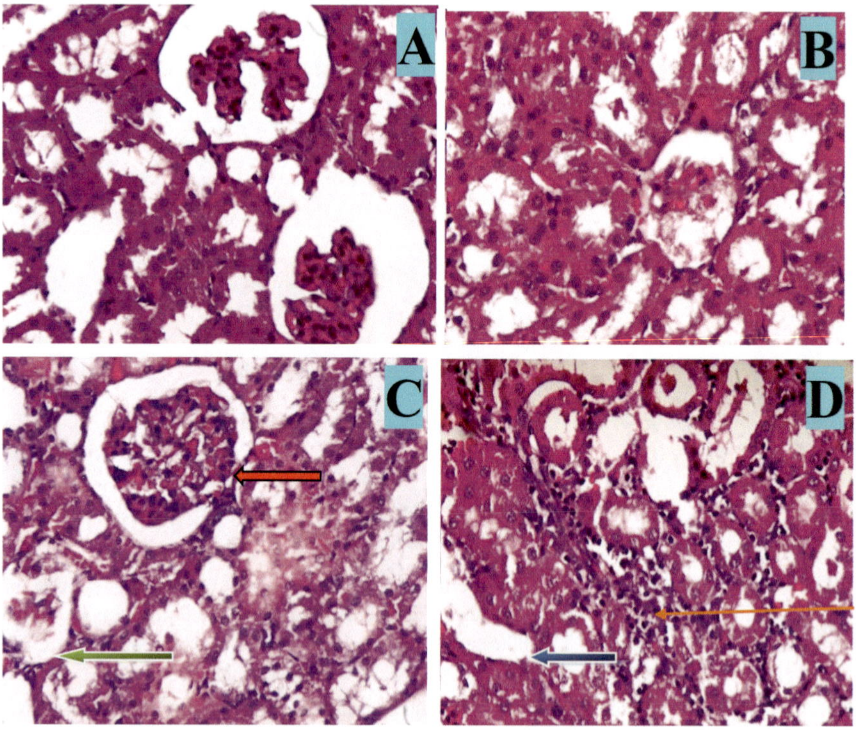

FIGURE 10.5 Photomicrograph of the kidney section of Male Wistar rats stained by Haematoxylin and Eosin. (A) Control; (B) 500 mg/kg bodyweight of the *S. aromaticum* extract; (C) 1000 mg/kg bodyweight of the *S. aromaticum* extract; (D) 1500 mg/kg bodyweight of the *S. aromaticum* extract; Magnification ×400.

Note: Red arrow, Mild tubular necrosis; Green arrow, mild vascular congestion; Slender arrow, mild infiltration of inflammatory cells; Blue arrow, loss of luminar space.

10.4 DISCUSSION AND CONCLUSION

There has been an increasing interest in herbs and spices as safe and non-toxic sources of numerous health benefits. The antioxidants that they possess are known to confer protection to the body against the negative effects of free radical or reactive oxygen species generation. These radicals affect various physiological processes resulting in diseases such as cancers and atherosclerosis.[22,32] The clove tree is a native plant, originally found in Indonesia, but is now widely cultivated all around the world.[7,4] The buds are the most highly prized part of this important tree, even though many other parts of the tree are also very beneficial. Due to its antioxidant activities,[35] cloves are believed to be able to prevent degenerative diseases such as diabetes, hypertension, and impotency.[1]Many herbal preparations are made with the inclusion of *S. aromaticum* because of these numerous advantages. However, not much thought is given to the possible toxic effect in cases of overdose or prolonged use.

Phytochemical screening of the *S. aromaticum* extract showed the presence of alkaloids, flavonoids, saponins, tannins, cardiac glycosides, and anthraquinones. Other spices such as Ajwain *(Trachyspermum ammi*), Aniseed (*Pimpinella anisum*), Fenugreek seeds *(Trigonella foenum-graecum),* Poppy seed (*Papaver somniferum*), turmeric (*Curcuma longa*), cinnamon (*Cinnamomum verum*), and black pepper (*Piper nigrum*) have also been shown to contain similar phytochemicals profile has seen in *S. aromaticum.*[7,10,18] Flavonoids are important because of their antioxidant properties. Shan et al.[35] established the relationship between the presence of flavonoids and antioxidant levels using various spices. In many antibiotics, alkaloids, saponins, and tannins are useful in the treating of common pathogenic strains as reported by Kubmarawa et al.[17] and Mensah et al.[23]

The proximate analysis of cloves (Figure 10.1) revealed the carbohydrate value to be highest at 61.85% and moisture content to be the lowest at 0.96%. Khathir et al.[15] revealed that depending on the variation of temperatures, air velocity, and humidity in the drying chamber and method of drying, the final moisture content of dry cloves may vary from 0.2 to 19.2%. Hence, the value obtained in this study is within the range described. This study also showed the crude fat content of the extract to be 10.77%. Cloves contain 3.64% fiber while crude protein was significantly high at 14.36%. Dietary fiber is known to aid digestion and reducing the risk of constipation, diabetes, and cancers of the stomach.[12]

According to Pearson,[31] plants that provide an excess 12% of its calorific value are good source of protein. The ash content of plant is usually the source of mineral and these minerals are essential in cell growth and sustenance of organisms. *S. aromaticum* has an ash content of 8.42% and contains many essential minerals.

The mineral composition of *S. aromaticum* is shown in Table 10.3. Magnesium and Phosphorus were shown to have the highest value of 22.04 ± 0.15 mg/100 g and 15.04 ± 0.12 mg/100 g of the extract, respectively. This is highly beneficial, as the presence of Mg in human diet is known to protect against hypertension and various forms of heart diseases.[29] Lead and chromium were of the lowest value. Lead presence at high quantity in plant products is usually a contaminant and can be toxic.[28]

Hematological parameters are a widely acceptable means of determining the toxicity of a substance. In this study, *S. aromaticum* was shown to pose no toxicity potential at the concentration considered from the analysis done in the blood. Vijayasteltar et al.[42] using a standardized polyphenolic extract of clove buds also reported this. The study reported that when the standardized extract was used for 2 weeks at up to 1000 mg/kg b.wt./day, no adverse effect was seen in hematology and biochemical parameters in Wistar rats.

However, upon histological evaluation, it was observed that at concentrations higher that 500 mg/kg body weight, there was increasing congestion of the portal vein and the hepatocytes showed cytoplasmic vacuolation and lipid infiltration of cytoplasm. In a case reported by Hartnoll et al.[11] ingestion of oil of cloves in a 2-year-old boy resulted in coma, fits, a coagulopathy, and acute liver damage.[21] Thompson et al.[41] demonstrated that when rat hepatocytes were incubated with eugenol, the major component of *S. aromaticum*, there was a resultant loss of cellular antioxidant and ~85% cell death within 5 h.

Janes et al.[13] also reported a case of hepatic failure resulting from clove oil in a 15-month old. Herbal supplements have been known to result in liver toxicity because of reasons ranging from contamination, poor labeling, interaction between the various ingredient and parts of the plants used.[34] Clove essential oil can be toxic to the nervous system and irritating to the skin and digestive mucous if taken in higher than recommended doses. Tammannavar et al.[40] described a case of a dental patient who developed allergic contact urticarial following the use of clove essential oil as a mild anesthesia during dental treatment. Reactive haptens may be formed in the presence of proteins leading to hypersensitive reactions.[40]

10.5 SUMMARY

The ethanolic extract of *S. aromaticum* has been shown to possess phytochemicals such as alkaloids, flavonoids, tannins, and cardiac glycosides. Others are saponins and anthraquinones. These phytochemicals are important in conferring the beneficial properties associated with Cloves. Mineral contents in *S. aromaticum* include important and essential elements like Zinc, Copper, Iron, Manganese, Molybdenum, and Phosphorus. This present study evaluated the safety of *S. aromaticum* in Wistar rats at up to 1500 mg/kg bwt for 2 weeks. Hematological parameters revealed no significant difference between control group and the treatment groups. Conversely, the use of *S. aromaticum* was observed to have deleterious effect on the kidney and liver at concentrations greater than 1000 mg/kg bodyweight. The prolonged or use of clove buds at high dosage is therefore discouraged. The use of multiple herbal preparations containing *S. aromaticum* should be avoided.

KEYWORDS

- **cloves**
- **hematological parameters**
- **histopathological effects**
- **phytochemicals**
- ***Syzygium aromaticum***
- **toxicological effect**

REFERENCES

1. Anuj, G.; Sanjay, S. Eugenol: A Potential Phytochemical With Multifaceted Therapeutic Activities. *Pharmacol. Online* **2010,** *2*, 108–120.
2. AOAC. In *Official Methods of Analysis*, 17th ed.; Association of Official Analytical Chemists (AOAC): Washington, USA, 2003.
3. Avwioro, O. G. *Histochemistry and Tissue Pathology: Principle and Techniques*, 1st ed.; Claverianum Press: Ibadan, Nigeria, 2010; pp 134–260.
4. Batiha, G. E.; Alkazmi, L. M.; Wasef, L. G.; Beshbishy, A. M.; Nadwa, E. H.; Rashwan, E. K. *Syzygium Aromaticum*: Traditional Uses, Bioactive Chemical Constituents, Pharmacological and Toxicological Activities. *Biomolecules* **2020,** *10*(2), 202–210.

5. Briozzo, J.; Núñez, L.; Chirife, J.; Herszage, L.; D'Aquino, M. Antimicrobial Activity of Clove Oil Dispersed in A Concentrated Sugar Solution. *J. Appl. Bacteriol.* **1989,** *66*(1), 69–75.
6. Butt, M. S.; Naz, A.; Sultan, M. T.; Qayyum, M. M. Anti-Oncogenic Perspectives of Spices/Herbs: A Comprehensive Review. *EXCLI J.* **2013,** *12*, 1043–1065.
7. Cortés-Rojas, D. F.; de Souza, C. R.; Oliveira, W. P. Clove (*Syzygium aromaticum*): A Precious Spice. *Asian Pac. J. Trop. Biomed.* **2014,** *4*(2), 90–96.
8. Dorman, H. J. D.; Figueiredo, A. C.; Barroso, J. G.; Deans, S. G. In Vitro Evaluation of Antioxidant Activity of Essential Oils and Their Components. *Flav. Frag. J.* **2000,** *15,* 12–16.
9. Ekor M. The Growing Use of Herbal Medicines: Issues Relating to Adverse Reactions and Challenges in Monitoring Safety. *Front. Pharmacol.* **2014,** *4*, 177–182.
10. Harsha, N.; Sridevi, V.; Chandana-Lakshmi, M. V. V.; Rani, K.; Vani, N. D. S. P hytochemical Analysis of Some Selected Spices. *Int. J. Innov. Res. Sci. Eng. Technol.* **2013,** *2*(11), 6618–6621.
11. Hartnoll, G; Moore, D.; Douek, D. Near Fatal Ingestion of Oil of Cloves. *Arch. Dis. Childh.* **1993,** *69*, 392–393
12. Ishida, H.; Suzuno, H.; Sugiyama, N.; Innami, S.; Tadokoro, T; Maekawa, A. Nutritive Evaluation on Chemical Components of Leaves, Stalks and Stems of Sweet Potatoes (*Ipomoea batatas*). *Food Chem.* **2000,** *68,* 359–367.
13. Janes, S. E.; Price, C. S.; Thomas, D. Essential Oil Poisoning: N-Acetylcysteine for Eugenol-Induced Hepatic Failure and Analysis of a National Database. *Eur. J. Pediatr.* **2005,** *164*(8), 520–522.
14. Jimoh, S.; Arowolo, L.; Alabi, K. Phytochemical Screening and Antimicrobial Evaluation of *Syzygium aromaticum* Extract and Essential oil. *Int. J. Curr. Microbiol. Appl. Sci.* **2017,** 6, 4557–4567.
15. Khathir, R.; Luecke, W.; Hartulistyoso, E.; Nelwarq L. O. Abstract: Study on Drying Cloves (*Syzygium Aromaticum*) Using Greenhouse Effect Solar Dryer Integrated by Biomass Energy. In *Competition for Resources in a Changing World: New Drive for Rural Development*; Cuvillier Verlag: Germany, 2008; pp 164–168.
16. Krejpcio, Z.; Krol, E; Sionkowski, S. Evaluation of Heavy Metals Contents in Spices and Herbs Available on the Polish Market. *Pol. J. Environ. Stud.* **2007,** *16*(1), 97–100.
17. Kubmarawa, D.; Ahunna, A.; Enwerem, N.; Okorie, D. Preliminary Phytochemical and Antimicrobial Screening of 50 Medicinal Plants from Nigeria. *Afr. J. Biotechnol.* **2007,** *6*(14), 1690–1696
18. Lawal, A. R.; Olayinka, B. U.; Murtadha, R. A. Ayinla, A.; Etejere, E. O. Comparative Analysis of Phytochemical and Proximate Composition of *Allium sativum* L. and *Zingiber officinale*. *Nig. J. Basic Appl. Sci.* **2018,** *26* (2), 82–87.
19. Lin, C. G.; Schaider, L. A.; Brabander, D. J.; Woolf, A. D. Pediatric Lead Exposure From Imported Indian Spices and Cultural Powders. *Pediatrics* **2010,** *125*(4), 828–835.
20. Liu, Q.; Meng, X.; Li, Y.; Zhao, C.; Tang, G.; Li, H. Antibacterial and Antifungal Activities of Spices *Int. J. Mol. Sci.* **2017,** *18*, 1283–1345. DOI: 10.3390/ijms18061283
21. LiverTox: Clinical and Research Information on Drug-Induced Liver Injury [Internet]. Bethesda (MD): National Institute of Diabetes and Digestive and Kidney Diseases; 2012-. Eugenol (Clove Oil) [Online], [Updated Oct 28, 2019]. https://www.ncbi.nlm.nih.gov/books/NBK551727/; Accessed on June 10, 2020

22. Lobo, V.; Patil, A.; Phatak, A.; Chandra, N. Free Radicals, Antioxidants and Functional Foods: Impact on Human Health. *Pharmacogn. Rev.* **2010,** *4*(8), 118–126. https://doi.org/10.4103/0973-7847.70902
23. Mensah, J. K.; Okoli, R. I.; Ohaju-Obodo, J. O.; Eifediyi, K. Phytochemical, Nutritional and Medical Properties Of Some Leafy Vegetables Consumed by Edo people of Nigeria. *Afr. J. Biotechnol.* **2008,** *7*, 2304–2309.
24. Mittal, M.; Gupta, N.; Parashar, P.; Mehra, V.; Khatri, M. Phytochemical Evaluation and Pharmacological Activity of *Syzygium aromaticum*: A comprehensive review. *Int. J. Pharm. Pharm. Sci.* **2014,** *6*, 67–72.
25. Obadoni, B. O.; Ochuko, P. O. Phytochemical Studies and Comparative Efficacy of the Crude Extracts of Some Homeostatic Plants in Edo and Delta States of Nigeria. *Glob. J. Pure Appl. Sci.* **2001,** *8*, 203–208.
26. Ogbunugafor, H.; Ugochukwu, C.; Kyrian-Ogbonna, A. The Role of Spices in Nutrition and Health: A Review of Three Popular Spices Used in Southern Nigeria. *Food Qual. Saf.* **2017,** *1*(3), 171–185.
27. Ogunwande, I.; Olawore, N.; Ekundayo, O.; Walker, T. M.; Schmidt, J. M.; Setzer, W. N. Studies on the Essential Oils Composition, Antibacterial and Cytotoxicity of *Eugenia uniflora* L. *Int. J. Aromather.***2005,** *15*, 147–152.
28. Oloyede O. I. Chemical Profile of Unripe Pulp of *Carica papaya*. *Pak. J. Nutr.* **2005,** *4*(6), 379–381.
29. Otunola, G. A.; Oloyede, O. B.; Oladiji, A. T.; Afolayan, A. J. Comparative Analysis of the Chemical Composition of Three Spices – *Allium sativum* L., *Zingiber officinale Rosc.* and *Capsicum frutescens* L. Commonly Consumed in Nigeria. *Afr. J. Biotechnol.* **2010,** *9*(41), 6927–6931.
30. Parle, M.; Deepa, K. Clove: A Champion Spice. *Int. J. Res. Ayurveda Pharm.* **2010,** *2*(1), 47–54.
31. Pearson, D. In *Chemical Analysis of Foods*, 7th ed.; Church Hill Livingstone: London, UK, 1976; pp 138–143.
32. Phaniendra, A.; Jestadi, D. B.; Periyasamy, L. Free Radicals: Properties, Sources, Targets, and Their Implication in various diseases. *Indian J. Clin. Biochem. IJCB* **2015,** *30*(1), 11–26. https://doi.org/10.1007/s12291-014-0446-0
33. Prasad, R. C.; Herzog, B.; Boone, B.; Sims, L.; Waltner-Law, M. An Extract of *Syzygium aromaticum* Represses Genes Encoding Hepatic Gluconeogenic Enzymes. *J. Ethnopharmacol.* **2005,** *96*(1–2), 295–301.
34. Seeff, L.; Stickel, F.; Navarro, V. J. Hepatotoxicity of Herbals and Dietary Supplements. In *Drug-Induced Liver Disease*; Kaplowitz, N., DeLeve, L. D., Eds.; Elsevier: Amsterdam, 2013; pp 631–58.
35. Shan, B.; Cai, Y. Z.; Sun, M.; Corke H. Antioxidant Capacity Of 26 Spice Extracts and Characterization of Their Phenolic Constituents. *J. Agric. Food Chem.* **2005,** *53*(20), 7749–7759. DOI: 10.1021/jf051513y
36. Sofowora, A. *Phytochemical Screening of Medicinal Plants and Traditional Medicine in Africa*. Spectrum Books Ltd: Ibadan, Nigeria, 1993, p 238.
37. Srinivasan, K. Spices as Influencers of Body Metabolism: An Overview of Three Decades of Research. *Food Res. Int.* **2005,** *38*, 77–86.
38. Srinivasan, K. Role of Spices Beyond Food Flavoring: Nutraceuticals with Multiple Health Effects. *Food Rev. Int.* **2005,** *21*(2), 167–188.

39. Tajuddin, T; Ahmad, S; Latif, A; Qasmi, I. Effect Of 50% Ethanolic Extract of *Syzygium aromaticum* (Clove) on Sexual Behavior of Normal Male Rats. *BMC Complement. Altern. Med.* **2004,** *4*(1), 17–22.
40. Tammannavar, P.; Pushpalatha, C.; Jain, S.; Sowmya, S. V. Unexpected Positive Hypersensitive Reaction to Eugenol. *BMJ Case Rep.* [Online] 2013. https://doi.org/10.1136/bcr-2013-009464 (accessed Aug 31, 2021).
41. Thompson, D. C.; Constantin-Teodosiu, D.; Moldéus, P. Metabolism and Cytotoxicity of Eugenol in Isolated Rat Hepatocytes. *Chem. Biol. Interact.* **1991,** *77*(2), 137–147.
42. Vijayasteltar, L.; Nair, G. G.; Maliakel, B.; Kuttan, R.; Im, K. Safety Assessment of a Standardized Polyphenolic Extract of Clove Buds: Subchronic Toxicity and Mutagenicity Studies. *Toxicol. Rep.* **2016,** *3*, 439–449.

CHAPTER 11

Therapeutic Mechanisms of Bioactive Compounds in Lipopolysaccharide-Induced Liver Injury with Murine Models

Ayokanmi Ore, Ebenezer Tunde Olayinka, Babajide Oluwaseun Ajayi, and Oluseyi Adeboye Akinloye

ABSTRACT

Lipopolysaccharide (LPS) is an endotoxin-glycolipid found in Gram-negative bacteria, and is implicated in inflammatory liver injury. Animal models of LPS-induced liver injury have proven to be vital to the study of pathogenesis as well as therapy against liver diseases. LPS (gut-derived or exogenously administered) is transported in circulation to the liver, bound to LPS binding protein (LBP). The liver is involved in the clearance of LPS and as a result, is exposed to its toxic effects. LPS elicits its action via the LPS signaling, following its binding to toll-like-receptor-4 (TLR4) on Kupffer cells surface. These events lead to release of cytokines like interleukin-1β, tumor necrosis-factor-alpha (TNFα), including several other inflammatory proteins. Other effects caused by LPS include oxidative stress and altered lipid profiles. Several studies on animal models have demonstrated the ability of plant-based extracts and compounds to protect the liver against LPS-related injuries. These plant-based substances protect the liver via multiple mechanisms including LPS clearance, anti-inflammatory, antioxidant, antiapoptosis, immunoregulatory, etc.

Curative and Preventive Properties of Medicinal Plants: Research on Disease Management and Animal Model Studies. Megh R. Goyal, Junaid Ahmad Malik, & Ademola Olabode Ayeleso (Eds.)

11.1 INTRODUCTION

Liver diseases constitute an important health challenge worldwide and it has a global distribution.[54] Liver disease is estimated to cause about 2 million deaths annually.[9] Affected individuals are often asymptomatic for a long period of time, making it very difficult to generate accurate data on disease incidence and prevalence.[27,54] Recent data imply that acute and chronic hepatic diseases are prevalent globally, with significant morbidity and mortality.[9,27,54] Factors responsible for hepatic diseases include diet patterns, viral infection, exposure to toxicants, drugs, toxins, etc.[50]

Acute liver failure (ALF) has about 80% mortality rate.[6] Virus-induced hepatitis is the most important cause of ALF, and is closely followed by drugs and toxins including endotoxins. ALF is a serious clinical condition, distinguished by hepatocellular damage and hepatocellular malfunction.[6] It is characterized by progressive loss of hepatic function within a short period of time. Despite the availability of specific therapies and the success rate of liver transplantation, the overall mortality rate of ALF patients still remains high.[10,11] However, there are continuous efforts in place to find therapeutic options other than liver transplantation, which is generally expensive and outside the reach of most patients.

Use of plant-based therapies is rapidly gaining attention worldwide due to their multiple therapeutic mechanisms, therapeutic efficacies, and relatively minimal toxicities or side effects. It is a preferred therapeutic option due to a number of reasons including accessibility and affordability.[26] A number of studies have highlighted the efficacy of plant-derived extracts in the treatment of various diseases including liver diseases.[26,36]

This chapter describes murine models of lipopolysaccharide (LPS)-induced acute liver injury. LPS model has proven to be vital in the study of both pathophysiology and therapy of ALF. This is due to the ability of LPS to produce relevant pathological features within a short interval. The pathophysiology of LPS-induced liver injury is described, as well as mechanistic view of the therapeutic effects exerted by plant-derived compounds, extracts, or their active fractions in murine models.

11.2 PATHOPHYSIOLOGY OF LPS-INDUCED ACUTE LIVER INJURY

LPS is an endotoxin-glycolipid present on the surface of Gram-negative bacteria. Endogenously derived LPS from gut-based bacteria has been identified as a major player in the pathophysiology of acute liver injury. LPS-induced hepatic injury involves inflammatory response, release of oxidants, due to the liver's role in elimination of LPS during exposure.[58] LPS from gut-based microbes and exogenous sources are transported in circulation bound to LPS binding protein (LBP). Systemic inflammation and hepatic damage from LPS exposure occur due to release of mediators of inflammation like IL1β, TNFα, and IL6. These are cytokines released due to the attachment of LPS to the toll-like receptor-4 (TLR4) on the membrane of Kupffer cells.[13,59] Subsequently, there is an upsurge in production of c-reactive-protein (CRP), and other inducible proteins/enzymes like cyclooxygenase-2-(COX2) and-inducible-nitric-oxide-synthase (iNOS). These events cause an increase in generation free radicals and oxidants such as superoxide radical (O_2^-), and nitric oxide (NO).[12,34] These contribute to LPS-induced hepatotoxicity characterized by hepatocellular damage and subsequent elevation of plasma activities-of-liver-based enzymes like aspartate-aminotransferase- (AST) and alanine-aminotransferase- (ALT).[21]

11.2.1 LPS SIGNALING

The active receptor for LPS has been identified as the CD14/TLR4 complex located on Kupffer cells surface[41] (Figure 11.1). Following the interaction of LPS with TLR4, a series of signal cascades that activates-nuclear factor kappa light chain enhancer of activated-B-cells (NF-kB). TLR4 has a cytoplasmic domain that is necessary for downstream signaling.[47]

Both IL1 receptor and TLR4 interact with the TIR-containing cytosolic adaptor known as myeloid differentiation-protein-88 (MyD88) resulting in activation of IL-1R–associated kinase (IRAK) especially IRAK4, and TNF-Receptor-associated factor-6 (TRAF6).[7] Subsequently, TRAF6 activates NF–kB either through the IKK-complex or via the mitogen-activated-protein-kinases- (MAPKs). A regulatory protein called IkB controls the activity of NF-kB. Stimulation of Ik-B kinases (IKK) leads to the phosphorylation of IkB and NF-kB is subsequently transported into the nuclear compartment. In the nucleus, NF-kB initiates the transcription of relevant genes for cytokine production and inflammatory response.

11.2.2 INFLAMMATORY RESPONSE

During LPS toxicity, NF-kB facilitates the release of inflammatory-cytokines comprising TNFα, IL1β, and IL6 from activated Kupffer cells.[59] The actions of NF-kB are subjected to regulation. In the nucleus, NF-kB interact with specific DNA binding sites to initiate expression (or transcription) of pro-inflammatory genes.[19] These events result in expression of mRNA of several mediators of inflammation including cytokines, enzymes, and acute phase proteins involved in LPS-induced inflammatory response. Additional cytokines are produced from leukocytes and vascular endothelial cells due to stimulation by TNF-α. These mediators of inflammation are released into circulation causing endothelial permeability and further inflammation of other tissues.[18,19]

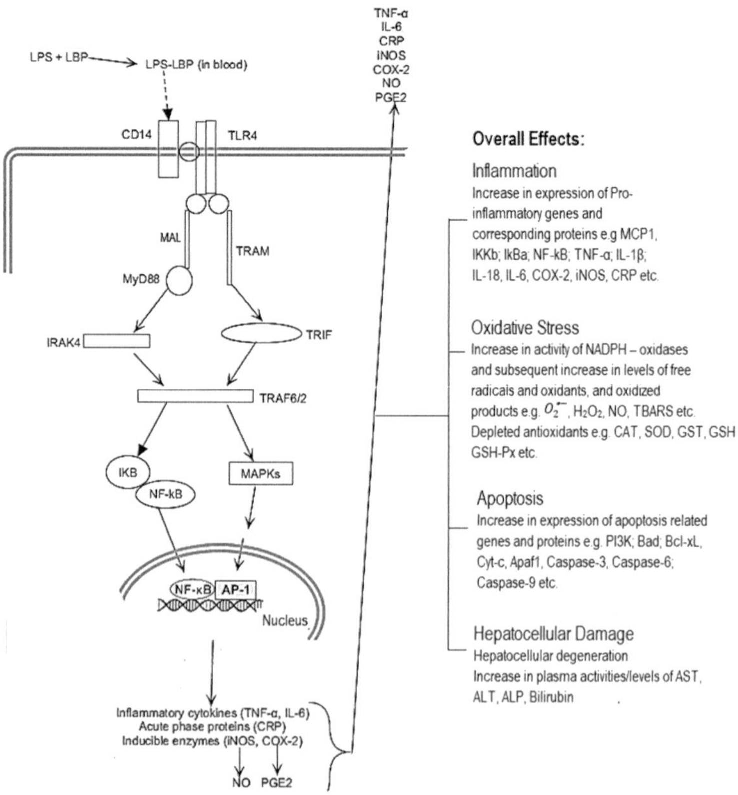

FIGURE 11.1 LPS signaling, inflammatory response, oxidative stress, and liver injury.

11.2.3 OXIDATIVE STRESS

LPS signaling is linked to the induction of COX2 and iNOS.[12] These proteins are known to mediate inflammatory processes in the liver. COX2 catalyzes the biosynthesis of prostaglandins and LPS exposure has been linked to elevated COX2 expression and stimulation of prostaglandin production, including prostaglandin E_2 (PGE_2), a pro-inflammatory lipid product.[19] Inducible nitric oxide synthase (iNOS) catalyzes NO production from L-arginine; expressed in endothelium, epithelium, and inflammatory cells. When iNOS is induced by LPS via cytokines, it produces large quantities of NO, leading to inflammatory injury.[19]

LPS model of acute liver failure is accompanied by production of reactive species resulting in depletion of hepatic reduced glutathione (GSH).[12,32] The NADPH-Oxidases (NOXs) are important enzymes in the activation of immune response to LPS.[19,23] As a destructive process, phagocytic NOXs generate superoxide when exposed to bacteria.[40] In addition to these, NOXs are also stimulated by substances such as TNFα, IFNγ, and IL1β resulting in O_2^- generation.[19] O_2^- is converted to H_2O_2 via the action of SOD, while the H_2O_2 produced is transformed to H_2O and O_2 by the action of catalase (CAT) and glutathione-peroxidase (GSH-Px).

11.2.4 HEPATIC INJURY

LPS model of acute liver injury is similar to fulminant hepatitis. In most reports, it is characterized by hepatocellular damage accompanied by elevated serum or plasma level of liver function enzymes ALT and AST, severe inflammation, and apoptosis.[44,61] It is well established that LPS induces liver injury and apoptosis via TNFα production in murine models and the lethality of injury depends on quantity of TNF-α secreted.[17,39] This model of TNF-α-induced hepatocyte apoptosis has been shown to involve p55 receptor-signaling in LPS-exposed animals.[29] Additionally, levels of Fas ligand (FasL) may rise in the liver during LPS exposure.[42] Thus, blocking FasL response with soluble Fas fusion protein has been proven to be useful against LPS-induced liver injury in mouse model.[22] IL18 produced from Kupffer cells also—play a-vital role-in liver injury and-apoptosis induced by LPS.[46]

11.3 PROTECTIVE ROLES OF PLANT-DERIVED COMPOUNDS IN LPS MODEL OF ACUTE LIVER INJURY

In this section, a total of 10 plant-derived compounds with significant protective effects against LPS-mediated liver injury have been examined. These plants include curcumin, hesperidin, carnosic acid, dioscin, emodin, quercetin, damaurone D, amygdalin, thymoquinone, and chlorogenic acid. Their chemical structures and common plant/food sources are presented in Figure 11.2 and Table 11.1, respectively. The mechanism behind the hepatoprotective activities exhibited by these compounds are highlighted in the following subsections and illustrated in Figure 11.3.

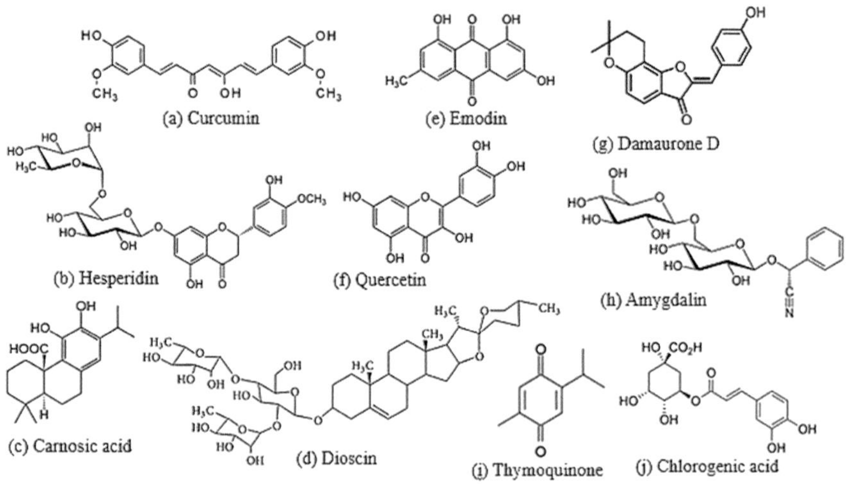

FIGURE 11.2 Chemical structures of phytochemicals with hepatoprotective roles in LPS-induced acute liver injury.

11.3.1 CURCUMIN (CUR)

CUR is a bright yellow phytochemical, a major curcuminoid found in turmeric (*Curcuma longa*). In a C57BL/6 and albino mouse model of LPS-induced liver injury,[44,61] curcumin (CUR), significantly protected the liver against inflammation, oxidative damage, apoptosis, and hepatocellular damage. In the study carried out by Zhong et al.,[61] CUR exhibited anti-inflammatory activity against LPS as indicated by decrease in hepatic level of pro-inflammatory proteins including IKKb; IkBa; NF-kB; TNFα; IL1β; IL18. It also displayed significant antioxidant properties via increase in hepatic activities-of-SOD, CAT, and-GSH-Px; increase in hepatic GSH concentration; and decrease in hepatic levels of free radicals like nitric oxide, hydroxyl radical, and superoxide radical. CUR also prevented apoptosis via decrease in hepatic expression of Caspase-3, Caspase-6, Caspase-9, Cyt-c, PI3K, Bad, Bcl-xL, and Apaf1. The overall hepatoprotective effect

TABLE 11.1 Plant-Derived Compounds and Their Protective Roles in Murine Models of LPS-Induced Acute Liver Injury.

Compound	Class/type	Common plant sources	Animal model	LPS dose (route)	Dosage of phyto-compound (route) and duration before LPS challenge	Activity exhibited by phyto-compound against LPS induced acute liver injury	Ref
Curcumin (CUR)	Phenols (curcuminoid)	Turmeric (*Curcuma longa*)	Male C57BL/6 mice	5 mg/kg bw (i.p.)	20, 40, and 80 mg/kg bw/d CUR (p.o.) for 28 days	**Anti-inflammatory:** ↓TNFα; ↓IL1β; ↓IL18; ↓IKKb; ↓IkBa; ↓NF-kB (in liver) **Antioxidant:** ↑GSH-Px; ↑SOD; ↑CAT; ↑GSH, ↓free radicals (in liver) **Antiapoptotic:** ↓ PI3K; ↓Bad; ↓Bcl-xL; ↓Cyto-c; ↓Apaf1; ↓Caspase-9; ↓Caspase-3; ↓Caspase-6) **Hepatoprotective:** ↓ALT; ↓AST; ↓ALP (in serum); improved liver ultrastructure.	61
			Male albino mice	3 mg/kg bw (i.p.)	20 mg/kg bw/d CUR (p.o.) for 7 days	**Hepatoprotective:** ↓ALT; ↓AST; ↓ALP (in serum); improved liver ultrastructure	44
Hesperidin (HDN)	Flavonoid (flavanone glycoside)	Citrus fruits	Male Wistar rat	1 mg/kg bw (i.p.)	50, 100, 200 mg/kg bw/d HDN (p.o.); for 7 days	**Antioxidant:** ↑SOD; ↑GSH; ↓free radicals/oxidative stress (in liver) **Hepatoprotective:** ↓ALT; ↓AST; ↓ALP (in serum); improved liver ultrastructure	21
			Female albino rat	7 mg/kg bw (i.p.)	50, 100, 200 mg/kg bw/d HDN (p.o.) for 15 days	**LPS scavenging:** ↓LPS (in serum) **Anti-inflammatory:** ↓IL1 receptor; ↓IL6; receptor;↓MPO; ↓IL6; ↓IL1α; ↓IL1β; ↓TNFα (in liver) **Antioxidant:** ↑SOD; ↑GSH-Px; ↑GST (in liver) ↑GSH; ↓free radicals/oxidative stress	36
Carnosic acid (CAN)	Phenolic diterpene	Rosemary *(Rosmarinus officinalis)* plus common sage *(Salvia officinalis)*	Male Sprague–Dawley rat	1 mg/kg bw (i.p.) every 5 days for 35 days	15,30, and 60 mg/kg bw/d CAN (p.o.) for 35 days	**Anti-inflammatory:** ↓iNOS (in liver); ↓TNFα; ↓IL6; ↓iNOS (in serum) **Antioxidant:** ↑GSH; ↑SOD; ↑GSH-Px (in liver); ↑SOD; ↑GSH (in serum); ↓free radicals/oxidative stress	53

TABLE 11.1 *(Continued)*

Compound	Class/type	Common plant sources	Animal model	LPS dose (route)	Dosage of phyto-compound (route) and duration before LPS challenge	Activity exhibited by phyto-compound against LPS induced acute liver injury	Ref
						Hepatoprotective: ↓ALT; ↓AST; ↓ALP (in serum); improved liver ultrastructure	
Dioscin (DIO)	Steroidal saponin	*Dioscorea nipponica* Makino	Male C57BL/6J mice	8 or 5 mg/kg bw (i.p.)	80,40, and 20 mg/kg bw/d DIO (p.o.) for 7 days	**Anti-inflammatory:** ↓iNOS (in liver); ↓TNFα; ↓IL6; ↓iNOS (in serum) **Antioxidant:** ↑GSH; ↑SOD; ↑GSH-Px (in liver); ↑SOD; ↑GSH (in serum); ↓free radicals/oxidative stress **Hepatoprotective:** ↓ALT; ↓AST (in serum); improved liver ultrastructure	57
Emodin (EMD)	Anthra-quinone	Rhubarb *(Rheum rhabarbarum)*	Male Balb/c mice	20 mg/kg bw (i.p.)	20,40, and 80 mg/kg/d EMD (i.g.) for 7 days	**Hepatoprotective**: ↓ALT; ↓AST (in serum); improved liver ultrastructure	14
Quercetin (QCT)	Flavonol (flavonoid)	Red onions *(Allium cepa)* and kale	Male Wistar rat	1 mg/kg bw (i.p.)	25 mg/kg bw/d QCT (i.p) for 14 days	**Antioxidants:** ↑GSH; ↑SOD; ↓TBARS (in liver) **Hepatoprotective:** ↓ALT; ↓AST, ↓BIL (in serum)	33
Damaurone D (DAD)	Isoprenylated aurone	*Rosa damascena*	Male ICR mice	10 mg/kg bw (i.p.) for 4/24 h	10 mg/kg bw DD (i.p.) twice with 18 h-interval	**Anti-inflammatory:** ↓TNFα; ↓MCP1; ↓IL1β (in plasma); ↓F4/80; ↑IkB-α; ↑p-ACC; ↑p-AMPK (in liver); ↓TNFα; ↓MCP1; ↓Il1β, ↓Ho1 (in liver) **Hepatoprotective:** ↓ALT; ↓AST (in serum); improved liver ultrastructure	52
Amygdalin (AMG)	Cyanogenic glycoside	Seeds apples, apricots, plums, peaches, bitter almonds	Male Sprague–Dawley rat	2.5 mg/kg bw (i.p.) every 3 day for 30 days (from day 30 of study)	0.5, 1, and 1.5 mg/kg bw AG (p.o.) for 60 days.	**LPS scavenging:** ↓LPS (in serum and liver) **Anti-inflammatory:** ↓IL1β; ↓IL6; ↓IL8; ↓IL18; ↓TNFα (in liver); ↓PI3K/AKT & JAK2.STAT3 pathway	56

TABLE 11.1 *(Continued)*

Compound	Class/type	Common plant sources	Animal model	LPS dose (route)	Dosage of phyto-compound (route) and duration before LPS challenge	Activity exhibited by phyto-compound against LPS induced acute liver injury	Ref
						Hepatoprotective: ↓ALT; ↓AST; ↓ALP (in serum); improved liver ultrastructure	
Thymoquinone (TMQ)	Monoterpene	*Nigella sativa*	Male Wistar rats	1 mg/kg bw/d (i.p) for 3 weeks	2, 5, 10 mg/kg bw/d TMQ for 3 weeks	**Antioxidant:** ↑SOD; ↑CAT; ↓MDA (in liver) **Hepatoprotective:** ↓ALT; ↓AST; ↓ALP (in serum); improved liver ultrastructure	8
Chlorogenic acid (CGA)	Ester of quinic acid and caffeic acid	*Coffee beans*	C57BL/6J mice	30 mg/kg bw,(i.p.)	50 mg/kg bw/d CGA (i.p.) daily for 5 days	**Anti-inflammatory:** ↓TNF-α; ↓PGC-1α; ↓NF-kB p65; ↓TLR4 (in liver). **Hepatoprotective:** ↓ALT; ↓AST (in serum); improved liver morphology	55

↓, *inhibition of/decrease in concentration, expression or activity;* ↑, *activation of/increase in concentration, expression or activity* of CUR was revealed by diminished serum activities of ALT, AST, and ALP (hepatic function biomarkers) and improvement in histological structure of the liver.

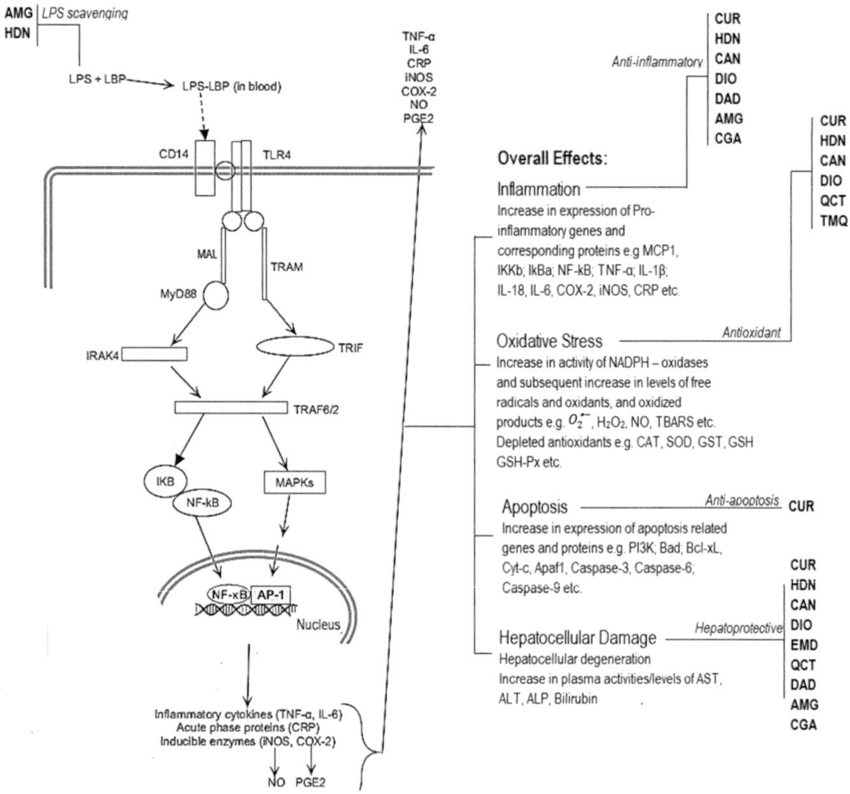

FIGURE 11.3 Effects of plant-based bioactive compounds in LPS models.

In another related study conducted by Thriveni et al.[44] on male albino mice, 20 mg/kg bw CUR alleviated LPS-induced liver-injury (see Table 11.1).

11.3.2 HESPERIDIN (HDN)

HDN is a flavonoid (flavanone glycoside) found in abundance in citrus fruits. In a male Wistar rat,[21] HDN alleviate the hepatic damage and oxidative stress induced by LPS. These were achieved via antioxidant activity as suggested by increase in hepatic GSH level and activity of SOD, decrease in hepatic levels of free radicals and oxidative stress markers including nitrites and TBARS. In addition, decline in serum-activities of ALT, AST, ALP, and bilirubin level with significant protection of hepatocytes was observed.

In a related research conducted by Rotimi et al.,[36] using the same set of doses against LPS-induced liver injury in female albino rat,[36] HDN caused a decrease in circulating LPS. In addition, HDN exhibited anti-inflammatory activity against LPS-induced inflammation as indicated by decrease in myeloperoxidase (MPO) activity, decrease in hepatic expressions of IL1α, IL1β, and IL1 receptor, IL6 and IL6 receptor, and TNFα concentration. HDN also prevented oxidative damage via upsurge in liver SOD, GST, GSH-Px, and level of-GSH, as well as decrease in hepatic levels of nitric oxide and lipid oxidation products.

11.3.3 CARNOSIC ACID (CAN)

CAN is a phenolic-diterpene present in *Rosmarinus-officinalis* (rosemary) and *Salvia officinalis* (also called common sage). CAN displayed anti-inflammatory and antioxidant activities in male Sprague–Dawley rat model of LPS-induced liver injury.[53] CAN alleviated the inflammatory response from LPS endotoxicity via decrease in hepatic iNOS level and serum TNFα, IL6, and iNOS levels. CAN also exhibited antioxidant activity via increase in hepatic GSH level and increase in activities of-GSH-Px and-SOD with decline in hepatic levels-of carbonylated proteins, MDA,-and serum NO level. Hepatoprotective activity of CAN was also indicated by decrease in serum ALT, AST, and ALP along with improved liver morphology.

11.3.4 DIOSCIN (DIO)

DIO is a steroidal saponin found in *Dioscorea nipponica* Makino. DIO alleviate LPS-induced liver injury in male C57BL/6J mice model.[57] DIO at 80, 40, and 20 mg/kg bw exhibited hepatoprotective role in LPS endotoxicity. It causes a significant decrease in hepatic iNOS, serum TNF- α, IL-6, and iNOS levels. DIO also caused an improvement in hepatic GSH level, activities of hepatic GSH-Px, hepatic SOD and serum SOD activities, GSH level, decreased hepatic levels of carbonylated proteins and MDA, as well as serum NO level. Hepatoprotective effects of DIO were indicated via decrease in serum ALT, AST, and improved liver morphology.

11.3.5 EMODIN (EMD)

EMD is an anthraquinone abundant in Rhubarb (*Rheum rhabarbarum*). EMD at 20, 40, and 80 mg/kg bw alleviated liver injury induced by LPS in mouse model.[14] This was indicated by decrease in serum ALT, AST as well as improved liver morphology in EMD-treated mice.

11.3.6 QUERCETIN (QCT)

QCT is a flavonol (flavonoid) abundant in red onions (*Allium cepa*) and kale. QCT exerted significant hepatoprotective effect in male Wistar rat model of LPS-induced liver injury.[33] QCT at 25 mg/kg bw alleviated LPS-induced oxidative stress as implied by increase in hepatic GSH level, and SOD levels and decline in hepatic TBARS level. Furthermore, QCT caused decline in-serum activities of AST and-ALT and bilirubin level.

11.3.7 DAMAURONE D (DAD)

DAD is an isoprenylated aurone present in *Rosa damascene*. DAD exhibited a significant protection against LPS-induced inflammatory liver injury in male ICR mice.[52] DAD at 10 mg/ kg bw showed anti-inflammatory effect against LPS-induced inflammatory response decrease in plasma TNFα, IL1β, and MCP-1 concentrations, decrease in hepatic expression of F4/80, increase in expression of IkB-α, p-ACC; p-AMPK; TNFα, Mcp1, Il1β, and Ho1 genes. Hepatoprotective effects of DAD also involved decrease in plasma AST and ALT activities as well as improvement in liver morphology.

11.3.8 AMYGDALIN (AMG)

AMG is a cyanogenic glycoside present in kernels (or seeds) of plums, almonds, peaches, apricots, and apples. AG displayed hepatoprotective effects in a male Sprague–Dawley rat model.[56] AMG at 2.5 mg/kg bw protected against liver injury caused by LPS via LPS scavenging and anti-inflammatory activities. In-this-study, AMG caused a reduction in serum and hepatic LPS concentration. Anti-inflammatory action of AMG was exhibited via decrease in hepatic expression of TNFα, IL1β; IL6, IL8, and IL18, and inhibition of expression of JAK2/STAT3 and PI3K/AKT signaling pathway proteins. Moreover, AMG is also protected against hepatocellular damage.

11.3.9 *THYMOQUINONE (TMQ)*

TMQ is a monoterpene abundant in *Nigella sativa.* TMQ displayed antioxidant and hepatoprotective effects in male Wistar rat model of LPS-mediated liver injury.[8] TMQ at 2, 5, and 10 mg/kg bw caused an increase in hepatic total thiol level and activities of enzymic antioxidants, CAT, and SOD and decrease in hepatic MDA level TMQ also caused a reduction in plasma activities of ALP, AST, and ALT with improved liver morphology following LPS-mediated hepatic damage.

11.3.10 *CHLOROGENIC ACID (CGA)*

In a C57BL/6J mouse model, CGA exhibited anti-inflammatory and hepatoprotective effects against LPS-induced inflammation and hepatic damage.[55] CGA at 50 mg/kg bw showed anti-inflammatory action against LPS-induced liver damage as indicated by decrease in hepatic mRNA expression of TLR4 and TNFα, increase in hepatic expression of PGC-1α and decrease in hepatic NF-kB-p65 mRNA expression. Hepatoprotective effects of CGA were indicated by decline in serum activities of ALT and AST coupled with improved liver morphology, following LPS-mediated hepatic damage.

11.4 EFFECTS OF PLANT EXTRACTS ON LPS-INDUCED LIVER INJURY

This section describes plant extracts that have shown significant protection against LPS-induced liver injury. These plant extracts include:

- Active fraction of ethanol extract of *Scirpus yagara* tuber (AF-EESYT),
- Aqueous green tea extract (AGTE),
- Aqueous rooibos extract (ARE),
- *B. vulgaris* ethanolic extract (BVEE),
- *C. vulgaris* ethanolic extracts (CVEE),
- Danhong injection (DHI),
- Ethanol extract of (defatted) *Buchholzia coriacea* seed (HEBCS),
- Ethanol extract of *Illicium henryi* (EEIH),

- Ethanolic extracts of *S. baicalensis* root (EESBR),
- *Forsythia suspensa* extract (FSE),
- *Gardenia jasminoides* water extract (GJWE),
- Grape seed procyanidin extract (GSPE),
- Green tea polyphenols (GTPs),
- *Lonicera caerulea* berry extract (LCBE),
- Lotus seedpod extract (LSE),
- *Origanum majorana* Extract (OME),
- *Pulicaria petiolaris* methanol extract (PPME),
- Red ginseng alcohol extract (RGAE),
- Red ginseng water extract (RGWE),
- *Rhus verniciflua* stokes (RVS),
- Stevoside extract (STE-EXT),
- Sun ginseng methanol extract (SGME),

Their phytoconstituents and mechanism of hepatoprotection in LPS-induced liver injury are presented in Table 11.2 and Figure 11.4 and highlighted in the following section.

11.4.1 GRAPE SEED PROCYANIDIN EXTRACT (GSPE)

GSPE is a standardized extract containing procyanidins and phenolic acids. GSPE alleviated LPS-related liver injury in female Wistar rat model. GSPE at 50, 75, 100, and 200 mg/kg bw, exhibited anti-inflammatory actions against LPS-induced inflammation via decreased in expression of TNFα, iNOS, Il10, Il6, and CRP genes in the liver.[31] It also shows antioxidant actions via increase in hepatic GSH level and decrease in hepatic nitrate and nitrite levels.

11.4.2 GREEN TEA POLYPHENOLS (GTPS)

Wang et al.[49] characterized green tea polyphenols as gallic acid, catechin, epicatechin, gallo-catechin gallate, epigallocatechin, epicatechin gallate, epigallocatechin gallate, anthocyanins, leuco-anthocyanins, and phenolic acids as characterized by HPLC. They demonstrated the hepatoprotective effects of GTPs against LPS-related inflammation, oxidative distress, and hepatic damage in male ICR mice. GTPs at 200 mg/kg bw show anti-inflammatory activity via decreased plasma levels of TNFα, IL1β, IL6, and IL18,

TABLE 11.2 Plant-Based Extracts and Their Protective Roles in Murine Models of LPS-Induced Acute Liver Injury.

Plant source (part)	Solvent extract (abbreviation)	Animal model	LPS dose (route)	Dosage of phyto-compound (route) and duration before LPS challenge	Activity exhibited by extract against LPS induced liver injury	Ref
Grape (seed)	Grape seed procyanidin extract (GSPE)	Female Wistar rats	7 mg/kg bw (i.p.)	50 mg/kg/d, 75 mg/kg/d, 100 mg/kg/d, or 200 mg/kg/d (p.o.) for 7 days	Anti-inflammatory: ↓Crp; ↓Tnf-α; ↓iNos; ↓Il-10; ↓Il-6 (in liver) Antioxidant: ↑GSH; ↓NO (in liver)	31
Green tea	Green tea polyphenols (GTPs)	Male ICR mice	15 mg/kg bw (i.p.)	200 mg/kg bw/d (i.g) for 7 days	Anti-inflammatory: ↓IL1β; ↓IL6; ↓IL18; ↓TNFα; ↓iNOS; ↓COX-2 (in liver) Antioxidant: ↑SOD; ↑GSH; ↓MDA (in liver) Hepatoprotective: ↓ALT; ↓AST (in plasma); improved liver ultrastructure	49
Grean tea	Aqueous green tea extract (AGTE)	Male spague-dawley rats	10 μg/kg bw (i.p.) day 1–15	10 g extracted in 750 mL water; (p.o.) in drinking water; day 16–30	Antioxidant: ↑SOD; ↑GSH-Px; ↑GSH-Rd; ↑GSH ↓NO; ↓MDA (in liver) Antiapoptosis: ↓Bcl-2; ↓casp 8; ↓cyt c Antifibrosis: ↓Hyp; ↓HA Hepatoprotective: ↓ALT; ↓AST; ↓BIL (in serum); improved liver ultrastructure	38
Grean tea	Aqueous green tea extract (AGTE)	Wistar albino rats	1 g/kg bw (i.p.) at day 7 of gestation, postnatal day 1 and postnatal day 15	AGTE 100 mg/kg/d and 200 mg/kg/d p.o. from day 0 of gestation to postnatal day 30	Antioxidant: ↓NO; ↑TAC Antiappoptosis: ↓ssDNA; ↓sFas Hepatoprotective: ↓ALT; ↓AST; ↓BIL (in serum); improved liver ultrastructure	5
Sun ginseng (SG) *Panax ginseng*	Sun ginseng methanol extract (SGME)	Male Wistar rats	5 mg/kg bw (i.v.)	50, 100 mg/kg bw/d for 15 days	Anti-inflammatory: ↓NF-kBp65; ↓IkB-a; ↓iNOS (in liver) Antioxidant: ↓TBARS (in liver)	20
Buchholzia coriacea (Seeds)	Hydroethanolic extract of	Albino rat	4 mg/kg bw (i.p.)	125, 250, 500 mg/kg (p.o.) for 7 days	Anti-inflammatory: ↓TNF-α (in serum) ↓CRP (in liver)	30

TABLE 11.2 *(Continued)*

Plant source (part)	Solvent extract (abbreviation)	Animal model	LPS dose (route)	Dosage of phyto-compound (route) and duration before LPS challenge	Activity exhibited by extract against LPS induced liver injury	Ref
	(defatted) *B. coriacea* seeds (HEBCS)				Antioxidant: ↑CAT; ↑SOD; ↑GSH; ↓NO; ↓MDA (in liver) Hepatoprotective: ↓ALT; ↓AST; ↓ALP (in serum); improved liver ultrastructure	
Berberis vulgaris (roots); *Calluna vulgaris* (shoots)	*B. vulgaris* and *C. vulgaris* ethanolic extracts BVEE, CVEE	Male rats	LPS 4 mg/kg bw (i.p.)	BVEE or CVEE at 200 mg/kg bw/d for 3 weeks	Anti-inflammatory: ↓COX-2; ↓iNOS; ↓TNF-α (in liver) Antioxidant: ↓NO; ↓TBARS; ↓XO (in liver) Hepatoprotective: improved liver ultrastructure	1
Fermented rooibos *(Aspalathus linearis)* leaves plus stem	(aqueous rooibos extract, ARE)	Male Wistar rats	0.5 mg/kg bw, (i.p.)	ARE (2%, w/v) in drinking fluid 27 days	Anti-inflammatory: ↓IL1β; ↓IL6; ↓IL10; ↓TNFα (in liver) Antioxidant: ↑CAT; ↑SOD; ↑GSH-Px; ↑GR; ↑GSH; ↑GSSG (in liver) Hepatoprotective: ↓ALT; ↓AST; ↓LDH (in serum)	4
Scutellaria baicalensis (root)	ethanolic extracts of *S. baicalensis* root (EESBR)	Male C57BL/6J mice	5 mg/kg (i.p.)	200 mg/kg/ bw/d for 7 days (p.o.)	Anti-inflammatory: ↓NF-KB; ↓COX-2; ↓iNOS; ↓TNFα; ↓IL1β; ↓IL6; ↓IL1β (in liver) Antioxidant: ↓MDA	43
Forsythia suspensa (Fruits)	Forsythia suspensa extract (FSE)	Male Sprague—Dawley rats	20 mg/kg BW LPS (i.p.)	100 mg/kg bw/d for 7 days	Anti-inflammatory: ↓IL1β; ↓IL6; ↓TNFα (in serum & liver); ↓NF-κB (in liver) Antioxidant: ↑SOD; ↑GSH-Px Hepatoprotective: ↓ALT; ↓AST (in serum); improved liver ultrastructure	60

TABLE 11.2 *(Continued)*

Plant source (part)	Solvent extract (abbreviation)	Animal model	LPS dose (route)	Dosage of phyto-compound (route) and duration before LPS challenge	Activity exhibited by extract against LPS induced liver injury	Ref
Scirpus yagara Ohwi (tuber)	Active fraction of H_2O/EtOH extract of *Scirpus yagara* tuber (AF-EESYT)	Female BALB/c mice	15 mg/kg bw (i.p.)	50, 300 mg/kg bw/twice/d for 3 days	Anti-inflammatory: ↓TNF-α; ↓IL6 (in serum and liver)	24
Lonicera caerulea (Berry)	Extracted in acidified methanol: *Lonicera caerulea* berry extract (LCBE)	Male Sprague—Dawley rats	200 μg /kg bw/d (i.p.) for 4 weeks	50, 100, and 200 mg/kg bw/d (i.g.) for 4 weeks.	Anti-inflammatory: ↓CRP; ↓IL6; ↓TLR2; ↓TLR4; ↓MAPKs (in liver) Antioxidant: ↑GSH; ↓ROS Hepatoprotective: ↓ALT; ↓AST (in serum); improved liver ultrastructure	51
Rhus verniciflua stokes (dried branches)	*Rhus verniciflua* stokes (RVS) extracted at 90–110°C for 4 h; with 3% fustin	Male Wistar rats	5 mg/kg bw (i.p.)	100, 250, or 500 mg/kg bw/d (i.g.) for 14 days.	Anti-inflammatory: ↓IL1β; ↓IL6; ↓TNFα; ↓iNOS; ↓COX-2 (in liver) Antioxidant: ↓NO; ↓MDA (in liver) Hepatoprotective: ↓ALT; ↓AST (in serum); improved liver ultrastructure	28
Stevia rebaudiana (leaves)	Extracted using ethyl alcohol and water (1:1). *(STE-EXT)*	Male Wistar rats	5 mg/kg) i.p. on day 0	250, 500 mg/kg bw/d p.o. for 7 days post LPS challenge	Anti-inflammatory: ↓IL1β; ↓IL6; ↓TNFα (in liver) Antioxidant: ↓Nitrite; ↓H_2O_2; ↓MDA; ↑SOD; ↑GSH Hepatoprotective: ↓ALT; ↓AST (in serum); improved liver ultrastructure	37
Danhong injection (DHI)	30% ethanol-extract the rhizome and root of *Salvia-*	Male C57BL/6J mice	16 mg/kg bw (i.p.)	0.75, 3 g/kg bw, for 30 min	Anti-inflammatory: ↓TNF-α; ↓IL-6 (in liver) Antioxidant: ↓GST; ↓MDA (in liver) Antiappoptosis: ↓casp-3; ↓Bax; ↓Bcl-2;	15

TABLE 11.2 *(Continued)*

Plant source (part)	Solvent extract (abbreviation)	Animal model	LPS dose (route)	Dosage of phyto-compound (route) and duration before LPS challenge	Activity exhibited by extract against LPS induced liver injury	Ref
	miltiorrhiza plus water-extract of flowers of *Carthamus-tinctorius* (3:1)				↓TUNEL +ve-cells Hepatoprotective: ↓ALT; ↓AST; ↓BIL (in serum); improved liver ultrastructure	
Origanum majorana (sweet marjoram) leaves	extracted in 95% ethanol (1:20 w/v) *Origanum majorana* Extract (OME)	Male albino rats	10 mg/kg bw (i.p.)	0.5 mg/kg bw/d for 10 days	Antioxidant: ↓NO; ↓MDA; ↑GSH (in liver) Hepatoprotective: ↓ALT; ↓AST (in serum);	48
Red ginseng (RG)	80% alcohol extract (RGAE) and water extract (RGWE)	Male Sprague-Dawley rats	5 mg/kg bw (i.p.)	2 mL (0.66 mg/mL) of RG for 4 weeks	Antioxidant: ↓NO; ↑SOD; ↑CAT (in liver) Hepatoprotective: ↓ALT; ↓AST; ↓ALP; ↓LDH (in serum)	2
Pulicaria petiolaris (aerial parts)	*Pulicaria petiolaris* (PP) extracted with methanol (PPME)	Male Swiss albino mice	10 mg/kg bw (i.p.)	50, 100 mg/kg bw/d) for 5 days before	Anti-inflammatory: ↓TNFα; ↓IL6; ↓NF-κB (in liver) Antioxidant: ↓MDA; ↑GSH; ↑SOD (in liver) Hepatoprotective: ↓ALT; ↓AST; ↓ALP; ↓LDH (in serum); improved liver structure	3
Root bark of *Illicium henryi*	*Ethanol extract of I. henryi (EEIH)*	Male BALB/c mice	8 mg/kg bw (i.p.)	1.25, 2.5, or 5.0 mg/kg bw/d for 5 days	Anti-inflammatory: ↓TLR4; ↓NF-KB; ↓IL1β; ↓IL6; ↓TNFα; ↓COX2; ↓iNOS (in liver) Antioxidant: ↓NO; ↓TBARS; ↑SOD; ↑GSH (in liver) Hepatoprotective: ↓ALT; ↓AST (in serum); improved liver ultrastructure	16

TABLE 11.2 *(Continued)*

Plant source (part)	Solvent extract (abbreviation)	Animal model	LPS dose (route)	Dosage of phyto-compound (route) and duration before LPS challenge	Activity exhibited by extract against LPS induced liver injury	Ref
Gardenia jasminoides Ellis (GJ)	*GE water extract (GJWE)*	Male Sprague-Dawley rats	50 μg/kg bw (i.p.)	10, 30 mg/kg bw/d for 3 days	Hepatoprotective: ↓ALT; ↓AST (in serum); improved liver ultrastructure	15
Lotus seedpod	Lotus seedpod extract (LSE)	Male ICR mice	LPS 20 mg/kg bw i.p.	1, or 2 g in (99 or 98 g) diet for 8 weeks	Anti-inflammatory: ↓IL-1β; ↓IL-6; ↓TNF-α (in liver)	45
					Antioxidant: ↓CAT; ↑SOD; ↑GSH-Px; ↑GSH (in liver)	
					Hepatoprotective: ↓ALT; ↓AST (in serum); improved liver ultrastructure	

↓, *inhibition of/decrease in concentration, expression or activity;* ↑, *activation of/increase in concentration, expression or activity.*

decreased hepatic expression of IL1β, TNFα, IL6, and IL18 and decrease in hepatic expression of COX2 and iNOS. GTPS also show antioxidant effects via increase in hepatic SOD activity, GSH concentration, and drop in MDA content. Hepatoprotective actions of GTPs also include decrease in plasma ALT, AST activities, and improved hepatic ultrastructure as seen in histopathological images.

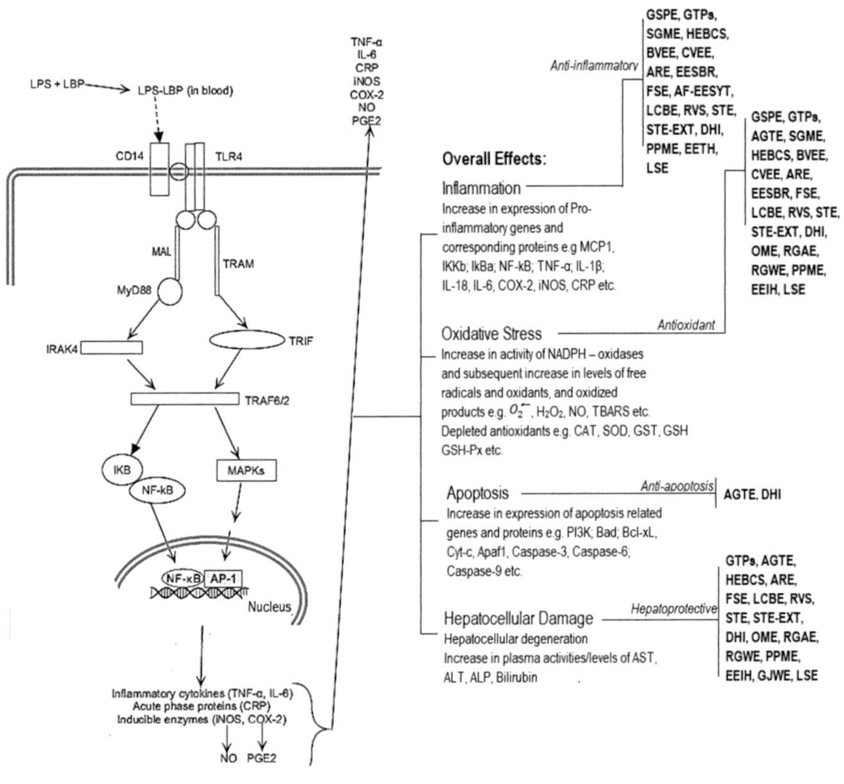

FIGURE 11.4 Therapeutic mechanisms of plant-derived extracts on LPS-induced liver injury.

11.4.3 AQUEOUS GREEN TEA EXTRACT (AGTE)

Sami et al.[38] prepared and characterized AGTE using HPLC. Compounds found in AGTE include tannins, chlorogenic acid, triterpinoids, vanillic and synergic acid, coumarin, naringenin, etc. In a male Sprague–Dawley rat model of LPS-induced liver injury,[38] AGTE supplemented in drinking water demonstrated antioxidant effects against LPS-induced oxidative stress via increase in hepatic GSH level and activities GSH-Px, GSH-Rd, and SOD; decrease in hepatic levels of NO and MDA. AGTE also showed antiapoptosis effect as indicated by decrease in pro-apoptotic markers (Bcl-2, caspase 8, cyt c). AGTE is also protected against LPS-induced fibrosis via decrease in fibrosis markers, hydroxyproline (Hyp), and hyaluronic acid (HA). Furthermore, there was decrease in serum ALT, AST activities, and bilirubin level with improved liver ultrastructure as indicated by histopathological images of hepatic tissues.

Using HPLC technique, Allam et al.[5] identified gallic acid, ferulic acid, pyrogallol, iso-ferulic, catechein, alpha-coumaric, chlorogenic, benzoic, catechol, 3,4,5-methoxycinnamic acid, epicatechein, coumarin, caffeic acid, caffeine, salycilic acid, cinnamic acid, vanillic, naringenin, tannin, synergic acid, etc., in AGTE.

Allam et al.[5] reported the hepatoprotective effects of AGTE in Wistar albino rat model of liver injury induced by LPS. AGTE at 100 and 200 mg/kg/bw showed antioxidant activity by causing decrease in hepatic NO level and increase in total antioxidant capacity. In this study, AGTE protected against LPS-induced apoptosis via decrease in hepatic single-stranded DNA (ssDNA) and soluble fas antigen (sFas). AGTE also showed hepatoprotective effects as indicated by decrease in hepatic function markers (serum ALT, AST activities, and bilirubin level) and improved liver ultrastructure.

11.4.4 SUN GINSENG METHANOL EXTRACT (SGME)

Kang et al.[20] prepared SGME, characterize it by RP-LC, and demonstrated its hepatoprotective effects against LPS in Male rat model. SGME was found to contain ginsenosides (see Table 11.2). SGME at 50 and 100 mg/kg bw showed anti-inflammatory effects via decrease in hepatic expression iNOS, NF-kBp65, IkB-a, and antioxidant activity as indicated by a significant decrease in hepatic TBARS level.

11.4.5 HYDROETHANOLIC EXTRACT OF BUCHHOLZIA CORIACEA SEED (HEBCS)

Seeds of *Buchholzia coriacea* Engl. is commonly referred to as wonderful kola because of its numerous medicinal applications in folk medicine. Ore et al.[30] described the hepatoprotective effects of HEBCS in LPS model. In this study, HEBCS demonstrated anti-inflammatory, antioxidant, and hepatoprotective effects against LPS. HEBCS alleviated the LPS-induced inflammatory response via decrease in serum TNF-α and CRP levels. It also alleviated oxidative distress as shown by increase in hepatic activities of CAT, SOD, and level of GSH as well as decrease in hepatic levels of NO, MDA. HEBCS also showed hepatoprotective effects against LPS-related injury as implied by decrease in serum ALT, AST, and ALP activities and improved liver ultrastructure.

11.4.6 BERBERIS VULGARIS ETHANOLIC EXTRACTS (BVEE) AND CALLUNA VULGARIS ETHANOLIC EXTRACTS (CVEE)

Abdel-Latif et al.[1] showed the ability BVEE and CVEE to defend against liver injury caused by LPS in rat model. Both BVEE and CVEE at 200 mg/kg bw show anti-inflammatory activity against LPS-related inflammatory response as implied by decrease in COX2, iNOS, and TNFα expressions in rat hepatocytes. Both BVEE and CVEE also show antioxidant actions via decrease in hepatic levels of NO, TBARS as well as decrease in hepatic xanthine oxidase activity. Both extracts also alleviate LPS-induced hepatocellular damage as indicated by improved liver ultrastructure in the BVEE and CVEE treated rats.

11.4.7 AQUEOUS ROOIBOS EXTRACT (ARE)

Rooibos (*Aspalathus linearis*) is a legume indigenous to South Africa. It is used in making herbal beverage, which possesses antioxidant and other health-promoting properties. Aqueous rooibos extract is prepared from hot water extraction of fermented rooibos leaves and stem. HPLC evaluation of ARE by Ajuwon et al.[4] revealed the presence of flavonol, flavanol, aspalathin, rutin, quercetin, luteolin, etc. (see Table 11.2). In a male Wistar rat's model,[4] ARE administered as drinking fluid defended against LPS-related hepatic injury. ARE exhibited significant anti-inflammatory and antioxidant activities against LPS-induced hepatic inflammation/oxidative stress. ARE also alleviates LPS-induced oxidative stress as implied by increase in hepatic antioxidants. There was also decrease in serum ALT, AST, and LDH activities in response to ARE treatment.

11.4.8 ETHANOLIC EXTRACTS OF SCUTELLARIA BAICALENSIS ROOT (EESBR)

Scutellaria baicalensis Georgi (*S. baicalensis*) is widely used in traditional medicines, with anti-inflammatory, anticancer, antiviral activities among others. In a study carried out by Thanh et al.[43] in male C57BL/6J mice, EESBR protected against LPS-induced liver injury. 200 mg/kg/ bw EESBR exhibited anti-inflammatory actions via decrease in hepatic levels of mediators of inflammation (see Table 11.2). EESBR also alleviate LPS-induced oxidative stress as implied by decrease in hepatic MDA level.

11.4.9 FORSYTHIA-SUSPENSA-EXTRACT-(FSE)

Forsythia suspensa is a popular long-established medicine in China used as a therapy for infections and numerous disease conditions. In a study carried out by Zhao et al.,[60] FSE attenuated liver injury caused by LPS rat model, via anti-inflammatory and antioxidant actions. FSE displayed anti-inflammatory actions by inducing decrease in serum and hepatic concentrations of TNFα, IL1β, and IL6, as well as decrease in hepatic NF-κB expression. FSE also improve antioxidant actions in LPS-intoxicated rats via increase in serum and hepatic activities of GSH-Px and SOD. FSE also protected the hepatocytes as implied by decrease in serum ALT and AST activities as well as improvement in liver morphology.

11.4.10 ACTIVE FRACTION OF AQUEOUS ETHANOL EXTRACT OF SCIRPUS YAGARA TUBER (AF-EESYT)

Li et al.[24] evaluated the active fraction (petroleum ether-ethylacetate) from ethanol extract of *Scirpus yagara* tuber (AF-EESYT) in female mouse model of LPS-induced liver injury.[24] HPLC analysis of AF-EESYT revealed the presence of mainly phenolic acids. AF-EESYT demonstrated significant anti-inflammatory effect in this model by causing decreased serum and hepatic levels of inflammatory cytokines.

11.4.11 LONICERA CAERULEA BERRY EXTRACT (LCBE)

Wang et al.[51] reported the ability of LCBE to suppress LPS-induced inflammation in male Sprague–Dawley rats. In this study, LCBE at 50, 100, and

200 mg/kg bw protected against LPS-induced liver injury via its influence on TLRs and oxidative stress-associated MAPK signaling. LCBE alleviated LPS-induced inflammatory response via decrease in expression of TLR2, TLR4, and MAPKs as well as decrease in hepatic concentrations of CRP and IL-6. Moreover, LCBE also improved hepatic antioxidants against LPS-induced oxidative stress as implied by increase in GSH level and decrease in hepatic ROS level. LCBE also show protection against hepatocellular damage as indicated by decrease in serum ALT and AST activities as well as restoration of normal histological appearance of the hepatocytes.

11.4.12 RHUS-VERNICIFLUA-STOKES (RVS)

RVS is an edible plant used in folk medicine as remedy against metabolic diseases, gastritis, and a number of cancers. Moon et al.[28] prepared RVS water extract by extraction at 90–110°C for 4 h and standardized with 3% fustin. RSV at 100, 250, or 500 mg/kg bw RVS exhibited anti-inflammatory, antioxidant, and hepatoprotective effects in male Wistar rat model. In this study, RVS showed anti-inflammatory actions and antioxidant action by causing decrease in hepatic NO and MDA levels. It also protected the hepatocytes as shown by decrease in serum ALT and AST activities as well as improved liver histological structure.

11.4.13 STEVIA EXTRACT (STE-EXT)

Stevia rebaudiana is a noncaloric sweet herb called "honey herb" used in the treatment of heart burn and other conditions. In addition, it has been used to manage obesity, diabetes mellitus, and inflammation. Sa et al.[37] prepared *Stevia rebaudiana* leaf extract (STE-EXT) using water and ethyl alcohol (1:1). They showed that STE-EXT ameliorates LPS-induced liver injury in rat model. HPTLC analysis of STE-EXT shows eight phytoconstituents including stevioside (STE). STE-EXT shows anti-inflammatory actions against LPS-induced inflammatory response by causing decrease in hepatic concentrations of IL1β, IL6, and TNFα. It is also protected against oxidative stress via decrease in hepatic nitrite, H_2O_2, and MDA concentrations and rise in hepatic activity of SOD and GSH level. Hepatoprotective effect of STE-EXT was further indicated by decline in activities of ALT and AST in serum coupled with improvement in hepatocellular structure.

11.4.14 DANHONG INJECTION (DHI)

DHI is made up of 30% ethanol-extract the rhizome and root of *Salvia-miltiorrhiza* plus water-extract of flowers of *Carthamus-tinctorius* (3:1). Gao et al.[15] reported that DHI alleviated LPS-induced liver injury in male C57BL/6J mouse model. In this study, DHI displayed significant hepatoprotective effects against LPS toxicity via anti-inflammatory, antioxidant, and antiapoptotic mechanisms. DHI exacted anti-inflammatory action by causing decrease in hepatic expression of TNFα and IL6 as well as hepatic concentrations of TNF-α and IL-6. DHI also alleviated oxidative stress induced by LPS by causing decrease in hepatic MDA level. Anti-apoptosis effects of DHI were exhibited via decrease in hepatic concentration of casp3, Bax, and Bcl2 as well as causing decrease in number of TUNEL positive cells. DHI protected against hepatocellular damage as indicated by decrease in serum hepatic markers, as well as improvement in liver histological structure.

11.4.15 ORIGANUM MAJORANA EXTRACT (OME)

Wahby et al.[48] reported the antioxidant and hepatoprotective effects of OME in male albino mouse model of LPS-induced hepatotoxicity. They prepared OME by extraction of *Origanum majorana* leaves with 95% ethanol (1:20 w/v). HPLC evaluation of OME indicated the presence of phloridzin, 3,4-dicaffeoylquinic acid, rutin, 4,5-dicaffeoylquinic acid, quercetin, gallic acid, tannic acid, cinnamic acid, and chlorogenic acid. Antioxidant mechanism of OME involves increase in hepatic GSH level and decrease in hepatic NO and MDA level. OME alleviated LPS-induced hepatocellular damage as indicated by decrease in serum ALT and AST activities.

11.4.16 RED GINSENG ALCOHOL EXTRACT (RGAE) AND RED GINSENG WATER EXTRACT (RGWE)

Red ginseng (RG) is one of the Korean medicinal herbs with numerous medicinal importance including antioxidant effect. Adam et al.[2] reported the antioxidant and hepatoprotective effects of RGAE and RGWE in LPS-induced liver injury in rat model. Bothe RGAE and RAWE show antioxidant effects as indicated by decrease in hepatic NO level, increase in hepatic SOD and CAT activities and hepatoprotective effects as indicated by decrease in serum ALT, AST, ALP, and LDH activities.

11.4.17 PULICARIA PETIOLARIS METHANOL EXTRACT (PPME)

The genus Pulicaria is widely distributed in Europe, Asia, and Africa. They have been used for treating various diseases and have also shown a wide range of bioactivities. Ahmed et al.[3] reported the suppression of hepatotoxic effects induced by LPS by PPME in male swiss albino mouse model. In this report, PPME at 50 and 100 mg/kg bw alleviated LPS-induced inflammatory response via NF-κB dependent mechanism. This was evident by decreased activation of hepatic NF-kB and decrease in hepatic TNFα and IL6 levels. PPME also alleviate LPS-induced oxidative stress as indicated by decrease in hepatic MDA level and increase in hepatic GSH level and SOD activity. PPME also protected against LPS-induced hepatocellular damage.

11.4.18 ETHANOL-EXTRACT-OF ILLICIUM-HENRYI (EEIH)

Illicium-henryi known as hardwood crab, is used in traditional Chinese medicine. Islam et al.[16] evaluated the protective effect of EEIH in LPS model. Characterization of EEIH by HPLC–QTOF–MS revealed the presence of (+)-catechin, 30,40,7-trihydroxyflavone, anisatin, majucin, taxifolin, illihenryionese, illifrognone, burmanicumol, vanillin, illioliganones, henrylactone, etc. EEIH at 1.25, 2.5, or 5.0-mg/kg-bw show anti-inflammatory and antioxidant actions against LPS. It also showed hepatoprotective activity on LPS-induced hepatocellular damage as indicated serum markers and histological appearance of the liver.

11.4.19 GARDENIA JASMINOIDES WATER EXTRACT (GJWE)

Gardenia jasminoides is a shrub of Rubiaceae, applied in folk treatment of hepatitis, aphthous ulcer, and trauma in Asia. Lin et al.[25] studied the hepatoprotective effects of GJWE in LPS-induced liver injury. In their report, GJWE at 10 or 30 mg/kg bw exhibited hepatoprotective effect against LPS-related hepatocellular damage as indicated by significant decrease in serum ALT and AST activities and improved liver histological appearance.

11.4.20 LOTUS SEEDPOD EXTRACT (LSE)

Tseng et al.[45] explored the protective properties of LSE in LPS model of hepatic injury in male IRC mice. HPLC evaluation of LSE showed the presence of catechin, procyanidin B2, ρ-coumaric acid, and epigallocatechin. In this study,[45] LSE showed anti-inflammatory, antioxidant, and hepatocytes protection. LSE showed anti-inflammatory action via decrease in serum and hepatic levels of IL1β, IL6, and TNFα. LSE caused an upsurge in hepatic activity/level of SOD, GSH-Px, and GSH. LSE further shows hepatoprotective effect as indicated by decrease in serum ALT and AST activities as well as improved liver histological appearance.

11.5 SUMMARY

In this chapter, authors have described the mechanistic roles of some plant extracts and plant-based compounds in murine models of LPS-induced inflammatory liver injury. Important protective mechanisms exhibited by these extracts and compounds have also been highlighted. They include LPS scavenging, anti-inflammatory, antiapoptosis, antioxidant, and free radical scavenging activities among others. It is important to infer that herbal extracts and plant-derived compounds have important roles to play in the clinical management and treatment of liver diseases especially acute liver disease.

KEYWORDS

- **antioxidants**
- **apoptosis**
- **inflammation**
- **Kupffer cells activation**
- **lipopolysaccharide**
- **liver injury**
- **murine models**
- **oxidative stress**
- **plant derived compounds**
- **plant extracts**

REFERENCES

1. Abdel-Latif, M. S.; Hegazy, A.; Khalil, S.; Ghareeb, D. Prophylactic Effect of Herbal Extracts on LPS-induced Inflammatory Response in Rat Hepatocytes. *Int. J. Phytomed.* **2017,** *9*, 20–28.
2. Adam, G. O.; Kim, G. B.; Lee, S. J.; Lee, H.; Kang, H. S.; Kim, S. J. Red Ginseng Reduces Inflammatory Response via Suppression MAPK/P38 Signaling and p65 Nuclear Proteins Translocation in Rats and Raw 264.7 Macrophage. *Am. J. Chinese Med.* **2019,** *47*(7), 1589–1609.
3. Ahmed, N.; El-Agamy, D. S.; Mohammed, G. A.; Abo-Haded, H.; Elkablawy, M.; Ibrahim, S. R. M. Suppression of LPS-Induced Hepato- and Cardiotoxic Effects by *Pulicaria petiolaris* via NF-κB Dependent Mechanism. *Cardiovasc. Toxicol.* **2020,** *20*(2), 121–129.
4. Ajuwon, O. R.; Oguntibeju, O. O.; Marnewick, J. L. Amelioration of Lipopolysaccharide Induced Liver Injury by Aqueous Rooibos (*Aspalathus Linearis*) Extract via Inhibition of Pro-Inflammatory Cytokines and Oxidative Stress. *BMC Complement. Altern. Med.* **2014,** *14*, 392–399.
5. Allam, A. A.; Gabr, S. A.; Ajarem, J.; Alghadir, A. H.; Sekar, R.; Chow, B. K. C. Geno Protective and Anti-Apoptotic Effect of Green Tea Against Perinatal Lipopolysaccharide-Exposure Induced Liver Toxicity in Rat Newborns. *Afr. J. Tradit. Complement. Altern. Med.* **2017,** *14*(2), 166–176.
6. Altinbas, A.; Bechmann, L. P.; Akkiz, H.; Gerken, G.; Canbay, A. Acute Liver Failure, Chapter 27. In *Hepatology - A Clinical Textbook*, 10th ed.; Mauss, S., Berg, T., Rockstroh, J., Sarrazin, C., Wedemeyer, S., Eds.; Free Online Book, 2020; pp 712–726. https://www.natap.org/2014/HCV/052214_01.htm
7. Andreakos, E.; Sacre, S. M.; Smith, C.; Lundberg, A.; Kiriakidis, S.; Stonehouse, T.; Monaco, C.; Feldmann, M.; Foxwell, B. M. Distinct Pathways of LPS-induced NF-kB Activation and Cytokine Production in Human Myeloid and Non-myeloid Cells Defined by Selective Utilization of MyD88 and Mal/TIRAP. *Blood* **2004,** *103*(6), 2229–2237.
8. Asgharzadeh, F.; Bargi, R.; Beheshti, F.; Hosseini, M.; Farzadnia, M.; Khazaei, M. Thymoquinone Restores Liver Fibrosis and Improves Oxidative Stress Status in a Lipopolysaccharide-Induced Inflammation Model in Rats. *Avicenna J. Phytomed.* **2017,** *7*(6), 502–510.
9. Asrani, S. K.; Devarbhavi, H.; Eaton, J.; Kamath, P. S. Burden of Liver Diseases in the World. *J. Hepatol.* **2019,** *70*, 151–171.
10. Bernal, W.; Auzinger, G.; Dhawan, A.; Wendon, J. Acute Liver Failure. *Lancet* **2010,** *376*, 190–201.
11. Bernal, W.; Wendon, J. Liver Transplantation in Adults With Acute Liver Failure. *J. Hepatol.* **2004,** *40*, 192–197.
12. Choudhury, M. G.; Saha, N. Induction of Inducible Nitric Oxide Synthase by Lipopolysaccharide and the Influences of Cell Volume Changes, Stress Hormones and Oxidative Stress on Nitric Oxide Efflux from the Perfused Liver of Air-Breathing Catfish, Heteropneustes fossilis. *PLoS One* **2016,** *11*(3), 1–21.
13. Cronin, J. G.; Turner, M. L.; Goetze, L.; Bryant, C. E.; Sheldon, I. M. Toll-Like Receptor 4 and MYD88-Dependent Signaling Mechanisms of the Innate Immune System are Essential for the Response to Lipopolysaccharide by Epithelial and Stromal Cells of the Bovine Endometrium. *Biol. Reprod.* **2012,** *86*(2), 51–56.

14. Ding, Y.; Liu, P.; Chen Z. L.; Zhang, S. J.; Wang, Y. Q.; Cai, X.; Luo, L.; Zhou, X.; Zhao, L. Emodin Attenuates Lipopolysaccharide-Induced Acute Liver Injury via Inhibiting the TLR4 Signaling Pathway In Vitro and In Vivo. *Front. Pharmacol.* **2018,** *9,* 1–16.
15. Gao, L. N.; Yan, K.; Cui, Y. L.; Fan, G. W.; Wang, Y. F. Protective Effect of *Salvia miltiorrhiza* and *Carthamus tinctorius* Extract Against Lipopolysaccharide-Induced Liver Injury. *World J. Gastroenterol.* **2015,** *21*(30), 9079–9092.
16. Islam, M. S.; Yu, H.; Miao, L.; Liu, Z.; He, Y.; Sun, H. Hepatoprotective Effect of the Ethanol Extract of *Illicium henryi* Against Acute Liver Injury in Mice Induced by Lipopolysaccharide. *Antioxidants* **2019,** *8*(10), 1–21.
17. Jirillo, E; Caccavol, D.; Magrone, T.; Piccigallo, E.; Amati, L.; Lembo, A.; Kalis, C.; Gumenscheimer, M. The Role of the Liver in the Response to LPS: Experimental and Clinical Findings. *J. Endotoxin Res.* **2002,** *8*(5), 319–327.
18. Josephs, M. D.; Bahjat, F. R.; Fukuzuka, K.; Ksontini, R.; Solorzano, C. C.; Edwards, C. K.; Tannahill, C. L.; MacKay, S. L.; Copeland, E. M.; Moldawer, L. L. Lipopolysaccharide and D-galactosamine-Induced Hepatic Injury is Mediated by TNF-Alpha and Not by Fas Ligand. *Am. J. Physiol. Regul. Integr. Comp. Physiol.* **2000,** *278*, R1196–R1201.
19. Kallapura, G.; Pumford, N. R.; Hernandez-Velasco, X.; Hargis, B. M.; Tellez, G. Mechanisms Involved in Lipopolysaccharide Derived ROS and RNS Oxidative Stress and Septic Shock. *J. Microbiol. Res. Rev.* **2014,** *2*(1), 6–11.
20. Kang, K. S.; Yamabe, N.; Kim, H. Y.; Yokozawa, T. Effect of Sun Ginseng Methanol Extract on Lipopolysaccharide-Induced Liver Injury in Rats. *Phytomedicine* **2007,** *14*, 840–845.
21. Kaur, G.; Tirkey, N.; Chopra, K. Beneficial Effect of Hesperidin on Lipopolysaccharide-Induced Hepatotoxicity. *Toxicology* **2006,** *226*, 152–160.
22. Kondo, T.; Suda, T.; Fukuyama, H.; Adachi, M.; Nagata, S. Essential Roles of the Fas Ligand in the Development of Hepatitis. *Nat. Med.* **1997,** *3*, 409–413.
23. Lee, I. T.; Shih, R.; Lin, C.; Chen, J.; Yang, C. Role of TLR4/NADPH Oxidase/ROS-Activated p38 MAPK in VCAM-1 Expression Induced by Lipopolysaccharide in Human Renal Mesangial Cells. *Cell Commun. Signal.* **2012,** *10*, 1–15.
24. Li, P.; Liang, Q.; Cui, X.; Li, J.; Zou, N.; Wu, Q.; Duan, J. Protective Effects of the Active Fraction from the Tuber of *Scirpus yagara* in Mouse Endotoxin Shock Model. *J. Ethnopharmacol.* **2014,** *158*, 331–337.
25. Lin, W. H.; Kuo, H. H.; Ho, L. H.; Tseng, M. L.; Siao, A. C.; Hung, C. T; Jeng K. C.; Hou, C. W. *Gardenia jasminoides* Extracts and Gallic Acid Inhibit Lipopolysaccharide-Induced Inflammation by Suppression of JNK2/1 Signaling Pathways in BV-2 Cells. *Iran. J. Basic Med. Sci.* **2015,** *18*, 555–562.
26. Mintah, S. O.; Asafo-Agyei, T; Archer, M.; Atta-Adjei, P.; Boamah, D.; Kumadoh, D.; Appiah, A.; Ocloo, A.; Boakye, Y. D.; Agyare, C. Medicinal Plants for Treatment of Prevalent Diseases. In *Pharmacognosy: Medicinal Plants*; Perveen, S., Al-Taweel, A., Eds.; InTechOpen Online, 2019; pp 1–19.
27. Mokdad, A. H.; Forouzanfar, M. H.; Daoud, F.; Mokdad, A. A.; El Bcheraoui, C.; Moradi-Lakeh, M. Global Burden of Diseases, Injuries, and Risk Factors for Young People's Health During 1990–2013: A Systematic Analysis for the Global Burden of Disease Study 2013. *Lancet* **2016,** *387*, 2383–2401.
28. Moon, J. E.; Shin, J.; Kwon, O.; Kim, J. Y. A Standardized Extract of *Rhus verniciflua* Stokes Protects Wistar Rats Against Lipopolysaccharide-Induced Acute Inflammation. *J. Med. Food* **2015,** *18*(11), 1223–1230.

29. Oberholzer, A; Oberholzer, C.; Bahjat, F. R.; Edwards, C. K.; Moldawer, L. L. Genetic Determinants of Lipopolysaccharide and D-Galactosamine Mediated Hepatocellular Apoptosis and Lethality. *J. Endotoxin Res.* **2001,** *7*, 375–380.
30. Ore, A.; Ugbaja, R. N.; Adeogun, A. I.; Akinloye, O. A. Hydroethanolic Extract of *Buchholzia coriacea* Seeds Alleviates LPS Induced Liver Injury in Rat via Antioxidant and Anti-inflammatory Actions. *J. Complement. Altern. Med. Res.* **2019,** *8*(2), 1–15.
31. Pallarès, V.; Fernández-Iglesias, A.; Cedó, L.; Castell-Auví, A.; Pinent, M.; Ardévol, A.; Salvadó, M. J.; Garcia-Vallvé, S.; Blay, M. Grape Seed Procyanidin Extract Reduces The Endotoxic Effects Induced by Lipopolysaccharide in Rats. *Free Radic. Biol. Med.* **2013,** *60*, 107–114.
32. Pandian, R. P.; Kutala, V. K.; Liaugminas, A.; Parinandi, N. L.; Kuppusamy, P. Lipopolysaccharide-induced Alterations in Oxygen Consumption and Radical Generation in Endothelial Cells. *Mol. Cell. Biochem.* **2005,** *278*, 119–127.
33. Pilkhwal, S.; Tirkey, N.; Kuhad, A.; Chopra, K. Effect of Bioflavonoid Quercetin on Endotoxin-Induced Hepatotoxicity and Oxidative Stress in Rat Liver. *IJPT* **2010,** *9*, 47–53.
34. Qin, L.; Li, G.; Qian, X.; Liu, Y.; Wu, X.; Liu, B.; Hong, J. S.; Block, M. L. Interactive Role of the Toll-Like Receptor-4 and Reactive Oxygen Species in LPS-Induced Microglia Activation. *Glia* **2005,** *52*(1), 78–84.
35. Rajaratnam, M.; Prystupa, A.; Lachowska-Kotowska, P.; Załuska, W.; Filip, R. Herbal Medicine for Treatment and Prevention of Liver Diseases. *J. Pre-Clin. Clin. Res.* **2014,** *8*(2), 55–60.
36. Rotimi, S. O.; Bankole, G. E.; Adelani, I. B.; Rotimi O. A. Hesperidin Prevents Lipopolysaccharide- Induced Endotoxicity in Rats. *Immunopharmacol. Immunotoxicol.* **2016,** *38*(5), 364–371.
37. Sa, L.; Chaudharya, S.; Sa, R. R. Hydroalcoholic extract of *Stevia rebaudiana* bert. Leaves and Stevioside Ameliorates Lipopolysaccharide Induced Acute Liver Injury in Rats. *Biomed. Pharmacother.* **2017,** *95*, 1040–1050.
38. Sami, A.; Gabr, N. S.; Elsaed, W. M. Aqueous Green Tea Extract and Prediction of Fibrosis in Lipopolysaccharide Intoxicated Rats. *Int. J. Pharmacol.* **2019,** *15*, 403–417.
39. Su, G. L. Lipopolysaccharides in Liver Injury: Molecular Mechanisms of Kupffer cell activation. *Am. J. Physiol. Gastrointest. Liver Physiol.* **2002,** *283*, G256–G265.
40. Sumimoto, H. Structure, Regulation and Evolution of Nox-family NADPH Oxidases that Produce Reactive Oxygen Species. *FEBS J.* **2008,** *275*, 3249–3277.
41. Takeda, K.; Kaisho, T.; Akira, S. Toll-like receptors. *Ann. Rev. Immunol.* **2003,** *21*, 335–376.
42. Tannahil, C. L.; Fukuzuka, K.; Marum, T.; Abouhamze, Z.; MacKay, S. L. D.; Copeland, E. M.; Moldawal, L. L. Discordant Tumor Necrosis Factor-Alpha Superfamily Gene Expression in Bacterial Peritonitis and Endotoxemic Shock. *Surgery* **1999,** *126*, 349–357.
43. Thanh, H. N.; Minh, H. P. T.; Le, T. A.; Ly, H. D. T.; Huu, T. N.; Duc, L. V.; Kim, T. D.; Thanh, T. B. Ethanol Extracts of *Scutellaria baicalensis* Protect Against Lipopolysaccharide Induced Acute Liver Injury in Mice. *Asian Pac. J. Trop. Biomed.* **2015,** *5*(9), 761–767.
44. Thriveni, V.; Manjunatha, H.; Santhosh, K. H. Hepatoprotective Effect of Curcumin and Capsaicin against Lipopolysaccharide Induced Liver Damage in Mice. *Pharmacogn. J.* **2017,** *9*(6), 947–951.

45. Tseng, H. C.; Tsai, P. M.; Chou, Y. H.; Lee, Y. C.; Lin, H. H.; Chen, J. H. *In Vitro* and *In Vivo* Protective Effects of Flavonoid-Enriched Lotus Seedpod Extract on Lipopolysaccharide-Induced Hepatic Inflammation. *Am. J. Chinese Med.* **2019,** *47*(1), 153–176.
46. Tsutsui, H.; Matsui, K.; Okamura, H.; Nakanishi, K. Pathophysiological Roles of Interleukin-18 in Inflammatory Liver Diseases. *Immunol. Rev.* **2000,** *174*, 192–209.
47. Ulevitch, R. J.; Tobias, P. S. Receptor-Dependent Mechanisms of Cell Stimulation by Bacterial Endotoxin. *Ann. Rev. Immunol.* **1995,** *13*, 437–457.
48. Wahby, M. M.; Yacout, G.; Kandeel, K.; Awad, D. LPS-induced Oxidative Inflammation and Hyperlipidemia in Male Rats: The Protective Role of Origanum Majorana Extract. *Beni-Suef Univ. J. Basic Appl. Sci.* **2015,** *4*, 291–298.
49. Wang, D.; Zhang, M.; Wang, T.; Cai, M.; Qian, F.; Sun, Y.; Wang, Y. Green Tea Polyphenols Prevent Lipopolysaccharide-Induced Inflammatory Liver Injury in Mice by Inhibiting NLRP3 Inflammasome Activation. *Food Funct.* **2019,** *10*(7), 3898–3908.
50. Wang, K. Molecular mechanisms of hepatic apoptosis. *Cell Death Dis.* **2014,** *5*, 1–10.
51. Wang, Y.; Li, B.; Zhu, J. Y.; Zhang, Q.; Zhang, X.; Li, L.; Ma, Y.; Meng, X. *Lonicera caerulea* Berry Extract Suppresses Lipopolysaccharide-Induced Inflammation via Toll-Like Receptor and Oxidative Stress-Associated Mitogen-Activated Protein Kinase Signaling. *Food Funct.* **2016**, 24, 1–10.
52. Wang, Z.; Ka, S. O.; Han, Y. T.; Bae, E. J. Dihydropyranoaurone Compound Damaurone D Inhibits LPS-induced Inflammation and Liver Injury by Inhibiting NF-κB and MAPK Signaling Independent of AMPK. *Arch. Pharm. Res.* **2017,** *41*(3), 314–323.
53. Xiang, Q.; Liu, Z.; Wang, Y.; Xiao, H.; Wu, W.; Xiao, C.; Liu, X. Carnosic Acid Attenuates Lipopolysaccharide-Induced Liver Injury in Rats via Fortifying Cellular Antioxidant Defense System. *Food Chem. Toxicol.* **2013,** *53*, 1–9.
54. Xiao, J.; Wang, F.; Wong, N. K.; He, J.; Zhang, R.; Sun, R.; Xu, Y.; Liu, Y.; Li, W.; Koike, K.; He, W.; You, H.; Miao, Y.; Liu, X.; Meng, M.; Gao, B.; Wang, H.; Li, C. Global Liver Disease Burdens and Research Trends: Analysis from a Chinese Perspective. *J. Hepatol.* **2019,** *71*(1), 212–221.
55. Xu, Y.; Chen, J.; Yu, X.; Tao, W.; Jiang, F.; Yin, Z.; Liu, C. Protective Effects of Clorogenic Acid on Acute Hepatotoxicity Induced by Lipopolysaccharide in Mice. *Inflamm. Res.* **2010,** *59*(10), 871–877.
56. Yang, Y.; Zhao, J.; Song, X.; Li, L.; Li, F.; Shang, J.; Wang, W. W. Amygdalin Reduces Lipopolysaccharide-induced Chronic Liver Injury in Rats by Down-Regulating PI3K/AKT, JAK2/STAT3 and NF-κB Signaling Pathways. *Artif. Cells Nanomed. Biotechnol.* **2019,** *47*(1), 2688–2697.
57. Yao, H.; Hu, C.; Yin, L.; Tao, X.; Xu, L.; Qi, Y.; Han, X.; Xu, Y.; Zhao, Y.; Wang, C.; Peng, J. Dioscin Reduces Lipopolysaccharide-Induced Inflammatory Liver Injury via Regulating TLR4/MyD88 Signal Pathway. *Int. Immunopharmacol.* **2016,** *36,* 132–141.
58. Yao, Z.; Mates, J. M.; Cheplowitz, A. M.; Hammer, L. P.; Maiseyeu, A.; Phillips, G. S.; Wewers, M. D.; Rajaram, M. V.; Robinson, J. M.; Anderson, C. L.; Ganesan, L. P. Blood-borne LPS is Rapidly Eliminated by Liver Sinusoidal Endothelial Cells via HDL. *J. Immunol.* **2016,** *197*(6), 2390–2399.
59. Yoshioka, M.; Ito, T.; Miyazaki, S.; Nakajima, Y. The Release of Tumor Necrosis Factor-alpha, Interleukin-1, Interleukin-6 and Prostaglandin E2 in Bovine Kupffer Cells Stimulated with Bacterial Lipopolysaccharide. *Vet. Immunol. Immunopathol.* **1998,** *66*(3–4), 301–307.

60. Zhao, P.; Piao, X.; Pan, L.; Zeng, Z.; Li, Q.; Xu, X.; Wang, H. *Forsythia suspensa* Extract Attenuates Lipopolysaccharide Induced Inflammatory Liver Injury In Rats Via Promoting Antioxidant Defense Mechanisms. *Anim. Sci. J.* **2017,** *88*(6), 873–881.
61. Zhong, W.; Qian, K.; Xiong, J.; Mad, K.; Wang, A.; Zou, Y. Curcumin Alleviates Lipopolysaccharide Induced Sepsis and Liver Failure by Suppression of Oxidative Stress-Related Inflammation via PI3K/AKT and NF-Kb Related Signaling. *Biomed. Pharmacother.* **2016,** *83*, 302–313.

CHAPTER 12

Telfairia occidentalis for the Alteration of Expression Ratio of Bax/Bcl-2 mRNA in Lead-Exposed Rat Testicular Tissues

Adefolaju Gbenga Anthony, Falana Benedict Abiola, Odukoya Samson Ayodeji, and Mwakikunga Anthony

ABSTRACT

This study was aimed at estimating the levels of important proapoptotic and antiapoptotic genes expressed in the testicular tissue of rats fed with *Telfairia occidentalis* after being exposed to lead nitrate. Eighteen healthy Sprague–Dawley rats were randomly allocated into three groups containing six animals per group. Group A served as control. Group B animals were administered 75 mg/kg/b.w. of lead nitrate only. Group C animals were also administered 75 mg/kg/b.w. of lead nitrate as well as 300 mg/kg/b.w. of an alcoholic (methanol) extraction of *T. occidentalis* (*METo*). Lead exposure and *METo* administration lasted 2 weeks, and the rats were sacrificed on the 15th day. The testes and epididymis were removed and RNA was isolated from paraffin-embedded testicular tissues. Total RNA was reverse transcribed and cDNA was amplified by Real-Time quantitative PCR to determine the mRNA expression levels of antiapoptotic B-cell lymphoma 2 (*Bcl2*) and proapoptotic Bcl-2-associated X (*Bax*) gene. The 75 mg/kg lead nitrate significantly ($p < 0.05$) led to an increment in the amount of expressed *Bcl-2* mRNA as well as a significant downgrading in the amount of expressed *Bax* mRNA. Administration of 300 mg/kg *METo* significantly ($p < 0.05$) prevented the alteration of *Bcl-2* mRNA and *Bax* mRNA expression levels.

Curative and Preventive Properties of Medicinal Plants: Research on Disease Management and Animal Model Studies. Megh R. Goyal, Junaid Ahmad Malik, & Ademola Olabode Ayeleso (Eds.)

These results demonstrate that *T. occidentalis* has anticancer effects due to its effects on the expression characteristics of important genes of the anti-/pro-apoptotic pathway.

12.1 INTRODUCTION

The initiation and progression of cancer is a multiplex process, which involves hereditary as well as epigenetic changes that are modulated by several environmental factors.[4] As a result of the worldwide problem of malignancies, number of epidemiological research studies have been conducted to determine the main risk factors to identify ubiquitous chemical substances and heavy metals as potential culprits, though they have been considered as potentially amendable factors.[8,12,27] WHO depicted 19 of such substances as being vital for people's wellbeing. These components are normally discovered in different environments, for example, in the air, in water, even in foodstuff.[25] Deficiency of these substances is considered as basic for homeostatic equilibrium; and the build-up of some of the heavy metals may interfere with the host's defense against malignancies. For instance, Arsenic, Nickel (Ni), Cadmium reportedly functions in propagating the oxidative stress, and have been labeled as 1st-Class carcinogenic agents in human beings by International Agency for Research on Cancer and have been identified in numerous malignant specimens.[19] Lead is likewise characterized as plausibly poisonous for humans. Emissions and effluents from homes, industries, and factories most often contain countless metals that defile the topsoil as well as groundwater and waterways.[8]

These heavy metals are not biodegradable and have been isolated in various stages of the food chain, for example, leafy foods as regularly eaten nourishments and even in unanticipated sources, for example, topsoil, airborne, road dusts that exist in the roads that we use daily. These substances can then get into the human body via respirational and digestive routes and later amass in various organ systems, prompting their interference with different body organs.[8,12]

Telfairia occidentalis comprises of two different species.[16] Because the vegetable's leaves and seeds are both edible,[22] it is extensively cultivated and eaten in many parts of West Africa. It is commonly known as fluted pumpkin, fluted gourd, and "Ugu" in the Nigerian Igbo dialect.[6,22] It is rich in protein, with the leaf having as much as 37% protein per 100 g. It also contains fats, nutrients (for example riboflavin, Vitamin C, and nicotinamide) and minerals (for example, calcium, and magnesium), polyphenols, and flavonoids, tannins, saponins, and alkaloids.[5]

Several research reports have documented the therapeutic effects of *T. occidentalis* on different tissues and organs of animal models of human diseases. For example, *T. occidentalis* is known to have antidiabetic, antibacterial and anti-inflammation capacities,[23] the treatment of hypercholesterolemia,[24] and improvement of the effect of quinine-induced testicular injury.[23] Primary studies screening for cancer chemopreventive have demonstrated that *T. occidentalis* contains phenols, phytosterols, chlorophyll, glycosides, alkaloids, tannins, flavonoids, and saponins.[20] Adedosu et al.[1] reported that *T. occidentalis* protected rats from oxidative damage, inflammatory reactions, and hematologic imbalances related to prostate cancer induced by atrazine.

One of the main features of the development or progression of cancers is apoptosis, also known as physiological cell death. Apoptosis is moderated by dual mechanisms: the inherent and the extraneous.[9,26] The mitochondrial track is facilitated by proapoptosis genes like *Bax* and antiapoptosis genes like *Bcl-2*.[10] Therefore, the proportion of proapoptosis and antiapoptosis mRNA-expression controls the signalling of cells and tissues to die or to live[7]; as a result, increased *Bcl-2* gene expression and/or decreased expressions of *Bax* have been named as one crucial feature of cancer development or progression.[2,17]

In this study, examining the effects of *T. occidentalis* on this crucial intrinsic mechanism could offer a better elucidation of their anticancer capabilities.

12.2 MATERIALS AND METHODS

12.2.1 KITS AND REAGENTS

Bouin's Fluid, Ethanol, and Methanol were obtained from Sigma (St. Louis, MO). Agarose D-1 Low EEO-GQT was purchased from Conda Laboratories (Madrid, Spain). The items from Life technologies (California, USA) were: Power SYBR® Green PCR Master Mix, High Capacity cDNA Reverse Transcription Kit, Trizol reagent. Thermo Scientific (Pittsburgh—PA, USA) supplied Proteinase K, O'GeneRuler Low Range DNA Ladder and DNase I, RNase-free kit, and water lacking RNase.

12.2.2 PLANT PREPARATION

T. occidentalis leaves were obtained, identified, and authenticated (at Botany Department, Osun State University, Osogbo) and prepared.[16] Briefly, the leaves were pulverized and 200 g of the triturate were soaked in methanol for 72 h, and mixed. The mince was filtered and poured into another container for evaporation in a water bath at 40°C. Repeating this procedure for three days produced 32.54 g of the extract, from which a stock solution was prepared.

12.2.3 ANIMALS AND TREATMENT

The upkeep and nurture of research animals conforms to the procedures by US National Institute of Health for the appropriate utilization of experimental animals.[16] Ethical authorization for the study was obtained from the College of Health Sciences Research Committee, Osun State University, Oshogbo. Nigeria. The rats were housed at the animal unit of the Department of Anatomy, in properly ventilated enclosures under normal photoperiodic specifications and provided rat chow (Bendel Feeds, Oshogbo) and water with constant availability.[16] Sprague–Dawley (SD) rats were used for this study with weights ranging between 120 and 140 g. Eighteen healthy SD rats were arbitrarily assigned into three groups having six animals per group:

- Group A was designated as the control;
- Group B animals were given 75 mg/kg/b.w. of lead nitrate only; and
- In group C, the rats were also given 75 mg/kg/b.w. of lead nitrate as well as 300 mg/kg/b.w. of an alcoholic (methanol) extract of *T. occidentalis* (*METo*).

Lead exposure and *METo* administration lasted 2 weeks, and the rats were sacrificed after receiving anesthesia on 15th day. The testes were removed from the rats and were fixed in Bouin's fluid and embedded in paraffin.

12.2.4 ISOLATION OF RNA AND REAL-TIME QUANTITATIVE PCR

Isolation of whole RNA from tissue slices embedded in paraffin was done.[3] With the aid of rotary microtome, 25 μm slices were cut from the embedded tissue blocks. To remove the embedding medium, the cuts were transferred into RNase-free microcentrifuge tubes containing one ml of xylene and the tubes were put in the oven for 3 min at 50°C to liquefy the paraffin, after which the tubes were centrifuged for one minute at the highest speed to pellet the tissue, and re-washed with xylene without tampering with the tissue pellet. The tissue was then rinsed two times with 1 mL of absolute alcohol and was air dried.

Protease K digestion buffer (150 μL) 1× was added to each sample and incubated at 55°C for 3 h. Trizol (1 mL) was added to each tissue sample and incubated at 15 to 30°C for at least 5 min to disengage the nucleo-protein complexes. Chloroform (0.2 mL) was added and the tubes vortexed for 15 s and incubated at 30°C for 3min. The tissue samples were centrifuged at 12,000 × *g* for 15 min at 4°C. The aqueous phase was pipetted into new tubes and 10 μg of glycogen was added and mixture was stirred.

Whole RNA was obtained by mixing with 0.6 mL of isopropyl alcohol, and placed at −20°C for 1 h and then centrifuged at 12,000 × *g* for 10 min at 4°C. RNA pellet was washed with absolute alcohol, briefly air-dried, and was dissolved in RNase-free water. The concentration and purity of the whole RNA were analyzed by the Nanodrop-1000 spectrophotometer.[21] In line with the manufacturer's manual, genomic DNA was removed from total RNA using the DNase I, RNase-free kit. The DNAase I treated RNA was purified once more with the Gene-JET RNA Purification kit, re-quantified, and stored at −80°C until used. In accordance with the manufacturer's directives, cDNA synthesis was performed using the MultiScribe™ Reverse Transcriptase from 700 ng RNA.[3]

The reverse transcriptase reaction was carried out in a GeneAmp® PCR System 9600 Thermal Cycler for 10 min at 25°C, 120 min at 37°C and then the enzyme was deactivated for 5 min at 85°C. The cDNA aliquots were then used in real-time polymerase chain reactions for *Bax* and *BCl-2* with *Sdha, Actb,* and *B2m* used as the housekeeping reference genes. PCR reactions were amplified for 40 cycles prior to which the AmpliTaq Gold® DNA polymerase was activated for 10 min at 95°C. Each cycle consisted of a denaturing step for 15 s at 95°C, and annealing/extension step for 1 min at 60°C.

PCR amplification was performed in a final volume of 20 μL using the Power SYBR® Green PCR Master Mix with the QuantStudio® 5 system. All primers used in this study were manufactured by Inqaba Biotechnical Industries (Pty) Ltd. (Pretoria, RSA) The primers series employed, and the base-pair sizes of all amplified yields are shown in Table 12.1. To ascertain that nonspecified products were not amplified, amplification yields were run through 3% agarose gels, stained with ethidium bromide, and images were acquired with the BioRad Gel Doc® XR (Model 170-8170 Segrate, Milan. Italy). Melt curves were analyzed for all PCR products with the QuantStudio® 5 Real-Time PCR system software. The comparative mRNA levels of expression of each gene of interest in all samples were analyzed with the qbase-PLUS software (Biogazelle, Zulte, Belgium).[3]

TABLE 12.1 Real-Time PCR Primers.

Gene	NCBI accession no	Sequences (5'-3' direction)		Product size (bp)
Genes of interest				
Bax	NM_017059.2	**F**	GGGCCTTTTTGCTACAGGGT	106
		R	TTCTTGGTGGATGCGTCCTG	
Bcl-2	NM_016993.1	**F**	ACTCTTCAGGGATGGGGTGA	94
		R	TGACATCTCCCTGTTGACGC	
Reference genes [21]				
Sdha	NM_130428.1	**F**	CCCATTCCAGTCCTTCCCAC	102
		R	CACAATCTGATCCTGGCCGT	
Actb	NM_031144.3	**F**	CTGTGTGGATTGGTGGCTCT	134
		R	AGCTCAGTAACAGTCCGCCT	
B2m	NM_012512.2	**F**	CACACGCAGTCTGAAAACCC	170
		R	AAAGGAGCTTTGGGGACACA	

12.2.5 STATISTICAL ANALYSIS

Data analysis was carried out with JMP® (Version 15.2 SAS Institute Inc., Cary, NC). Data were recorded as mean ± standard error of the mean. The standard distribution and homogeneousness of the variance were verified with the aid of the F-test ($p < 0.05$); then, a one-way ANOVA ($p < 0.05$) was utilized to analyze the results. All the averages were also analyzed with the Tukey's honest significance test.

12.3 RESULTS

12.3.1 METHANOLIC EXTRACT OF *T. OCCIDENTALIS* SUSTAINS THE BCL2 ASSOCIATED X, APOPTOTIC MODULATOR (BAX) AND BCL2, APOPTOTIC MODULATOR (BCL-2) MRNA RATIO IN THE TESTICULAR TISSUE OF SD RATS

As illustrated in Figure 12.1 and after comparison with the control group ($p < 0.05$), treatment of the animals with 75 mg/kg/b.w. of lead nitrate led to a significant up-regulation of *Bcl-2*; but the levels of *Bax* mRNA was downregulated in rat testes. Administration of extracts of *T. occidentalis* (300 mg/kg) reversed this ratio so that the amounts of proapoptosis *Bax* and antiapoptosis *Bcl*-2 mRNA expressed were restored to similar levels in group C tissue samples, as the control group.

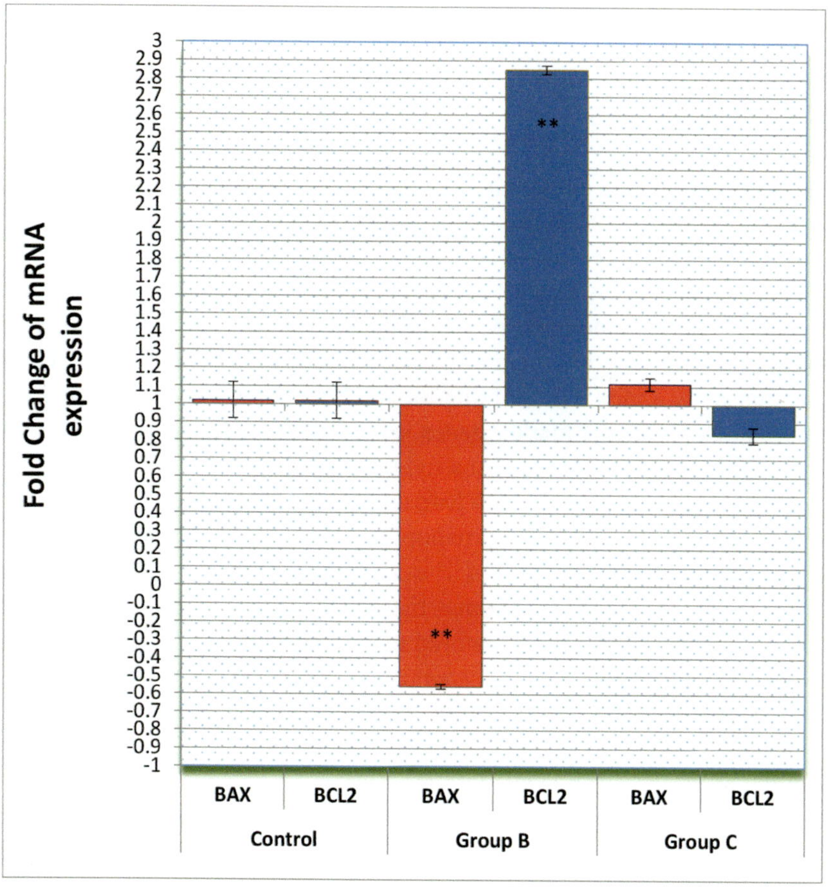

FIGURE 12.1 Effects of *Telfairia occidentalis* on proapoptosis *Bax* and antiapoptosis *Bcl*-2 gene expression in rats treated with 75 mg/kg/b.w. of lead nitrate. Data (mean ± SEM) are recorded as fold changes of mRNA expression comparative to values of the control group (designated as 1), which represents three separate reptations of Total RNA extraction. The group illustrated by ** shows statistically significantly ($p < 0.05$) difference.

These observed effects of *T. occidentalis* achieved statistical significance ($p < 0.05$) for both genes of interest. In addition, administration of the methanolic extract of *T. occidentalis* seemed to flip the mRNA expression of the *Bax* target gene from the negative levels of fold change of expression into the positive zones of fold change of expression while the mRNA expression of the *Bcl*-2 target gene seemed to change from the positive levels of fold change of expression into the negative zones of fold change of expression. However, the observed changes did not achieve statistical significance when compared to the control group.

12.4 DISCUSSION

This study is the first to examine the molecular mechanisms underlying the anticancer potentials of *T. occidentalis* thereby confirming preliminary data suggesting that it may contain anticancer agents.

The global burden of cancers has precipitated several research studies to evaluate the carcinogenic potential of environmentally ubiquitous trace elements and investigate the anticancer effects of medicinal plants. Therefore, this study aimed at estimating the mRNA expression characteristics of important carcinogenic marker genes in the testicular tissue of rats fed with *T. occidentalis* after being exposed to lead nitrate.

Treatment of the animals with lead nitrate significantly altered the normal ratio of antiapototic *Bcl-2* and proapoptotic *Bax* in testes of rat. Administration of the methanolic extract of *T. occidentalis* reversed this ratio so that in animals treated with the extract, the mRNA levels of proapoptosis *Bax* and antiapoptosis *Bcl*-2 were restored to similar levels as the control group. The methanolic extract of *T. occidentalis* restored the mRNA expression of Bax from the negative levels of fold change of expression into positive fold change of expression; while the mRNA expression of the *Bcl-2* changed from the overexpression levels into the near under-expression levels.

Several studies have shown that the raw extract of the seeds of *T. occidentalis* exhibited anticancer potential by significantly impeding oxidative action in blood's isolated polymorphonuclear cells and mononuclear cells and it also reduced serum prostatic acid phosphatase levels, and other parameters that appear to show its usefulness in the treatment of experimental nonmalignant prostatic hyperplasia in laboratory animals.[13–15]

While there are no in vitro or in vivo studies found in literature investigating the effects of leaf extract of *T. occidentalis*, yet several studies have reported the anticancer effects of some other members of the Cucurbitaceae family. For example, Cucurbitacin-I, a naturally occurring substance sequestered from Cucurbitaceae, showed strong anticancer activities against colon cancer cells by inducing apoptotic human colon cancer cell death, by blocking cell-cycle progress at Gap 2-M phase.[21] It was also reported that exposing the cells to Cucurbitacin-I considerably suppressed tumorigenic growth in the syngeneic Bagg-Albino-mice-Strain-C.

Güneş et al.[18] found that the ethanol extract of *Momordica charantia* L) demonstrated significant anticancer activities against human lung, breast, chronic myeloid leukemia, and T-cell leukemia cancer cell lines by cytotoxicity, apoptosis induction, and antiangiogenic effects.

Another similar study reported that a Cucurbitaceae-based formulation showed antiproliferative activity by inducing apoptotic death in A549 lung cells in a dose-dependent fashion.[11] The aforementioned reports show that members of the Cucurbitaceae family possess significant anticancer effects.

12.5 SUMMARY

This study shows that the anticancer activity of *T. occidentalis* can be attributed to its effects on the gene expression profile of vital antiapoptosis and proapoptosis genes *Bax* as well as *Bcl-2*. More studies are required to evaluate the effects of this extract on other members of the Bcl-2 family as well as other genes of the oncogenic pathways.

KEYWORDS

- **apoptosis**
- ***Bax***
- ***Bcl*-2**
- **Lead**
- **mRNA expression**
- ***Telfairia occidentalis***
- **testes**

REFERENCES

1. Adedosu, O.; Badmus, A.; Adeleke, G.; Raji, R. *Telfairia Occidentalis* Seed Extract Protects Against Oxidative Stress, Inflammation and Some Haematological Disorders

Associated With Atrazine-Induced Prostate Cancer in Rats. *Eur. J. Cancer* **2017,** *72*, S94–S98.
2. Adefolaju, G. A.; Mwakikunga, A. Lopinavir and Curcumin Directly Alters BAX/BCL2 and VEGF165b MRNA Levels to Suppress Human Squamous Cervical Carcinoma Cell Growth. *Int. J. Morphol.* **2019,** *37*, 584–591.
3. Adefolaju, G. A.; Falana, B. A.; Oyewopo, A. O.; Mwakikunga, A. Effects of *Talinum Triangulare* on Hepatic Antioxidant Gene Expression Profile in Carbon Tetrachloride-Induced Rat Liver Injury. In *Bioactive Compounds of Medicinal Plants: Properties and Potential for Human Health*; Goyal, M. R., Ayeleso, A. O., Eds.; Apple Academic Press: New Jersey, 2018; pp 87–100.
4. Adefolaju, G. A.; Theron, K. E.; Hosie, M. J. In-Vitro Effects of Protease Inhibitors on BAX, BCL- 2 and Apoptosis in Two Human Breast Cell Lines. *S. Afr. J. Sci.* **2015,** *111*, 1–9.
5. Aiyelaagbe, I. O.; Kintomo, A. A. Nitrogen Response of Fluted Pumpkin (*Telfairia Occidentalis Hook.* F) Grown Sole or Intercropped With Banana. *Nutr. Cycl. Agroecosys.* **2002,** *64*, 231–235.
6. Akoroda, M. O. Ethnobotany of *Telfairia occidentalis* (Cucurbitaceae) Among Igbos of Nigeria, [Online]. https://eurekamag.com/research/002/100/002100498.php (accessed Sep 28, 2021).
7. Bai, L.; Wang, S. Targeting Apoptosis Pathways for New Cancer Therapeutics. *Ann. Rev. Med.* **2014,** *65*, 139–155.
8. Canaz, E.; Kilinc, M.; Sayar, H.; Kiran, G.; Ozyurek, E. Lead, Selenium and Nickel Concentrations in Epithelial Ovarian Cancer, Borderline Ovarian Tumor, and Healthy Ovarian Tissues. *J. Trace Elem. Med. Biol.* **2017,** *43*, 217–223.
9. Chipuk, J. E.; Fisher, J. C.; Dillon, C. P.; Kriwacki, R. W.; Kuwana, T.; Green, D. R. Mechanism of Apoptosis Induction by Inhibition of the Anti-Apoptotic BCL-2 Proteins. *Proc. Natl. Acad. Sci. U. S. A.* **2008,** *105*, 20327–20332.
10. Chipuk, J. E.; Moldoveanu, T.; Llambi, F.; Parsons, M. J.; Green, D. R. The BCL-2 Family Reunion. *Mol. Cell.* **2010,** *37*, 299–310.
11. Dong, J.; Zhang, X.; Qu, C.; Rong, X.; Liu, J.; Qu, Y. Structural Characterization of *Momordica Charantia* L. (Cucurbitaceae) Oligopeptides and the Detection of Their Capability in Non-Small Cell Lung Cancer A549 Cells: Induction of Apoptosis. *RSC Adv.* **2019,** *9*, 8300–8309.
12. Ebrahimi, M.; Khalili, N.; Razi, S.; Keshavarz-Fathi, M.; Khalili, N.; Rezaei, N. Effects of Lead and Cadmium on the Immune System and Cancer Progression. *J. Environ. Health Sci. Eng.* **2020,** *18*, 335–343.
13. Ejike, C. E. C. C.; Ezeanyika, L. U. S. Inhibition of the Experimental Induction of Benign Prostatic Hyperplasia: A Possible Role for Fluted Pumpkin (*Telfairia Occidentalis* Hook f.) Seeds. *Urol. Int.* **2011,** *87*, 218–224.
14. Ejike, C. E. C. C.; Ezeanyika, L. U. S. Management of Experimental Benign Prostatic Hyperplasia in Rats Using a Food-Based Therapy Containing *Telfairia Occidentalis* Seeds. *Afr. J. Tradit. Complement. Altern. Med.* **2011,** *8*, 398–404.
15. 15. Eseyin, O. A.; Munavvar Zubaid, M.; Rathore, H. A Review of the Pharmacological and Biological Activities of the Aerial Parts of *Telfairia Occidentalis* Hook (Cucurbitaceae). *Trop. J. Pharm. Res.* **2014,** *13*, 1761–1769.
16. Falana, B. A.; Adekeye, W. M.; Adeleke, S. O. Amelioration of Lead Induced Changes in the Testes of Sprague-Dawley Rats, by Methanol Extract of *Telfairia occidentalis*. *Anat. J. Afr.* **2019,** *8*, 1606–1613.

17. Green, D. R.; Beere, H. M. Killers or Clean-Up Crew: Cancer Drug Discovery and Development. In *Apoptosis and Cancer Chemotherapy*; Hickman, J. A., Dive, C., Eds.; Humana Press: Totowa, NJ, 1999; pp 157–174.
18. Güneş, H.; Alper, M.; Çelikoğlu, N. Anticancer Effect of the Fruit and Seed Extracts of *Momordica Charantia* L. (Cucurbitaceae) on Human Cancer Cell Lines. *Trop. J. Pharm. Res.* **2019,** *18*(10), 2057–2065.
19. Hayes, R. B. The Carcinogenicity of Metals in Humans. *Cancer Causes Control* **1997,** *8*, 371–385.
20. Iweala, E. E. J. Preliminary Qualitative Screening for Cancer Chemopreventive Agents in *Telfairia Occidentalis*, *Gnetum Africanum*, *Gongronema Latifolium*, and *Ocimum Gratissimum* from Nigeria. *J. Med. Food Plants* **2009,** *1*, 58–63.
21. Kim, H. J.; Park, J. H. Y.; Kim, J. -K. Cucurbitacin-I, a Natural Cell-Permeable Triterpenoid Isolated From Cucurbitaceae, Exerts Potent Anticancer Effect in Colon Cancer. *Chem. Biol. Interact.* **2014,** *219*, 1–8.
22. Nwanna, E. E.; Oboh, G. Antioxidant and Hepatoprotective Properties of Polyphenol Extracts From *Telfairia Occidentalis* (Fluted Pumpkin) Leaves on Acetaminophen Induced Liver Damage. *Pak. J. Biol. Sci.* **2007,** *10*, 2682–2687.
23. Nwufo, M. I.; Emebiri, L. C. Pod Rots of Fluted Pumpkin (*Telfairia Occidentalis*) in Imo State, Nigeria. *Int. Biodeterior.* **1990,** *26*(1), 63–68.
24. Nwufo, M. I. Effects of Water Stress on the Post-Harvest Quality of Two Leafy Vegetables, *Telfairia Occidentalis* and *Pterocarpus Soyauxii* during Storage. *J. Sci. Food Agric.* **1994,** *64*, 265–269.
25. World Health Organization (WHO). *Report on Trace Elements in Human Nutrition and Health* [Online], International Atomic Energy Agency & Food and Agriculture Organization of the United Nations, Rome, 1996. https://apps.who.int/iris/handle/10665/37931 (accessed on Sept 30, 2021).
26. Youle, R. J.; Strasser, A. The BCL-2 Protein Family: Opposing Activities That Mediate Cell Death. *Nat. Rev. Mol. Cell Biol.* **2008,** *9*, 47–59.
27. Zaki, A.; Aldabah, F. H.; Mohammad, M. A.; Emran, T. M.; Amer, A. W. Effects of Environmental Exposure to Lead and Cadmium on Male Fertility. *Mansoura J. Forensic Med. Clin. Toxicol.* **2018,** *26*, 179–191.

PART IV

HEALTH BENEFITS OF MEDICINAL PLANTS

CHAPTER 13

Hydroethanolic Leaf Extract of *Parquetina nigrescens*: Hematological Indices and Biochemical Parameters in Male Wistar Rats

Lydia Oluwatoyin Ajayi, Ademola Olabode Ayeleso, and Temitope Adenike Oyedepo

ABSTRACT

This chapter is aimed at evaluating the adverse impacts of the mixture of water and ethanol (hydroethanolic) leaf extract of *Parquetina nigrescens* (HELEPN: Bullock plant) by assessing its effect on some biochemical parameters. The study evaluated the impact of HELEPN on the concentrations of hematological profile (white blood cells, hematocrit, red blood cells, platelet, hemoglobin, mean corpuscular volume, lymphocytes, red cell distribution width coefficient variation); lipid profile (high-density lipoproteins, low-density lipoproteins, total cholesterol, and triacylglycerol); protein profile (total protein, albumin, globulin); kidney parameters (urea and creatinine) and liver enzymes (alanine transaminase, aspartate transaminase, and alkaline phosphastase) using RayBio diagnostic kits. Histopathology was conducted on the liver, kidney, and heart as well as the phytochemical content, mineral content, and antioxidant capacity of the plant using parameters such as ferric reducing antioxidant ability and 2,2-diphenyl-1-picryhydrazyl scavenging ability of the plant extract. HELEPN administration at different doses in this study increased hematocrit levels, and decreased white blood cell count and

Curative and Preventive Properties of Medicinal Plants: Research on Disease Management and Animal Model Studies. Megh R. Goyal, Junaid Ahmad Malik, & Ademola Olabode Ayeleso (Eds.)

LDL but did not alter serum urea and creatinine, serum proteins, and liver enzymes. This study confirmed the use of this herb by traditional medical practice and inferred that the plant is not toxic when given a dose of 1500 mg/kg, which is the highest dose used in this study, also the plant could be said to be cardioprotective and these effects could be due to the phytochemicals and minerals present in the plant.

13.1 INTRODUCTION

Medicinal plants have gained acceptance worldwide for use as alternative therapeutic agents for numerous disease conditions. Interestingly, these plants contain different phytochemicals that are beneficial to human health.[27] One such medicinal plant is *Parquetina nigrescens*.

Parquetina nigrescens, sometimes called bullock plant, is a shrub found in equatorial West Africa. The plant is called African parquetina in English, Yorubas call it "Ewe ogbo," "Kwakwanin" in Hausa, and "Mbgidimgbe" in Igbo.[22] The plant grows on hills throughout the regions of Africa, specifically, it is abundant from Senegal to Nigeria, and from the Congo basin to south tropical Africa.[29] It has a twisted stem and a base that is woody, it is about 10–15 cm in height, while its breadth is about 6–8 cm wide with a long fine stem that contains the leaf part of the plant. Its flower grows from its side with branches having whitish outside and inner reddish coloration. The fruit, which has a featherlike seed,[7] is composed of the following two parts: An outer woody and an inner softer part.[29]

The plant has been in use as traditional medicine for many years. Available literature has shown the beneficial potential of *P. nigrescens*, such as pain relieving, anti-inflammatory, antipyretic effects,[25] and it prevents hemorrhagic anemia.[2] Extract of *P. nigrescens* has been revealed to be abundant in flavonoids, tannins, phlobatannins, saponins, cardiac glycosides, anthraquinones, and oils; and its abilities to scavenge reactive oxygen species (ROS) have been established.[8] The plant also has antidiabetic, antiulcerative cardiotonic, and hematinic properties.[8,26,29] It is being used to combat worm infestation, management of rheumatism,[25] and the treatment of menstrual disorders and gonorrhea.[24]

This chapter narrates the probable toxic activity of hydroethanolic leaf extract of *P. nigrescens* on the living system by assessing its impact on some biochemical parameters. The study was thus conducted to check and elucidate the effects that hydroethanolic leaf extract of *P. nigrescens* could have on hematological parameters, lipid, and parameters, liver enzymes as well as its effect on the architectural structure of the liver, kidney, and heart.

13.2 MATERIALS AND METHODS

13.2.1 COLLECTION AND EXTRACTION OF PARQUETINA NIGRECENS

The leaf of *P. nigrescens* was plucked at Orita Naira junction in Ogbomoso, Oyo State, Nigeria. The plant was recognized and confirmed by Ogunkunle, Department of Botany, Ladoke Akintola University of Technology, Ogbomoso, Nigeria, and assigned a Voucher No.: LHO 532. A specimen was kept in the herbarium at Department of Botany, Ladoke Akintola University of Technology, Ogbomoso, Nigeria. The leaves were spread inside a room and left to dry it was later ground into powder using an electric blender, a measure (50 g) of the sample was then fully immersed in 140 mL of ethanol and 60 mL of distilled water for 48 h to prepare the 70% mixture of water and ethanolic extract of *P. nigrecens*, the extract in ethanol was poured out and the remaining ethanol expelled using Soxhlet apparatus. The concentrated extract was then dried using a regulated oven at 45°C.[5]

13.2.2 CHEMICALS

Commercially purchased chemicals were used in this study and they were of standard analytical grade (Sigma Chemical Co, USA).

13.2.3 EXPERIMENTAL ANIMAL

Twenty male Wistar rats weighing between 180 and 250 g were bought from the animal facility of the Department of Anatomy, Adeleke University, Ede, Nigeria. The animals were kept in an enclosure made of woods under normal conditions of temperature and humidity in the animal holding facility of the Department of Biochemistry, Adeleke University after the ethical clearance was approved by the Adeleke University Ethical Review Committee (No. AUERC/FOS/01). They were provided with standard feed and given water freely with a 12 h light/dark cycle. The experiment was carried out according to NIH instructions for handling laboratory animals.

13.2.4 STUDY DESIGN

The animals were non-specifically separated into the following four groups (five animals assigned to each group):

Group A: received only DMSO (dimethyl sulfoxide) (normal control group)
Group B: received 500 mg/kg of 70% HELEPN
Group C: received 1000 mg/kg of 70% HELEPN
Group D: received 1500 mg/kg of 70% HELEPN

The rats were given HELEPN for 4 weeks at different doses, as indicated in the groupings. At the termination of the experimental duration, the rats were anesthetized using ketamine and xylazine (40–4 mg/kg) and then sacrificed.[3] Blood samples were collected into a plain bottle for hematology assay and EDTA bottles for biochemical assays. The liver, heart, and kidney were collected and kept in 10% formalin for histological examination.

13.2.5 PHYTOCHEMICAL SCREENING

The determination of alkaloid, flavonoid, and saponin was done using the method of harborne.[17] Tannin content of the plant was estimated by the method of Van–Burdan and Robinso.[31]

13.2.6 DETERMINATION OF MINERAL CONTENTS

Flame photometry was used to estimate sodium and potassium contents of the plants using the Jenway Flame Photometer (PFP7 model), while the vanadomolybate method was used for the determination of phosphorus.[6] In contrast, the other minerals were determined after wet digestion using an atomic absorption spectrophotometer.

13.2.7 IN-VITRO ASSAYS TO DETERMINE ANTIOXIDANT ABILITY OF PLANT

13.2.7.1 DPPH RADICAL SCAVENGING ACTIVITY [2,2-DIPHENYL-1-PICRYHYDRAZYL (DPPH) ASSAY]

The DPPH-radical ameliorating activity of the hydroethanolic extract of *P. nigrescens* was assayed according to the procedure of Blois,[11] as further modified by Brand-Williams *et al.*.[12]

13.2.7.2 FERRIC REDUCING ANTIOXIDANT POWER ASSAY (FRAP)

The FRAP estimation of the hydroethanolic extract of *P. nigrescens* was analyzed spectrophotometrically as described by Benzie and Strain[10] using ascorbic acid as standard.

13.2.8 TOXICITY STUDY

13.2.8.1 HEMATOLOGICAL ASSAY

The hematological parameters were evaluated with an automated hematological analyzer systemsKX-21 (Japan).

13.2.8.2 BIOCHEMICAL ASSAYS

The liver, kidney, and lipid profile levels of serum were done spectrophotometrically using Randox kits. Protein profile, which includes albumin; total protein, and globulin; and kidney parameters, such as urea and creatinine, were assayed for, lipid markers such as total cholesterol (TC), triglycerides (TG), high-density lipoproteins (HDL) were measured, and serum LDL-cholesterol was estimated using the method of Friedwald et al..[16] Liver enzymes [serum aspartate transaminase (AST), alanine aminotransferase, and (ALT), alkaline phosphatase (ALP)] were also quantified with Randox kit following the manufacturer's guidelines.

13.2.9 HISTOLOGY

Histopathological examination was done on the harvested organs. A section of the liver, kidney, and heart was weighed and dipped in 10% buffered formalin to ascertain proper fixation. It was processed and inserted into paraffin wax, and 5–6-μm thick sections of the selected parts were taken and covered with hematoxylin and eosin for staining. The photomicrographs of the organs were captured with a digital camera attached to a light microscope.

13.2.10 STATISTICAL ANALYSIS

Statistical analysis was done with SPSS (version 16). Mean ± standard deviations (SD), was used to express the results after which Tukey's post hoc test was adopted for pairwise comparison. Values of $p < 0.05$ were considered significant statistically.

13.3 RESULTS

13.3.1 YIELD OF EXTRACT

Relative to the plant material, HELEPN yielded 7.69% (w/w). The extract yielded was calculated by dividing powdered extracts (g) with the cumulative weight of the dried plant material (g), and multiplied by 100.

13.3.2 PHYTOCHEMICAL QUANTIFICATION

The quantitative phytochemical analysis of HELEPN showed the presence of flavonoids, alkaloids, saponins, tannins, cardiac glycosides, and anthraquinone (Table 13.1). Alkaloids had the highest quantity in the hydroethanol extract.

TABLE 13.1 Phytochemical Constituents of HELEPN.

Phytochemicals	Values
Alkaloids (mg/g)	72 ± 0.45
Saponin (mg/g)	18 ± 0.68
Flavonoids (mg/L)	0.02 ± 0.01
Tanins (mg/L)	4.12 ± 0.054
Cardiac glycosides (mg/L)	12.7 ± 0.12
Anthraquinones (mg/100 g)	15.2 ± 0.34

Values are the mean of three replicates ± standard deviation.

13.3.3 MINERAL COMPOSITION OF HELEPN

The results revealed that the most abundant mineral in the plant is potassium followed by chloride and then calcium. Toxic heavy metals (such as copper, cadmium, and lead) were not found in the extract (Table 13.2).

TABLE 13.2 Quantitative and Qualitative Mineral Content of HELEPN.

Mineral	Composition
Magnesium	3.58 ± 0.33
Manganese	1.38 ± 0.07
Iron	2.26 ± 0.03
Zinc	0.38 ± 0.09
Calcium	11.80 ± 0.29
Potassium	35.00 ± 1.28
Phosphorus	4.50 ± 0.16
Selenium	1.43 ± 0.01
Chromium	1.20 ± 0.08
Chloride	13.20 ± 0.08
Sodium	1.80 ± 0.08

Values are the mean of three replicates ± standard deviation.

13.3.4 IN-VITRO ANTIOXIDANT CAPACITY OF HELEPN

13.3.4.1 DPPH RADICAL SCAVENGING ACTIVITY AND FERRIC REDUCING ANTIOXIDANTS POTENTIAL OF HELEPN

The results revealed that the hydroethanolic extract of the plant possesses antioxidant properties as evident by their DPPH radical scavenging ability, and this was dependent on the concentration of the extract (Table 13.3). The extract also showed higher ferric reducing antioxidant power at a low concentration (Table 13.3).

TABLE 13.3 DPPH Radical Scavenging and Ferric Reducing Antioxidant Abilities of HELEPN.

Extract	Concentration (mg/mL)	%DPPH inhibition	%FRAP
P. nigrescens	0.2	20.4081	80.4878
	0.4	34.6938	70.7317
	0.6	40.8163	56.0975

13.3.5 PROTEIN PROFILE OF RATS FOLLOWING ADMINISTRATION OF HELEPN

The findings of this study showed that differences were not significant in the case of total protein of the groups administered HELEPN at 500, 1000, and 1500 mg/kg when compared across all groups. However, there was a noticeable increase between the HELEPN treated groups and the control, but it was not statistically significant.

Serum albumin levels of the experimental animals are shown in Figure 13.8. The study showed a reduction in the level of albumin in the group given 500 and 1000 mg/kg HELEPN and there was an upsurge in the albumin concentration of the group administered with 1500 mg/kg of HELEPN in comparison to the control at $p < 0.05$.

The study revealed statistically significant differences between the serum level of globulin in the groups given 500 and 1000 mg/kg doses in comparison to the control group and the group that was given 1500 mg/kg dose. There was a significant difference between the group that received 1500 mg/kg of hydroethanolic extract when compared to the control, but there was no convincing difference between the animals that received 500 and 100 mg/kg (Figure 13.1).

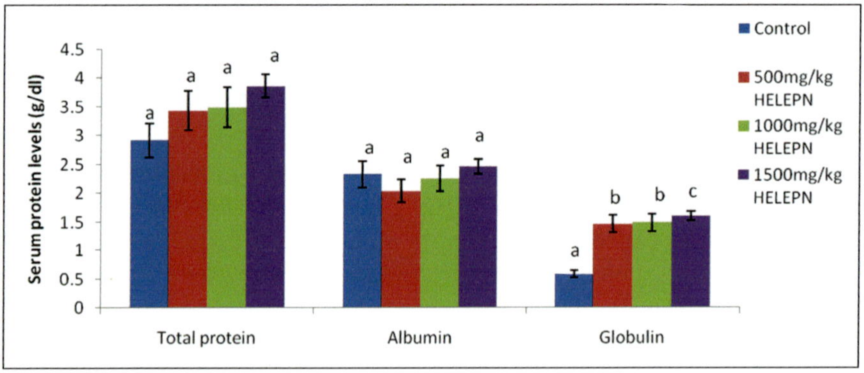

FIGURE 13.1 Effect of graded doses of HELEPN on protein profile: The letters a, b, and c on the bars of the same parameters indicate the statistical difference.

13.3.6 EFFECT OF GRADED DOSES OF HELEPN ON LIVER ENZYMES

The results revealed that there was no cogent difference ($p < 0.05$) in the serum levels of AST and ALT; and ALP in the rats given 500, 100, and 1500 mg/kg of the 70% hydroethanolic extract of *P. nigrescens,* in comparison to the control rats (Figure 13.2).

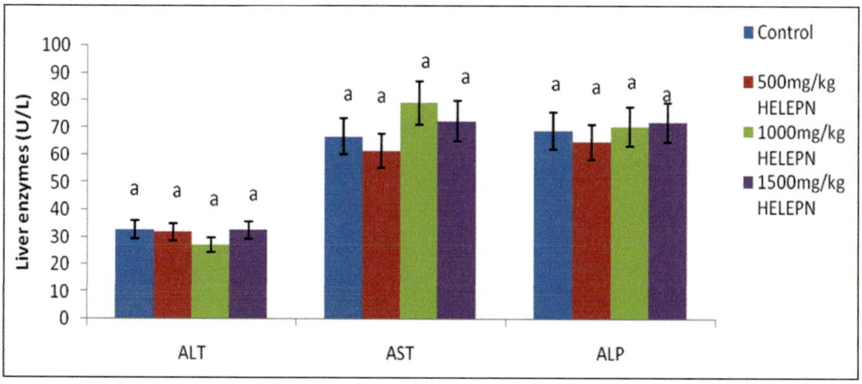

FIGURE 13.2 Implications of graded doses of HELEPN on liver enzymes: The letter a on the bars of the same parameters indicates the statistical difference.

13.3.7 THE EFFECT OF GRADED DOSES OF HELEPN ON LIPID PROFILE

There were no significant differences ($p < 0.05$) in the serum levels of TC, TG, and HDL at doses of 500, 1000, and 1500 mg/kg HELEPN in comparison to the normal control group. However, there was a cogent difference ($p < 0.05$) in the serum levels of LDL in all the groups that received HELEPN in comparison to the normal control group (Figure 13.3).

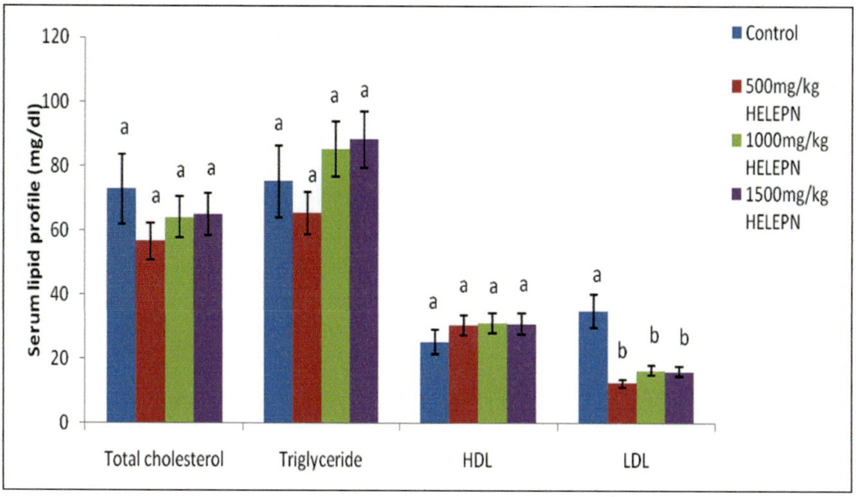

FIGURE 13.3 Effect of graded doses of HELEPN on lipid profile: The letters a, b, and c on the bars of the same parameters indicate the statistical difference.

13.3.8 THE EFFECT OF GRADED DOSES OF HELEPN ON ATHEROGENIC AND CORONARY INDICES

The results revealed significant decreases in both the atherogenic and coronary index of the groups given the extract at all doses relative to the control (Figure 13.4).

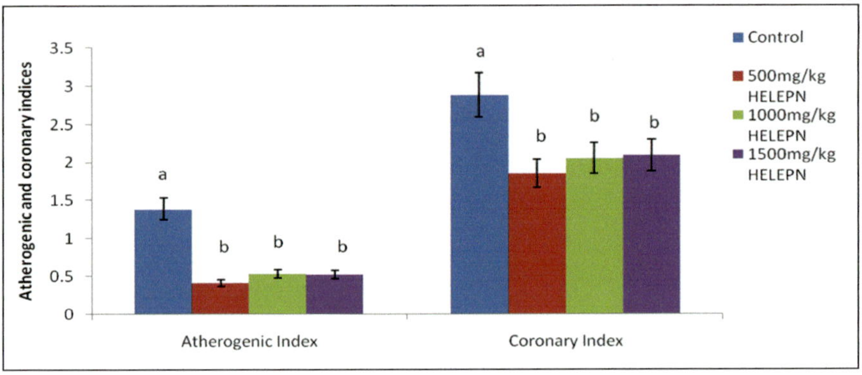

FIGURE 13.4 Effect of graded doses of HELEPN on atherogenic and coronary indices: The letters a, b, and c on the bars of the same parameters indicate the statistical difference.

13.3.9 EFFECT OF GRADED DOSES OF HELEPN ON SERUM UREA AND CREATININE

The serum level of urea was not different statistically from the normal control animals at $p < 0.05$ after the administration of different doses of HELEPN. Although a reduction was noticed at 1000 mg and a further decrease at 500 mg with an increase at 1500 mg (Figure 13.5).

Serum creatine was seen to be significantly lower in the groups that received 500 and 1000 mg/kg of HELEPN in comparison to the control. The animals that were administered 1500 mg/kg HELEPN were not different statistically from the normal control animals (Figure 13.5).

13.3.10 EFFECT OF GRADED DOSES OF HELEPN ON HEMATOLOGICAL PARAMETERS

Table 13.4 indicates the effects of HELEPN on hematological parameters indicated an upsurge in the level of red blood cell count (RBC), hematocrit

(HCT), and hemoglobin (Hb) in all groups that received HELEPN relative to the control group. Also, the values of platelets, lymphocytes, and lymphocyte count were not significantly different in all groups when compared to the control group. However, the amount of white cells in the blood increased substantially in the group that received 1000 mg/kg and decreased substantially in the group that received 1500 mg/kg in comparison to the normal control group.

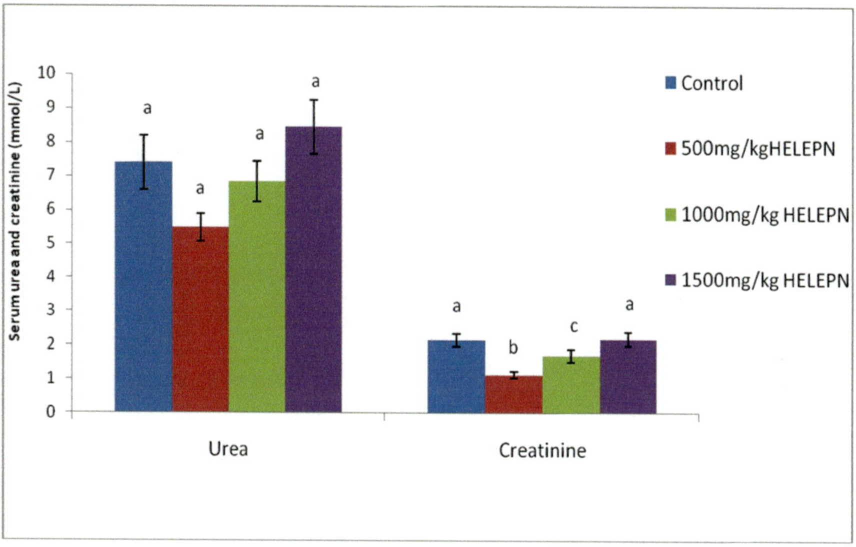

FIGURE 13.5 Effect of graded doses of HELEPN on serum urea and Creatinine: The letters a, b, and c on the bars of the same parameters indicate the statistical difference.

No significant differences were observed between the group that received 500 mg/kg and the normal control group. There were no cogent differences at $p < 0.05$ in the values of mean corpuscular volume (MCV), mean corpuscular hemoglobin (MCH), and mean corpuscular hemoglobin concentrations (MCHC) in all groups in comparison to the normal control group.

Also, no cogent difference was observed in the levels of red cell distribution width standard deviation, red cell distribution width coefficient of variation, platelet distribution width, mean platelet volume, and platelet coefficient ratio in all administered groups relative to the normal control group (Table 13.4).

TABLE 13.4 Effect of Graded Doses of HELEPN on Hematological Parameters.

Parameter	Control	500 mg/kg HELEPN	1000 mg/kg HELEPN	1500 mg/kg HELEPN
WBC	7.52 ± 0.88^{a}	7.04 ± 1.09^{a}	9.92 ± 0.82^{b}	4.36 ± 0.68^{c}
RBC	7.2 ± 0.47^{a}	8.79 ± 0.55^{b}	7.77 ± 0.47^{b}	8.5 ± 0.24^{b}
Hb	12.82 ± 0.13^{a}	15.64 ± 0.35^{b}	13.64 ± 1.18^{b}	15.46 ± 0.49^{b}
HCT	51.16 ± 0.99^{a}	60.34 ± 0.57^{b}	52.62 ± 2.84^{b}	59.54 ± 2.00^{b}
MCV	75.68 ± 11.84^{a}	68.84 ± 4.04^{a}	67.72 ± 1.52^{a}	69.78 ± 1.40^{a}
MCHC	24.94 ± 0.44^{a}	25.96 ± 0.66^{a}	26.02 ± 0.83^{a}	25.98 ± 0.62^{a}
MCH	16.68 ± 1.36^{a}	17.84 ± 0.93^{a}	17.64 ± 0.79^{a}	18.12 ± 0.44^{a}
Platelets	726.8 ± 16.2^{a}	651.4 ± 98.99^{a}	656.6 ± 146.07^{a}	778.6 ± 101.1^{a}
Lymphocytes%	77.4 ± 3.6^{a}	77 ± 1.2^{a}	73.06 ± 3.63^{a}	73.54 ± 5.05^{a}
Lymph count	5.82 ± 0.96^{a}	6.2 ± 0.14^{a}	7.6 ± 0.83	3.2 ± 0.57^{b}
RDWSD	56.18 ± 10.65^{a}	44.24 ± 5.27^{a}	46.08 ± 7.08^{a}	44.06 ± 1.6^{a}
RDWCV	22.3 ± 3.09^{a}	18.76 ± 1.19^{a}	19.94 ± 3.81^{a}	18.08 ± 0.55^{a}
PWD	12.4 ± 1.28^{a}	11.70 ± 0.41^{a}	11.74 ± 0.54^{a}	11.44 ± 0.68^{a}
MPV	9.12 ± 0.92^{a}	8.6 ± 0.2^{a}	8.76 ± 0.43^{a}	8.68 ± 0.26^{a}
PLCR	19.33 ± 4.90^{a}	8.6 ± 0.2^{a}	18.66 ± 2.29^{a}	17.72 ± 2.39^{a}

Superscripts a, b, and c against different values of the same parameters indicate that the values are statistically different.

13.3.11 HISTOLOGICAL SCREENING OF THE LIVER, KIDNEY, AND HEART (Figure 13.6)

Morphology of the liver given a varied dose of HELEPN

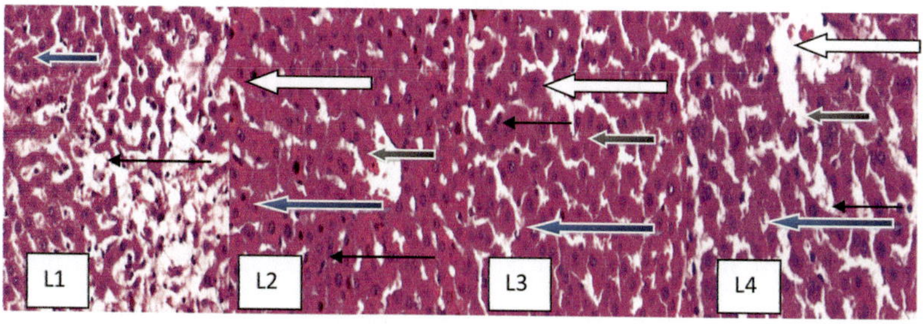

Control 500 mg/kg HELEPN 1000 mg/kg HELEPN 1500 mg/kg HELEPN

Group A–Control (L1): Photomicrograph of a liver section showing typical middle venules (white arrow) and normal portal tract (black arrow), the liver cells appear usual (blue arrow), the liver parenchyma, however, revealed a focal area of the dilated sinusoids with the moderate intrusion of inflamed cells (slender arrow).

Group B–500 mg/kg HELEPN (L2): Photomicrograph of a liver section revealing a typical central venule without profusion (white arrow) and moderately clogged portal vein (black arrow), the liver cells appear usual (blue arrow), the sinusoids appear typical and not perfused (slender arrow), no diseased lesion was observed.

Group C–1000 mg/kg HELEPN (L3): Photomicrograph of a liver section revealing typical central venules without clogging (white arrow) and lightly overcrowded portal vein (black arrow), the liver cells appear normal (blue arrow), the sinusoids appear orderly and not saturated (slender arrow), no diseased lesion was observed.

Group D–1500 mg/kg HELEPN (L4): Photomicrograph of a liver section revealing normal central venules (white arrow) and portal vein (black arrow), without overcrowding, the liver cells appear normal (blue arrow), the sinusoids appear typical and not infiltrated (slender arrow), no diseased lesion was observed.

Morpzhology of the kidney given a varied dose of HELEPN

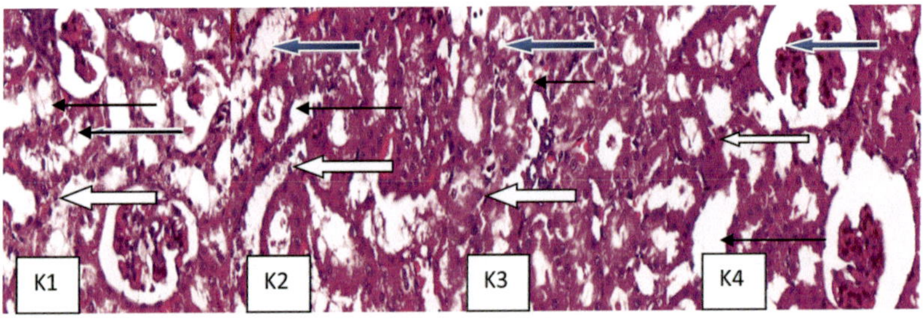

Control 500 mg/kg HELEPN 1000 mg/kg HELEPN 1500 mg/kg HELEPN

Group A–Control (K1): Photomicrographs of kidney sections revealing a typical kidney architecture. The cortex shows typical glomeruli with normal capsular spaces and mesangial cells (white arrow), the renal tubules, including PCT and DCT was seen to be typical as that of the normal control (blue arrow), and the interstitial spaces were observed to be normal (slender arrow).

Group B–500 mg/kg HELEPN (K2): Photomicrographs of kidney sections showing poor architecture. The kidney cortex revealed typical glomeruli with normal capsular spaces (white arrow). However, the tubules revealed mild epithelial degeneration and desquamation (blue arrow); there were no regular luminal spaces, and the interstitial spaces were observed to be normal (slender arrow).

Group C–1000 mg/kg HELEPN (K3): Photomicrographs of kidney sections revealing poor architecture. The cortex revealed typical glomeruli with normal capsular spaces (white arrow). However, the tubules were observed to have eosinophilic casts within the lumen of the tubules (blue arrow), and the interstitial spaces showed the main area of congestion of inflammatory cells (slender arrow).

Group D–1500 mg/kg HELEPN (K4): Photomicrographs of kidney sections revealing typical architecture. The renal cortex shows normal glomeruli with normal capsular spaces (white arrow). However, the tubules showed little eosinophilic casts within the lumen of some tubules (blue arrow), other normal tubules are seen (red arrow), and the interstitial spaces were observed to be normal (slender arrow).

Morphology of the heart given a varied dose of HELEPN

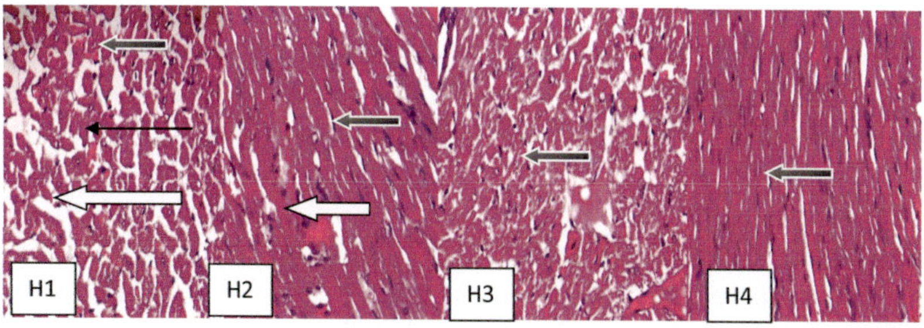

Control 500 mg/kg HELEPN 1000 mg/kg HELEPN 1500 mg/kg HELEPN

Group A–Control (H1): Photomicrograph of a heart section showing normal heart tissue with normal epicardial (white arrow) and with myocardial layer showing mild to moderate infarction (black arrow), the cardiac muscles generally are normal, no pathological lesion seen in the heart tissue.

Group B–500 mg/kg HELEPN (H2): Photomicrograph of a heart section showing normal heart tissue with normal epicardial (white arrow) and myocardial layer (black arrow), the cardiac muscles generally are normal, no pathological lesion seen in the heart tissue.

Group C–1000 mg/kg HELEPN (H3): Photomicrograph of a heart section stained by hematoxylin and eosin showing a normal heart with a normal myocardial layer (black arrow), the cardiac muscles generally are normal. No pathological lesion was seen.

Group D–1500 mg/kg HELEPN (H4): Photomicrograph of a heart section stained by hematoxylin and eosin showing a healthy heart tissue with a normal myocardial layer (black arrow), the cardiac muscles generally are normal. No pathological lesion was seen.

FIGURE 13.6 Histological screening at a varied dose of HELEPN.

13.4 DISCUSSION

Phytochemicals are the secondary metabolites of the plant which are indirectly involved in the growth and development of a plant and also function as a defensive agent against any pathogen. The results provided in this study indicated that HELEPN is rich in phytochemicals. Alkaloids being the most abundant phytochemical of HELEPN are known to serve as anti-feedants and also protect plants against microbes.[20] It has been established to have a vast-spectrum antibacterial property and is also used for pain relief. It was also observed to possess anti-inflammatory and antipyretic properties.[14,30]

Flavonoids, glycosides, and cardiac glycosides that are present in the HELEPN are indicative of their antioxidant ability. Flavonoids and glycosides have been observed to be antioxidants and are used as anti-inflammatory mediators in the treatment of capillary fragility.[19,18] Their presence in the extracts confers an antioxidant property to the plants and thus membrane-stabilizing properties.

Estimation of major dietary minerals showed that HELEPN has high concentrations of sodium and potassium, and this would enhance the maintenance of water balance in the body and nerve functions.[28] It also revealed the presence of iron, zinc, selenium, chromium, magnesium, and manganese in HELEPN. The presence of iron indicates that HELEPN could contribute to hemoglobin formation[9] and normal functioning of the central nervous system[1,9] while zinc, selenium, and chromium presence in HELEPN leaf implies that it would help in the normal operation of the immune system[4] and maintenance of cell and organ integrity.[15]

The ability of plants to scavenge free radicals is a crucial property underlying their various biological and pharmacological functions.[23] In this study, HELEPN extract showed radical scavenging activity and was seen to be concentration dependent. Also, HELEPN exhibited antioxidant activity, which can ameliorate oxidative stress with consequent positive health changes by accepting free radicals and breaking the oxidative damage chain. The phytochemicals such as flavonoids and alkaloids that confers the antioxidant ability on HELEPN disrupt lipid peroxidation and oxidation of macromolecules by donating a hydrogen atom to radicals.[13]

In this present study, HELEPN was able to significantly increase RBC, Hb, and PCV at all the different doses (500, 1000, and 1500 mg). This could have led to the activation of erythropoietin synthesis owing to the presence of a compound that may promote erythropoiesis in the extract. The active secondary metabolites such as alkaloids, cardiac glycosides, saponins, tannins, and phenols seen in HELEPN, may be indicative of its hematopoietic effects. These results agree with the findings of Agbor and Odetola,[2] showing similar effects of hydroethanolic leaf extract of the *P.*

nigrescens on erythrocyte parameters. In this study, increased erythrocyte indices at all doses imply that the erythropoietic potential of HELEPN leaf is not dependent on the dose of the extract. The study also showed that HELEPN might possess the ability to boost immunity at a moderate dose because the administration at 1000 mg/kg significantly increases the white blood cells and lymphocytes count in comparison to the normal control ($p < 0.05$).

In this study, the concentration of serum urea, total protein, albumin as well as liver enzymes such as ALT, ALP, and AST in rats after the administration of HELEPN were not statistically different from the normal control. This confirms the safety of HELEPN at 500, 1000, and 1500 mg/kg to the liver and the kidney and was evident by the histopathology findings of the liver and kidney tissues seen in all the groups administered HELEPN in comparison to the normal control.

Lipid profile analysis showed that the levels of TC, TG, and HDL were not convincingly different in the groups given HELEPN at 500, 1000, and 1500 mg/kg in comparison to the normal control. However, a cogent reduction was observed in the level of low-density lipoprotein cholesterol in comparison to that observed in the normal control group. This could be due to the presence of flavonoids in HELEPN. Chronic consumption of rutin has been established to be favorable to cardiovascular conditions,[21] as also seen in the histological examination of the heart.

13.5 SUMMARY

Parquetina nigrescens is a plant that has been used for medicinal purposes over the years, and it has gained acceptance in Africa. This study established that Hydroethanolic leaf Extract of *P. nigrescens* contains many phytochemicals and minerals which could be responsible for the many effects seen by the traditional users and confirms its use by the folklore. It also has antioxidative and anti-inflammatory potentials as evident in the constituent phytochemical and antioxidative capacity assay. Also, HELEPN was established to be safe for human consumption at the doses employed in this study; interestingly, it was observed to have lipid-lowering effect and could be used as a cardioprotective agent. The pharmacological potentials of HELEPN could be associated with a myriad of phytochemicals and minerals that are present in the extract.

KEYWORDS

- **antioxidants**
- **cardioprotective**
- ***parquetinanigrescens***
- **phytochemicals**
- **toxicity**

REFERENCES

1. Adeyeye, E. I.; Otokiti, M. K. O. Proximate Composition and Some Nutritionally Valuable Minerals of Two Varieties of *Capsicum Annum*. *Discov. Innov.* **1999,** *11* (1), 75–81.
2. Agbor, A. G.; Odetola, A. A. Effect of *Parquertina nigrecens* on Erythrocyte Indices and Serum Electrolytes of Rats Following Acute Blood Loss. *Pak. J. Biol. Sci.* **2005,** *8* (4), 527–531.
3. Ajayi, A. F.; Akhigbe, R. E. Assessment of Sexual Behavior and Fertility Indices in Male Rabbits Following Chronic Codeine Use. *J. Androl.* **2019,** *8* (2), 509–515.
4. Akubugwo, I. E.; Obasi, N. A.; Chinyere, G. C.; Ugbgu, A. E. Nutritional and Chemical Value of *Amaranthus hybridus* Leaves from Afikpo, Nigeria. *Afr. J. Biotechnol.* **2007,** *6* (24), 283–289.
5. Anita, D.; Garg, G.; Singh, B.; Ritu, M. Antimicrobial Properties of Methanolic Extract of Cumin (*Cuminum cyminum*) Seeds. *Int. J. Res. Ayurveda Pharm.* **2007,** *4* (1), 104–107.
6. AOAC International. *The official Methods of the AOAC International*, 16th ed.; Cunnif, P. A., Ed.; AOAC International: Arlington, VA, 1995.
7. Awobajo, F. O.; Olatunji-Bello, I. I.; Adegoke, O. A.; Odugbemi, T. O. Phytochemical and Antimicrobial Screening of *Hybanthusenne aspermus* and *Parquetina nigricense. Recent Res. Sci. Technol.* **2009,** *1* (4), 159–160.
8. Ayoola, A. O.; Akinloye, O.; Oguntibeju, O. O.; Oke, J. M.; Odetola, A. A. Antioxidant Activities of *Parquetina nigrescens*. *Afr. J. Biotechnol.* **2011,** *10*, 4920–4925.
9. Bakare, R.; Magbagbeola, O. A.; Akinwande, A. I.; Okunowo, W. Nutritional and Chemical Evaluation of *Momordica Charantia*. *J. Med. Plants Res.* **2010,** *4* (21), 2189–2193.
10. Benzie, I. F.; Strain, J. J. Ferric Reducing the Ability of Plasma (FRAP) a Measure of Antioxidant Power: The FRAP Assay. *Anal. Biochem.* **2010,** *239*, 70–76.
11. Blois, M. S. Antioxidant Determination using Stable Free Radicals. *Nature* **1985,** *26,* 1199–1200.
12. Brand-Williams, W.; Cuvelier, M. E; Berset, C. Use of a Free Radical Method to Evaluate Antioxidant Activity. *Lebensm. Wissu. Technol.* **1995,** *28*, 25–30.
13. Dai, J.; Mumper, R. J. Plant Phenolics: Extraction, Analysis, and Their Antioxidant and Anticancer Properties. *Molecules* **2010,** *15*, 7313–7352.

14. Elekwa, I.; Monanu, M. O.; Anosike, E. O. Effects of Aqueous Extracts of *Zanthoxylum Macrophylla* Roots on Membrane Stability of Human Erythrocytes of Different Genotypes. *Biokemistri* (Biochemistry) **2005,** *17* (1), 7–12.
15. Emebu, P. K.; Anyika, J. U. Proximate and Mineral Composition of Kale (*Brassica oleracea*) Grown in Delta State, Nigeria. *Pak. J. Nutr.* **2011,** *10* (2), 190–194.
16. Friedewald, W. T.; Levy, R. I.; Fredrickson, D. S. Estimation of the Concentration of Low-Density Lipoprotein Cholesterol in Plasma, Without the use of the Preparative Ultracentrifuge. *Clin. Chem.* **1972,** *18* (6), 499–502.
17. Harborne, J. B., Ed. *Phytochemical Methods.* Chapman and Hall, Ltd.: London, 1972; p 138.
18. Hošek, J.; Šmejkal, K. Flavonoids as Anti-inflammatory Agents. In *Encyclopedia of Inflammatory Diseases*; Parnham, M., Ed.; Springer: Basel, 2015; pp 1–17.
19. Iwu, M. M.; Obidoa, O.; Anazodo, M. Biochemical Mechanism of the Antimalarial Activity of *Azadirachta Indica* Leaf Extract. *Pharmacol. Res. Commun.* **1986,** *18* (1), 81–91.
20. Joanna, K. Introductory Chapter-Alkaloids—Their Importance in Nature and for Human Life. In *Alkaloids-Their Importance in Nature and Human Life*; Joanna, K., Ed.; IntechOpen Publishers, 2019; p 1–29.
21. Kalgaonkar, S.; Gross, H. B.; Yokoyama, W.; Keen, C. L. Effects of a Flavonol-Rich Diet on Select Cardiovascular Parameters in a Golden Syrian Hamster Model. *J. Med. Food.* **2010,** *13* (1), 108–115.
22. Kokwaro, J. O. *Medicinal Plants of East Africa*, 3rd ed.; University of Nairobi Press: Kenya, Nairobi, 2009; p 463.
23. Mayakrishnan, V.; Veluswamy, S.; Shanmuga, K.; Kannappan, P.; Abdullah, N. Free Radical Scavenging Potential of *Lagenariasicerari* (Molina) Standl Fruits Extracts. *Asia Pac. J. Trop. Med.* **2012,** *8*, 20–26.
24. Odetola, A. A.; Oluwole, F. S.; Adeniyi, B. A.; Olatiregun, A. M.; Ikupolowo, O. R.; Labode, O.; Busari, K. O.; Shorinola, J. A. Antimicrobial and Gastroprotective Properties of *Parquetinani Grescens*. *J. Biol. Sci.* **2006,** *6*, 701–705.
25. Owoyele, B. V.; Nafiu, A. B.; Oyewole, I. A.; Oyewole, L. A.; Soladoye, A. O. Studies on the Analgesic, Anti-Inflammatory, and Antipyretic Effects of *Parquetina nigrescens* Leaf Extract. *J. Ethnopharmacol.* **2011**, *122,* 86–90.
26. Ozaslan, M. *Parquetina nigrescens* Checks Ulceration and Oxidation. *Pakistan J. Biol. Sci.* **2011,** *14*, 1124–1125.
27. Romano, M.; Vitaglione, P.; Sellitto, S.; D'Argeio, G. Nutraceuticals for Protection and Healing of Gastrointestinal Mucosa. *Curr. Med. Chem.* **2012,** *19*, 109–117.
28. Roumelioti, M. E.; Glew, R. H.; Khitan, Z. J.; Rondon-Berrios, H.; Argyropoulos, C. P.; Malhotra, D.; Raj, D. S.; Agaba, E. I.; Rohrscheib, M.; Murata, G. H.; Shapiro, J. I.; Tzamaloukas, A. H. Fluid Balance Concepts in Medicine: Principles and Practice. *World J. Nephrol.* **2018,** *7* (1), 1–28.
29. Saba, A. B.; Oyagbemi, A. A.; Azeez, O. I. Antidiabetic and Hematinic Effects of *Parquetina nigrescens* on Alloxan-Induced Type-1 Diabetes and Normocytic Normochromic Anemia in Wistar Rats. *Afr. Health Sci.* **2010,** *10*, 276–282.
30. Souto, A. L.; Tavares, J. F.; Da Silva, M. S.; Diniz, M.; De Athayde-Filho, P. F.; Barbosa-Filho, J. M. Anti-inflammatory Activity of Alkaloids: an Update from 2000 to 2010. *Molecules (Basel, Switzerland).* **2011,** *16* (10), 8515–8534.
31. Van-Burden, T. P.; Robinson, W. C. Formation of Complexes between Protein and Tannin Acid. *J. Agric. Food Chem.* **1969,** *17* (4), 772–77.

CHAPTER 14

Role of Phytochemicals in *Allium sativum* on Obesity: A Review

Ayodeji Emmanuel Adepoju, Gregory Joseph,
Samuel Olufolarin Babarinde, and Ademola Olabode Ayeleso

ABSTRACT

The imbalance of energy metabolism over the last several decades has been a challenge for metabolic disorders. Obesity is caused by uncontrolled energy regulation, resulting in fat accumulation in the adipose tissue and the retraction of energy. Garlic *(Allium sativum)* phytoconstituents have been previously reported to regulate the process of energy metabolism. This review summarizes the implications of garlic phytoconstituents on energy homeostasis. Fructan polysaccharides, allicin, and a few polyphenols present in garlic are reviewed to better understand their health impacts on obesity. The metabolites of these phytoconstituents directly affect the organs responsible for energy homeostasis, as they activate and express the protein and genes responsible for satiation from the gut to the hypothalamus and thermogenesis from the adipose tissues.

14.1 INTRODUCTION

Garlic is a bulbous plant of central Asia origin of the Amaryllidaceae family.[3] It is widely used as a spice and for dressing in cooking.[46] The bulb has been used as a medicinal herb and used in traditional medicine across the world.[21,78] The stalk of garlic originates from the bulb and gives out white spherically arranged flowers; the bulb consists of a bulblet (cloves)

Curative and Preventive Properties of Medicinal Plants: Research on Disease Management and Animal Model Studies. Megh R. Goyal, Junaid Ahmad Malik, & Ademola Olabode Ayeleso (Eds.)

covered by white membranous scales.[46] Many studies have investigated the medicinal benefits of garlic.[49,59,96] Reduced risk for developing cardiovascular diseases and cancer, immunomodulatory, antifungal, antidote (for heavy metal poisoning), and hepatoprotective, anti-aging and antimicrobial potentials have been reported.[64,83,39] Garlic preparations, used as ethnomedicine, comprise garlic powder, garlic oil, extracts (aqueous and solvent) of raw or dried garlic, and aged garlic extracts. The medicinal potentials of garlic depend on the unique profile of the phytoconstituents[85,71] of garlic acting synergistically.

Obesity results from a complex interaction between the host genome and some factors contributing to an imbalance between energy intake and energy expenditure.[76] A total of 39% of adults above or equal to 18 years were overweight, and 13% of the world's adult population (1 out of 77 adults in the world) were obese,[90] as published on 1 April 2020 by WHO (https://www.who.int/news-room/fact-sheets/detail/obesity-and-overweight).

Obesity is linked with the risk of developing cardiovascular diseases,[62] diabetes, cancer[26], and musculoskeletal disorders, which result in increased mortality.[61] There are some emerging risk factors of obesity, including obstructive sleep apnea, gut microbiota, etc.[56,57] The obesity epidemic contributes to the reversal of the 20th century life expectancy gains. Sedentary lifestyles and intake of fat-rich and refined sugars diets[25] are the primary cause of the increased prevalence of obesity.[6]

Energy metabolism is mainly achieved in the fat body.[65] Animal fat tissues are of two types which include white adipose tissues (WAT), where the energy is stored, and brown adipose tissues (BAT), which work with energy dissipation[15,63,84] due to their involvement in thermoregulation stimulated by cold stress. Effective management of obesity over the years has been shown worthwhile. The treatments of obesity, including bariatric surgery, diet restrictions, and drug therapy, have displayed many limitations that at different times result in medical complications or the regain of weight.[91] Nutraceuticals are gaining much interest as they are used as adjunctive weight loss therapy.[9] Li et al.[44] reported that Intermittent Fasting promotes WAT browning by shaping the gut microbiota and exposure to cold temperatures. Still, to date, no effective strategies have been developed to increase BAT in adults substantially.

Garlic phytoconstituents have been researched to possess some biological activities and adjuvant therapy in some disease conditions. The review in this chapter was conceived from the thirst for expanding knowledge on the comparative impact of garlic phytoconstituents on obesity.

14.2 GARLIC AND ITS THERAPEUTIC POTENTIALS

The main chemicals in garlic include fructan carbohydrate and organo–sulfur (water and lipid-soluble) compounds such as S-allyl-cysteine (SAC), allicin, alliin, vinyldithiins, diallyl, ajoene, and poly-sulfides. Other garlic constituents include sterols and phenols such as saponins, flavonoids, and tannins.[1,23,49,50,87] The fructan–carbohydrate and organo–sulfur compounds are known chiefly for their typical smell and taste, indicative of their medicinal properties.[21] The alliin constituent of garlic (the amino acid derivative) is odorless; crushing the garlic releases the allinase enzyme that converts the alliin into allicin, further broken down into diallyl disulfide, which is responsible for the taste and smell of garlic.[93] The allyl group is added to the sulfur atom of cysteine amino acid to get another derivate of allicin, SAC. Research has established that SAC possesses antioxidant, anti-diabetic, anti-obesogenic, and anti-inflammatory properties.[7,72]

Garlic possesses medicinal potential, with its oral ingestion is proven to be safe both for rats and humans. Studies reported the clinical effectiveness of garlic as randomized, placebo-controlled. Double-blinded trials (Table 14.1), which revealed that garlic preparations significantly improve waist circumference and BMI in obesity,[5,73] reduce hypertension and hyperlipidemia in cardiovascular diseases,[77] and increase radical scavengers and normalize fasting blood sugar levels in diabetic patients.[29] However, reduced population size and short-term of garlic intervention in clinical trials make the efficacy of garlic stand to be questionable on a large scale and in long-term clinical trials.[29] Garlic might possess some side effects in chronic disease conditions related to bleeding, as it has been reported to interact with drugs like warfarin and aspirin.

14.3 PHYTOCONSTITUENTS OF GARLIC AND OBESITY

14.3.1 CARBOHYDRATE CONSTITUENT

Garlic (*Allium sativum* L.) has over 75% of its dry mass made up of carbohydrate,[35] with its main carbohydrate constituent being fructan-polysaccharide,[43,88,95] which possesses antioxidative, antiviral, probiotic potentials, among other medicinal benefits.[27,74,89] Fructans are carbohydrates that comprise fructose monomers with one D-glucose unit at its terminal.[68] Fructans are classified according to their varying degree of polymerization (DP) and monomeric bonding, with β-(2→1) fructosyl-fructose linkages

TABLE 14.1 Clinical Effectiveness of Garlic Preparations in Human Studies.

Sample size (Number of participants)	Garlic intervention	Duration	Outcome	Reference
		Obesity		
30	Garlic powder (6 g/day) once a day	8 weeks	Reduced BMI From 29.7 ± 3.7 kg/m² To 27. 2 ± 3.2 kg/m²	5
30	AGE packs (80 mL/pack) 5 times a week	12 weeks	Reduced BMI From 23.1 ± 3.1 kg/m² To 21. 8 ± 2.9 kg/m²	73
		Hyperlipidemia		
30	AGE packs (80 mL/pack) 5 times a week	12 weeks	Significantly lowered Triglycerides From 99.5 ± 39.6 mg/dL To 81.3 ± 20.8 mg/dL	73
50	3 g of raw garlic daily	90 days	Significantly reduced total cholesterol from 269.30 to 233.93 mg/dL	37
		Oxidative stress and diabetes		
25	Coated garlic powder tablets (900 mg with an alliin content of 1.3% and an allicin content of 0.6%)	2 months	Significant increase in reduced glutathione (GSH) concentration from 1.66 ± 0.84 to 2.32 ± 1.22	29
40	Garlic powder tablet allicor	4 weeks	Better metabolic control due to the lowering of FBS from 7.7 ± 0.6 mmol/L to 6.3 ± 0.4 mmol/L and serum fructosamine from 276 ± 16 µmol/L to 228 ± 14 µmol/L	77

Abbreviations: AGE – Aged garlic extract; BMI – Body mass index; FBS – Fasting blood sugar

(Figure 14.1a).[11,68] The classes include; Inulin-type fructan with short-chain composed of 2–9 DP, generally called fructooligosaccharides (FOS), and long-chain fructan (DP greater than 10) called inulins.[69]

Fructans are generally non-digestible by intestinal digestive enzymes as a result of the β-(2→1) linkage of fructose monomers, hence being called "non-digestible or fermentable or unavailable" carbohydrates.[21] Consumption of fructans has an advantage to human health, as they are being fermented in the colon and reduce the risk of metabolic diseases.[74] Undigestible carbohydrates have been reported to lower body fat storage, increase satiety and decrease food intake. The fermentation of the fructans by gut microbiota results in heat, gases, and short-chain fatty acid (SCFA) production.[33,48]

14.3.2 FRUCTAN–CARBOHYDRATE METABOLITE OF GARLIC AND OBESITY

Short-chain fatty acids are widely crucial in the pathophysiology of obesity via regulating energy intake.[45] The SCFAs are a colonic metabolite of fructans, consisting of acetate, propionate, and butyrate (Figures 14.1c–1e, respectively). Other fructans fermentation products may include lactate, succinate, and ethanol,[48] which are intermediates of carbohydrates fermentation by the microbiota. The production of SCFAs in the distal colon is in the ratio of 60:20:20 for acetate, propionate, and butyrate, respectively.[31]

Jorgensen et al.[35] reported that the total SCFAs concentration ranges between 70 and 130 $mmol^{-1}$. Increasing SCFAs synthesis creates an acidic environment in the gut, used as a defense mechanism in colonization resistance against pathogens.[13] There is an equal rate of absorption of major SCFAs in different colonic regions.[35,80] The SCFA metabolism occurs in the following three major sites of the body: Butyrate is used by the cells of the cecum–colon epithelium as a primary substrate for energy, and butyrate and propionate enter gluconeogenesis in the liver cells, the liver also takes up 50–70% of acetate for metabolism, and the muscle cells oxidizes residual acetate to generate energy.[31]

The SCFAs affect hepatic glucose and lipid metabolism. Acetate and butyrate increase gluconeogenesis and decrease lipogenesis, while propionate acts as a substrate for gluconeogenesis alone.[22,34] The SCFAs is involved in the expression of gut hormones; Peptide YY (PYY) and glucagon-like peptide-1 (GLP-1) mediated via G protein-coupled receptors FFA 2 and FFA 3.[13] The SCFAs have been associated with increased gut hormones, thereby down-regulating energy intake.

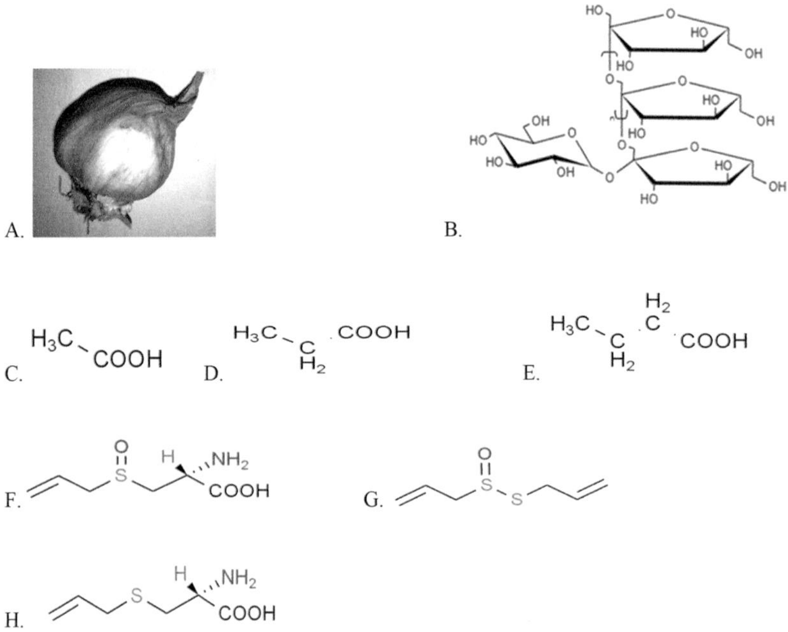

FIGURE 14.1 Chemical structures of biocompounds in garlic: A. Garlic bulb; B. β-(2→1) Frucosyl-fructose linked fructan polysaccharide (inulin); C. Acetate; D. propionate; E. butyrate; F. Allin; G. Allicin; H. S-allyl cysteine.

McNelis et al.[52] reported SCFA propionate to have a high affinity for FFA 2 receptor and protect animals fed with a high-fat diet from obesity and insulin resistance. Also, GLP-1 activates insulin secretion and decreases glucagon production in the pancreas,[45,86] while PYY increases glucose uptake in muscle and adipose tissue.[60,66] These gut peptides have been established to affect appetite and satiety via their effects on neuropeptide Y (NPY) and proopiomelanocortin (POMC), respectively, in the arcuate nucleus of the hypothalamus and also inhibiting gastric emptying and gastric acid secretion.[12,53]

Hernandez et al.,[32] reported the correlation between SCFA–FFA 3 and the plasma leptin in the adipose tissues, resulting in food intake regulation and SCFA-FFA 2 mediating insulin signaling, adipogenesis, and contributing to increased energy expenditure by promoting the activity of sympathetic neurons.[19,20] Leptin increases lipolysis by upregulating the sympathetic efferent signal to BAT[14,36] and WAT.[30] SCFAs, also regulate the proliferation of insulin biosynthesis and pancreatic β cells via FFA 2 or FFA 3.[19]

This role of SCFAs and GLP-1 of the entero-endocrine L-cells is found to be contrasting in irritable bowel syndrome (IBS) patients, as they are managed with a low FODMAP (fermentable oligosaccharides, disaccharides, monosaccharide, and polyols) diet, which suppresses the production of SCFAs and its secretagogues (PYY and GLP-1).[10]

14.3.3 AMINO ACID CONSTITUENTS

Garlic oil contains the amino acid alliin ($C_6H_{12}NO_3S$) (Figure 14.1f), which is odorless and can be converted to a compound called allicin.[8] Alliin (S-allyl cysteine sulfoxide) first was identified in 1947.[67,82] Allicin (Figure 14.1g), a sulfinyl compound that gives garlic its pungent odor and flavor, was discovered in 1944 by Cavallito et al.,[16] and noted for its potent antimicrobial activity. Allicin begins to break down quickly, especially if heated.[38] When raw garlic is either crushed or injured, there is a release of allinase that enzymatically converts allicin.[4] However, previous studies have suggested that allicin is unstable, and responsible for "garlic breath" which can cause adverse effects to the gastrointestinal tract lumen and has low bioavailability.[81] Allicin induces WAT and activates BAT, leading to the inhibition of obesity and maintaining energy homeostasis [energy expenditure (EE) and fat hypertrophy].[41] The stable form of garlic amino acid is SAC (Figure 14.1h). S-Allyl-cysteine is formed from γ-glutamyl-S-allyl cysteine catabolism by γ-glutamyl transferase and standardized commercial AGE.[18]

14.3.4 AMINO ACID METABOLITE OF GARLIC AND OBESITY

S-allyl-cysteine can be absorbed quickly in the gastrointestinal tract after oral administration without any changes. S-allyl-cysteine has been shown to possess various health benefits, including anticancer, antioxidant, antihyperlipidaemic, and anti-diabetic activity.[17,42] S-allyl-cysteine induced the expression of genes responsible for heat production/ thermogenesis and energy expenditure (UCP1, UCP2, UCP3, PRDM16, PGC1α/β, CPT1α, and MCAD) in the BAT.[24,47,79]

Zhang et al.[94] reported OXPHOS proteins' upregulation (ATP5A, UQCRC2, SDHB, and NDUFB8) post-allicin treatment. The AMP-activated protein kinase (AMPK) activity is being stimulated in BAT, Liver, WAT, and skeletal muscle,[55,92] as it relates to energy homeostasis. The thermogenic effect increases the rate of lipolysis, liberating fatty acids used as substrates for mitochondria electron transport, thereby utilizing accumulated energy (in the form of fats) in the adipose tissue to generate ATP.[41]

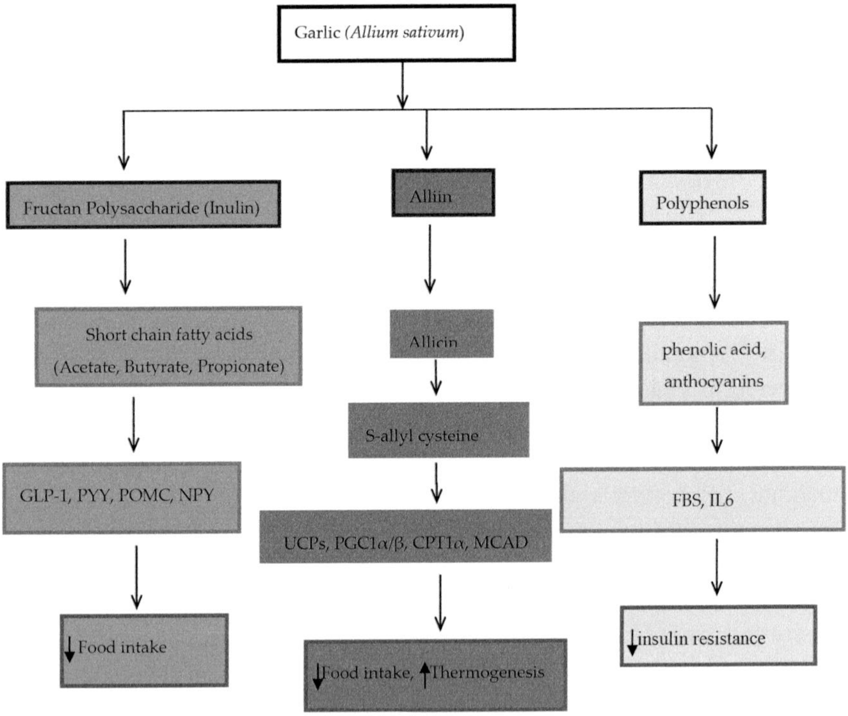

FIGURE 14.2 Chart illustrating the metabolic pathways of garlic phytoconstituents. Polysaccharides, amino acids, and antioxidants in pink, blue, and lilac-filled boxes, respectively. The phytoconstituents are represented in brown boxed outline, metabolites in blue boxed outline, genes/proteins expressed in red boxed outline, and biological effects in green boxed outline.

LEGEND: GLP-1–Glucagon-likepeptide-1; PYY–Peptide YY, POMC–Proopiomelanocortin; NPY–Neuropeptide Y; UCPs–Uncoupling proteins; PGC1α/β - proliferator-activated receptor gamma coactivator 1-alpha/beta; CPT1α–Carnitine palmitoyltransferase 1-alpha; TNFα–Tumor necrotic factor-alpha, IL-6–Interleukin 6.

Arrows: decreased, increased

14.3.5 POLYPHENOLS IN GARLIC AND OBESITY

Polyphenols also play a critical part in the health properties that garlic possesses. Some of the phenolic compounds present in garlic are phenolic acids[28,87] and anthocyanins.[2] Different classes of polyphenols identified in garlic bulb extract include flavonols (myricetin, quercetin, and kaempferol), hydroxycinnamic acid (caffeic acid, p-Coumaric acid, and ferulic acid), and hydroxybenzoic acid (gallic acid, protocatechuic acid, β-resorcylic acid, vanillic acid, and syringic acid).[39,58] Other phenolic compounds identified in garlic include; pyrogallol, rutin, vanillin, veratric acid, hesperidin, resveratrol, naringenin, biochanin, and formononetin.[58] The main anthocyanins detected in garlic skin was cyaniding-3-(6-malonyl)-glucoside.[23] Polyphenols possess strong biological activities. Polyphenols inhibit digestive enzymes implicated in the breakdown of lipids and carbohydrates and positively affect calorie intake, obesity, and blood glucose regulation.[51] Garlic polyphenols lower the blood glucose level either by preventing beta-cell damage, reducing insulin resistance (in the regulation of energy metabolism) (Figure 14.2), secretion and suppressing the activities of glucosidase enzymes.[54]

14.4 SUMMARY

The consumption of Garlic (*A. sativum* Linn) as food, condiments, flavoring/spice, and medicinal herb over the years has given positive results in suppressing inflammatory responses, activating thermogenesis, and regulating energy homeostasis. These potentials are peculiar for specific phytoconstituents. There is a need for further research on how garlic phyto-constituents regulate energy expenditure and intake.

KEYWORDS

- **allicin**
- **fructan**
- **garlic**
- **obesity**
- **polyphenol**

REFERENCES

1. Amagase, H. Clarifying the Real Bioactive Constituents of Garlic. *J. Nutr*. **2006,** *136*, 716S–725S.
2. Andersen, T. F. Malonated Anthocyanins of Garlic *Allium sativum* L. *Food Chem.* **1997,** *58*, 215–217.
3. *Angiosperm Phylogeny Group* (APG). An Update of the Angiosperm Phylogeny Group Classification for the orders and Families of Flowering Plants: AGP III. *Biol. J. Linn. Soc. Lond.* **2009,** *161*, 105–121.

4. Ankri, S.; Mirelman, D. Antimicrobial Properties of Allicin From Garlic. *Microbes Infect.* **1999,** *1*, 125–129.
5. Aslani, N.; Entezari, M. H; Askari, G.; Maghsoudi, Z.; Maracy, M. R. Effect of Garlic and Lemon Juice Mixture on Lipid Profile and Some Cardiovascular Risk Factors in People 30–60 Years Old with Moderate Hyperlipidaemia: A Randomized Clinical Trial. *Int. J. Prev. Med.* **2016,** *3*, 1–8.
6. Asrih, M.; Jornayvaz, F. R. Metabolic Syndrome and Nonalcoholic Fatty Liver Disease: Is Insulin Resistance the Link? *Mol. Cell Endocrinol.* **2015,** *418* (1), 55–65.
7. Ayeleso, A. O.; Lembede, B. W.; Nyakudya, T. T.; Adepoju, A. E.; Chegou, N. N.; Mukwevho, E. Administration of S-allyl Cysteine to Neonatal Rats Modulates Inflammatory Biomarkers in High-Fructose-Fed Rats in Adulthood. *Trop. J. Pharm. Res.* **2020,** *19* (5), 1053–1058.
8. Bagudo, B. U.; Acheme, O. D. Chemical Analysis of Garlic and it's Oil. *Der Chem. Sin.* **2014,** *5* (1), 128–134.
9. Barrea, L.; Altieri, B.; Polese, B.; De Conno, B.; Muscogiuri, G.; Colao, A. Nutritionist and Obesity: Brief Overview on Efficacy, Safety and Drug Interactions of the Main Weight-Loss Dietary Supplements. *Int. J. Obes. Suppl.* **2019,** *9* (1), 32–49.
10. Barrea, L.; Muscogiuri, G.; Annunziata, G.; Pugliesse, G.; Salzano, C.; Colao, A. From Gut Microbiota Dysfunction to Obesity: Could Short Chain Fatty Acids Stop This Dangerous Course. *Hormones* (*Athens*). **2019,** *18* (3), 245–250.
11. Baumgartner, S.; Thomas, G.; Dax, W. Characterisation of the High-Molecular Weight Fructan Isolated from Garlic (*Allium sativum* L.). *Carbohydr. Res.* **2000,** *328*, 177–183.
12. Beck, B. Neuropeptide-Y in Normal Eating and in Genetic and Dietary - Induced Obesity. *Philos. Trans. R. Soc. Lond. B Biol. Sci*. **2006,** *361*, 1159–1185
13. Byrne, C. S.; Chambers, E. S.; Morrison, D. J.; Frost, G. The Role of Short Chain Fatty Acids in Appetite Regulation and Energy Homeostasis. *Int. J. Obes.* **2015,** *39*, 1331–1338.
14. Caron, A.; Lee, S.; Elmquist, J. K.; Gautron, L. Leptin and Brain–Adipose Crosstalks. *Nat. Rev. Neurosci*. **2018,** *19* (3), 153–165.
15. Carpentier, R. C.; Blondin, D. P.; Virtanen, K. A.; Richard, D.; Haman, F.; Turcotte, E. E. Brown Adipose Tissue Energy Metabolism in Humans. *Front. Endocrinol.* **2018,** *9*, 447–452.
16. Cavallito, C. J.; Bailey, J. H. Allicin, the Antibacterial Principle of *Allium sativum*, Part I: Isolation, Physical Properties and Antibacterial Action. *J. Am. Chem. Soc.* **1944,** *66*, 1950–1951.
17. Chan, J. Y.; Yuen, A. C.; Chan, R. Y.; Chan, S. W. Review of the Cardiovascular Benefits and Antioxidant Properties of Allicin. *Phytother. Res. (PTR)* **2013,** *27*, 637–646.
18. Chen, Y. Y.; Lo, H. F.; Wang, T. F.; Lin, M. G.; Lin, L. L.; Chi, M. C. Enzymatic Synthesis of Γ-L-Glutamyl-S-Allyl-L-Cysteine: Naturally Occurring Organosulfur Compound from Garlic, by *Bacillus licheniformis* Γ-Glutamyltranspeptidase Enzyme. *Microb. Technol.* **2015,** *75*, 18–24.
19. Cignarelli, A.; Genchi, A. V.; Perrini, S.; Natalicchio, A.; Laviola, L.; Giorgino, F. Insulin and Insulin Receptors in Adipose Tissue Development. *Int. J. Mol. Sci.* **2019,** *20* (3), 759.
20. Colina, C.; Puhl, H. L.; Ikeda, S. R. Selective Tracking of FFAR3-Expressing Neurons Supports Receptor Coupling to N-Type Calcium Channels in Mouse Sympathetic Neurons. *Sci. Rep.* **2018,** *8*, 17379.
21. Corzo-Martìnez, M.; Corzo, N.; Villaiel, M. Biological Properties of Onions and Garlic. *Trends Food Sci. Technol.* **2007,** *18*, 609–625.

22. den Besten, G.; Lange, K.; Havinga, R.; van Dijk, T. H.; Gerding, A.; van Eunen, K. Gut Derived Short-Chain Fatty Acids are Vividly Assimilated into Host Carbohydrates and Lipids. *Am. J. Physiol. Gastrointest. Liver Physiol.* **2013,** *305*, G900–G910.
23. Du, C. T.; Francis, F. J. Anthocyanins of Garlic (*Allium sativum* L.). *J. Food Sci.* **1975,** *40* (5), 1101–1102
24. El-Sheakh, A. R.; Ghoneim, H. A.; Suddek, G. M.; Ammar, E. S. M. Attenuation of Oxidative Stress, Inflammation, and Endothelial Dysfunction in Hypercholesterolemic Rabbits by Allicin. *Can. J. Physiol. Pharmacol.* **2016,** *94*, 216–224.
25. Ferreira, Z.; Ayeleso, A.; Mukwevho, E. Control of Carbohydrate and Lipid Metabolism by NRF-1 and Sirtuins: Implications on Type 2 Diabetes and Obesity. *Chem. Biol. Lett.* **2014,** *1* (2), 66–76.
26. Gallagher, E. J.; LeRoith, D. Obesity and Diabetes: The Increased Risk of Cancer and Cancer-Related Mortality. *Physiol. Rev.* **2015,** *95* (3), 727–748.
27. Gibson, G. R.; Hu, X.; Li, P.; Lei, Z. (Eds.) *Development Application of Probiotics*; Chemical Industry Press: Beijing, China, 2008; pp 1–18.
28. Gorinstein, S.; Leontowicz, H.; Leontowicz, M.; Namiesnik, J.; Najman, K.; Drzewiecki, J. Comparison of the Main Bioactive Compounds and Antioxidant Activities in Garlic and White and Red Onions after Treatment Protocols. *J. Agric. Food Chem.* **2008,** *56*, 4418–4426.
29. Grune, T.; Scherat, T.; Behrend, H.; Conradi, E.; Brenke, R.; Siems, W. Influence of *Allium sativum* on Oxidative Stress Status a Clinical Investigation. *Phytomedicine* **1996,** *2*, 205–207.
30. Harms, M.; Seale, P. Brown and Beige Fat: Development, Function and Therapeutic Potential. *Nat. Med.* **2013,** *19*, 1252–1263.
31. Havenaar R. Intestinal Health Functions of Colonic Microbial Metabolites: A Review. *Benef. Microbes*. **2011,** *2*, 103–114.
32. Hernández, M. A. G.; Canfora, E. E.; Jocken, J. W. E.; Blaak, E. E. The Short-Chain Fatty Acid Acetate in Body Weight Control and Insulin Sensitivity. *Nutrients* **2019,** *11* (8), 1943.
33. Holmes, E.; Li, J. V.; Marchesi, J. R.; Nicholson, J. K. Gut Microbiota Composition and Activity in Relation to Host Metabolic Phenotype and Disease Risk. *Cell Metabol.* **2012,** *16*, 559–564.
34. Jin, E. S.; Szuszkiewicz-Garcia, M.; Browning, J. D.; Baxter, J. D.; Abate, N.; Malloy, C. R. Influence of Liver Triglycerides on Suppression of Glucose Production by Insulin In Men. *J. Clin. Endocrinol. Metab.* **2014,** *100*, 235–243.
35. Jørgensen, H.; Larsen, T.; Zhao, X. Q.; Eggum, B. O. The Energy Value of Short-Chain Fatty Acids Infused into the Caecum of Pigs. *Br. J. Nutr.* **1997,** *77*, 745–756.
36. Labbe, S. M.; Caron, A.; Lanfray, D.; Monge-Rofarello, B.; Bartness, T. J.; Richard, D. Hypothalamic Control of Brown Adipose Tissue Thermogenesis. *Front. Syst. Neurosci.* **2015,** *9*, 150–155.
37. Lachhiramka, P.; Patil, S. Cholesterol Lowering Property of Garlic (*Allium sativum*) on Patients with Hypercholesterolemia. *Int. J. Med. Sci. Public Health.* **2016,** *5*, 2249–2251.
38. Lai, K. C.; Hsu, S. C.; Kuo, C. L.; Yang, J. S.; Ma, C. Y.; Lu, H. F. Diallyl Sulfide, Diallyl Disulfide, and Diallyl Trisulfide Inhibit Migration and Invasion in Human Colon Cancer Colo 205 Cells Through the Inhibition of Matrix Metalloproteinase-2, -7, and -9 Expressions. *Environ. Toxicol.* **2013,** *28* (9), 479–88.
39. Lanzotti, V. The Analysis of Onion and Garlic. *J. Chromatogr.* **2006,** *1112*, 3–22.

40. Lee, C. G.; Rhee, D. K.; Kim, B. O.; Um, S. H.; Pyo, S. Allicin Induces Beige-like Adipocytes via KLF15 Signal Cascade. *J. Nutr. Biochem.* **2019,** *64*, 13–24.
41. Lee, J. H.; Park, A.; Oh, K. J.; Lee, S. C.; Kim, W. K.; Bae, K. H. The Role of Adipose Tissue Mitochondria: Regulation of Mitochondrial Function for the Treatment of Metabolic Diseases. *Int. J. Mol. Sci.* **2019,** *20* (19), 4924–4930.
42. Lembede, B. W.; Erlwanger, K. H.; Nkomozepi, P.; Chivandi, E. Effect of Neonatal Orally Administered S-Allyl Cysteine in High-Fructose Diet Fed Wistar Rats. *J. Dev. Orig. Health Dis.***2018,** *9* (2), 160–171.
43. Li, C.; Li, S.; Liu, K. Study on Isolation Purification and Antioxidation of Garlic Polysaccharide. *J. Hebei Univ. Sci. Technol.* **2007,** *3*, 243–246.
44. Li, G.; Xie, C.; Lu, S.; Nichols, R. G.; Tian, Y.; Li, L.; Patel, D.; Ma, Y.; Brocker, C. N.; Yan, T.; Krausz, K. W.; Xiang, R.; Gavrilova, O.; Patterson, A. D.; Gonzalez, F. J. Intermittent Fasting Promotes White Adipose Browning and Decreases Obesity by Shaping the Gut Microbiota. *Cell Metabol.* **2017,** *26* (4), 672–685.
45. Li, X.; Shimizu, Y.; Kimura, I. Gut Microbial Metabolite Short-Chain Fatty Acids and Obesity. *Biosci. Microbiota Food Health.* **2017,** *36* (4), 135–140.
46. Lisciani, S.; Gambelli, L.; Durazzo, A.; Marconi, S.; Camilli, E.; Rossetti, C. Carbohydrates Components of Some Italian Local Landraces: Garlic (*Allium sativum* L.). *Sustainability* **2017,** *9*, 1922–1927.
47. Liu, X.; Zheng, Z.; Zhu, X.; Meng, M.; Li, L.; Shen, Y. Brown Adipose Tissue Transplantation Improves Whole-Body Energy Metabolism. *Cell Res.* **2013,** *23*, 851–854.
48. Lunn, J.; Buttriss, J. Carbohydrates and Dietary Fiber. *Nutr. Bull.* **2007,** *32*, 21–64.
49. Majewski, M. *Allium sativum*: Facts and Myths Regarding Human Health. *Rocz. Panstw. Zakl. Hig.* **2014,** *65* (1), 1–8.
50. Martins, N.; Petropoulos, S.; Ferreira, I. C. F. R. Chemical Composition and Bioactive Compounds of Garlic (*Allium sativum* L.) as Affected by Pre- and Postharvest Conditions: A Review. *Food Chem.* **2016,** *211*, 41–50.
51. *McDougall, G. J.; Kulkarni, N. N.; Stewart, D.* Berry Polyphenols Inhibit Pancreatic Lipase Activity *In Vitro*. *Food Chemi.* **2009,** *115*, *193*–199.
52. McNelis, J. C.; Lee, Y. S; Mayoral, R.; van der Kant, R.; Johnson, A. M. F.; Wollam, J. GPR43 Potentiates Β-Cell Function in Obesity. *Diabetes* **2015,** *64*, 3203–3217.
53. Miller, G. D. Appetite Regulation: Hormones, Peptides, and Neurotransmitters and Their Role in Obesity. *Am. J. Lifestyle Med.* **2019,** *13* (6), 586–601.
54. Mughal, M. H. Garlic Polyphenols: A Diet Based Therapy. *Biomed. J. Sci. Technol. Res.* **2019,** *15* (4), 96–106.
55. Muise, E. S.; Guan, H. P.; Liu, J.; Nawrocki, A. R.; Yang, X.; Wang, C. Pharmacological AMPK Activation Induces Transcriptional Responses Congruent to Exercise in Skeletal and Cardiac Muscle, Adipose Tissues and Liver. *PLoS ONE.* **2019,** *14* (2), e0211568.
56. Muscogiuri, G.; Barrea, L.; Annunziata, G.; Di Somma, C.; Laudisio, D.; Colao, A. Obesity and Sleep Disturbance: The Chicken or the Egg? *Crit. Rev. Food Sci. Nutr.* **2019,** *59* (13), 2158–65.
57. Muscogiuri, G.; Cantone, E.; Cassarano, S.; Tuccinardi, D.; Barrea, L.; Savastano, S. Obesity Programs of Nutrition, Education, Research and Assessment (OPERA) Group: Gut Microbiota: A New Path to Treat Obesity. *Int. J. Obes. Suppl.* **2019,** *9* (1), 10–19.
58. Nagella, P.; Thiruvengadam, M.; Ahmad, A.; Yoon, J. Y.; Chung, I. M. Composition of Polyphenols and Antioxidant Activity of Garlic Bulbs Collected from Different Locations of Korea. *Asian J. Chem.* **2014,** *26* (3), 897–902.

59. Nasiri, A.; Ziamajidi, N.; Abbasalipourkabir, R.; Goodarzi, M. T.; Saidijam, M.; Behrouj, H. Beneficial Effect of Aqueous Garlic Extract on Inflammation and Oxidative Stress Status in the Kidneys of Type 1 Diabetic Rats. *Indian J. Clin. Biochem.* **2017,** *32*, 329–336.
60. Natarajan, N.; Hori, D.; Flavahan, S.; Steppan, J.; Flavahan, N.; Berkowitz, D. Microbial Short Chain Fatty Acid Metabolites Lower Blood Pressure via Endothelial G- Protein-Coupled Receptor 41. *Physiolol. Genomics*. **2016,** *48*, 826–834.
61. Onyemaech, N. O. C.; Anyanwu, G. E.; Obikil, E. N.; Onwuasoigwe, O.; Nwankwo, O. E. Impact of Overweight and Obesity on the Musculoskeletal System Using Lumbosacral Angles. *Patient Prefer. Adherenc.* **2016,** *10*, 291–296.
62. Ortega, F. B.; Lavie, C. J.; Blair, S. N. Obesity and Cardiovascular Disease. *Circ. Res.* **2016,** *118*, 1752–1770.
63. Peirce, V.; Carobbio, S.; Vidal-Puig, A. The different Shades of Fat. *Nature* **2014,** *510*, 76–83.
64. Phan, A. D. T.; Netzel, G.; Chhim, P.; Netzel, M. E.; Sultanbawa, Y. Phytochemical Characteristics and Antimicrobial Activity of Australian Grown Garlic (*Allium sativum* L.) *Cultivars Foods* **2019,** *8* (9), 358–362.
65. Piaggi, P.; Vinales, K. L.; Basolo, A.; Santini, F.; Krakoff, J. Energy Expenditure in the Etiology of Human Obesity: Spendthrift and Thrifty Metabolic Phenotypes and Energy-Sensing Mechanisms. *J. Endocrinol. Invest.* **2018,** *41* (1), 83–89.
66. Priyadarshini, M.; Villa, S.; Fuller, M.; Wicksteed, B.; Mackay, C.; Alquier, T. An Acetate-Specific GPCR, FFAR2, Regulates Insulin Secretion. *Mol. Endocrinol.* **2015,** *29*, 1055–1066.
67. Quintero-Fabián, S.; Ortuño-Sahagún, D.; Vázquez-Carrera, M.; López-Roa, R. I. Alliin, a Garlic (*Allium sativum*) Compound, Prevents LPS-induced Inflammation in 3T3-L1 Adipocytes. *Mediators Inflamm.* **2013,** *2013*, 381815.
68. Ritsema, T.; Smeekens, S. Engineering Fructan Metabolism in Plants. *J. Plant Physiol.* **2003,** *160* (7), 811–820.
69. Roberfroid, M.; Gibson, G. R.; Hoyles, L.; McCartney, A. L.; Rastall, R.; Rowland, I. Prebiotic Effects: Metabolic and Health Benefits. *Br. J. Nutr.* **2010,** *104* (Suppl 2), S11–S63.
70. Roberfroid, M. Prebiotics: The Concept Revisited. *J. Nutr.* **2007,** *137* (3), 830S–837S.
71. Santhosha, S. G.; Jamuna, P.; Prabhavathi, S. N. Bioactive Components of Garlic and Their Physiological Role in Health Maintenance: A Review. *Food Biosci.* **2013,** *3*, 59–74.
72. Saravanan, G.; Ponmurugan, P. Ameliorative potential of S-allylcysteine: Effect on Lipid Profile and Changes in Tissue Fatty Acid Composition in Experimental Diabetes. *Exp. Toxicol. Pathol.* **2012,** *64*, 639–644.
73. Seo, D. Y.; Lee, S. R.; Kim, H. K.; Baek, Y. H.; Kwak, Y. S.; Ko, T. H. Independent Beneficial Effects of Aged Garlic Extract Intake with Regular Exercise on Cardiovascular Risk in Postmenopausal Women. *Nutr. Res. Pract.* **2012,** *6*, 226–231.
74. Shoaib, M.; Shehzad, A.; Omar, M.; Rakha, A.; Raza, H.; Sharif, H. R. Inulin: Properties, Health Benefits and Food Applications. *Carbohydr. Polym.*, **2016,** *147*, 444–454.
75. Shrestha, D. K.; Sapkota, H.; Baidya, P.; Basnet, S. Antioxidant and Antibacterial Activities of *Allium sativum* and *Allium cepa*. *Bull. Pharm. Res.* **2016,** *6*, 50–55.
76. Sivamaruthi, B. S.; Kesika, P.; Suganthy, N.; Chaiyasut, C. A Review on Role of Microbiome in Obesity and Antiobesity Properties of Probiotic Supplements. *BioMedical Research International*, **2019,** *2019*, e3291367.

77. Sobenin, I. A.; Nedosugova, L. V.; Filatova, L. V.; Balabolkin, M. I.; Gorchakova, T. V.; Orekhov, A. N. Metabolic Effects of Time-Released Garlic Powder Tablets in Type 2 *Diabetes Mellitus*: The Results of Double-Blinded Placebo-Controlled Study. *Acta Diabetol.* **2008,** *45*, 1–6.
78. Soforowa, A. *Medicinal Plants and Traditional Medicine in Africa*; Chichester John Willey & Sons: London, UK, 1993; p 256.
79. Song, N. J.; Chang, S. H.; Li, D. Y.; Villanueva, C. J.; Park, K. W. Induction of Thermogenic Adipocytes: Molecular Targets and Thermogenic Small Molecules. *Exp. Mol. Med.* **2017,** *49*, 353–360.
80. Souza da Silva, C.; Bolhuis, J. E.; Gerrits, W. J.; Kemp, B.; Van den Borne, J. J. Effects of Dietary Fibers with Different Fermentation Characteristics on Feeding Motivation in Adult Female Pigs. *Physiol. Behav.* **2013,** *110*, 148–157.
81. Srivastava, R. A. K.; Pinkosky, S. L.; Filippov, S.; Hanselman, J. C.; Cramer, C. T.; Newton, R. S. AMP-Activated Protein Kinase: An Emerging Drug Target to Regulate Imbalances in Lipid and Carbohydrate Metabolism to Treat Cardio-Metabolic Diseases. *J. Lipid Res.* **2012,** *53* (12), 2490–2514.
82. Stoll, A.; Seebeck, E. *Alliin, die genuine Muttersubstanz des Knoblauchöls* (Allin: The Genuine Product of Garlic). *Experientia* **1947,** *3* (3), 114–115.
83. Tagoe, D. N. A.; Nyarko, H. D.; Akpaka, R. A comparison of Antifungal Properties of Onion (*Allium cepa*), Ginger (*Zingiber officinale*) and Garlic (*Allium sativum*) against *Aspergillus flavus*, *Aspergillus niger* and *Cladosporium herbarum*. *Res. J.Med. Plants.* **2011,** *5*, 281–287.
84. Townsend, K.; Tseng, Y. H. Brown Adipose Tissue: Recent Insights into Development, Metabolic Function and Therapeutic Potential. *Adipocyte* **2012,** *1* (1), 13–24.
85. Ugwu, C. E.; Suru, S. M. The Functional Role of Garlic and Bioactive Components in Cardiovascular and Cerebrovascular Health: What do We Know? *J. Biosci. Med.* **2016,** *4*, 28–42.
86. Vinke, P. C.; El Aidy, S.; van Dijk, G. The Role of Supplemental Complex Dietary Carbohydrates and Gut Microbiota in Promoting Cardiometabolic and Immunological Health in Obesity: Lessons from Healthy Non-Obese Individuals. *Front. Nutr.* **2017,** *4*, 34–40.
87. Vlase, L.; Parvu, M.; Parvu, E. A.; Toiu, A. Chemical Constituents of Three *Allium* Species from Romania. *Molecules* **2013,** *18*, 114–127.
88. Wang, W.; Huang, X.; Zeng, L. Advance on the Polysaccharide of Garlic (*Allium sativum* L.). *Guangzhou Food Sci. Technol.* **2004,** *4*, 144–146.
89. Whelan, K.; Abrahmsohn, O.; David, G. J. P.; Staudacher, H.; Irving, P.; Lomer, M. R. C. Fructan Content of Commonly Consumed Wheat, Rye and Gluten-Free Breads. *Int. J. Food Sci. Nutri.* **2011,** *62* (5), 498–503.
90. WHO (World Health Organisation). *Factsheet on Obesity*; 2015; p 12.
91. Wolfe, B. M.; Kvach, E.; Eckel, R. H. Treatment of Obesity: Weight Loss and Bariatric Surgery. *Circ. Res.* **2016,** *118* (11), 1844–1855.
92. Wu, L.; Zhang, L.; Li, B.; Jiang, H.; Duan, Y.; Xie, Z. AMP-Activated Protein Kinase (AMPK) Regulates Energy Metabolism through Modulating Thermogenesis in Adipose Tissue. *Front. Physiol.* **2018,** *9*, 122-127.
93. Yi, L.; Su, Q. Molecular Mechanisms for the Anti-Cancer Effects of Diallyl Disulphide. *Food Chem. Toxicol.* **2013,** *57*, 362–370.

94. Zhang, C.; He, X.; Sheng, Y.; Xu, J.; Yang, C.; Zheng, S.; Liu, J.; Li, H.; Ge, J.; Yang, M.; Zhai, B. Allicin Regulates Energy Homeostasis through Brown Adipose Tissue. *Iscience* **2020,** *23* (5), 101–113.
95. Zheng, Y.; Wang, H.; Guo, G. Effect of Garlic Polysaccharide on Cell Proliferation and Antioxidant Activities in PC12 Cells. *J. Jinan Univ. (Med. Ed.)* **2008,** *29*, 110–114.
96. Ziamajidi, N.; Behrouj, H.; Abbasalipourkabir, R.; Lotfi, F. Ameliorative Effects of *Allium Sativum* Extract on INOS Gene Expression and NO Production in Liver of Streptozotocin + Nicotinamide-Induced Diabetic Rats. *Indian J. Clin. Biochem.* **2018,** *33* (2), 147–153.

CHAPTER 15

Antidiabetic Potency of Tofu (Soybean Meal)

Modupe Olusola Adetayo, Maria Onomaghuan Ebesunun, Olugbenga Obajimi Adebawo, Adekunle Moses Adetayo, Ifeoluwa Opeyemi Alao-Owunna and Samuel Adegoke Adewole

ABSTRACT

This study investigated the effect of *Tofu* diet on lipoproteins and antioxidant status of alloxan-induced diabetic rats. Wistar rats (n = 28) were divided into four groups consisting of nondiabetic rats fed on rat chow (Group I), nondiabetic rats fed on *Tofu* diet (Group II), diabetes-induced rats fed on rat chow (Group III), and diabetes-induced rats that were fed on *Tofu* diet (Group IV). Diabetes was induced through intraperitoneal injection of alloxan (140 mg/kg body weight). By the end of 7 weeks of feeding, the animals were sacrificed. Blood, liver, brain, and heart organs were assayed for lipoproteins, lipid peroxidation, and catalase activity. Data obtained were analyzed using standard methods. Animals fed on *Tofu* had significantly lowered ($p < 0.05$) glucose, total cholesterol, triglyceride, lipid peroxidation levels, as well as significantly increased ($p < 0.05$) high-density lipoprotein cholesterol and catalase activity levels than the control groups. No significant difference was observed in the low-density lipoprotein cholesterol levels of all groups. Consumption of *Tofu* demonstrated effectiveness in reducing blood glucose levels, controlling lipid metabolism, reducing lipid peroxidation, and increasing antioxidant enzyme activity.

Curative and Preventive Properties of Medicinal Plants: Research on Disease Management and Animal Model Studies. Megh R. Goyal, Junaid Ahmad Malik, & Ademola Olabode Ayeleso (Eds.)

15.1 INTRODUCTION

Diabetes mellitus is a metabolic disease that is characterized by the high level of glucose in the blood and may arise as a result of a defect either in the secretion of insulin, or action of insulin or both.[14,44] There are two major types of diabetes that are clinically recognized; Type 1 and Type 2.[45] Type 1 diabetes, also known as insulin-dependent diabetes mellitus (IDDM), is caused by autoimmune destruction of β-cells which eventually leads to deficiency of insulin. Type 2 diabetes, also called non-IDDM, results from progressive loss of β-cell secretion of insulin commonly on the basis of resistance to insulin.[3]

Available statistics of 2014 by the World Health Organization reported that there are about 422 million people with diabetes mellitus worldwide and about 1.6 million deaths in the year 2016 were caused by diabetes.[44] There is an increasing prevalence of diabetes in the middle and low-income countries and it has been projected that 35.9% of the US population will be diagnosed with diabetes by the year 2050.[7] The cure for diabetes is still elusive and the main management strategy is focused on modification of the diet.

Tofu, also known as bean curd, is a soft white food that is prepared by soy milk coagulation to produce curds that are then pressed into blocks. It originates from China and is also a part of the cuisine of East Asia and Southeast Asia.[37] *Tofu* is rich in protein and some B-vitamins and also contains isoflavones as found in all whole soya products.[6]

Research into the nutritional composition of *Tofu* showed that it is low in fat, low in cholesterol, and low in calorie. In addition to being a great source of protein, *Tofu* has been found to be safe for the heart health and has been linked to a reduction in the risk of cancer.[32] It has been reported that when soy protein is taken regularly, it may help to decrease the levels of total cholesterol (TC), lower low-density lipoprotein (LDL) cholesterol, triglyceride, decrease the formation of blood clots by the platelets, and possibly increase HDL cholesterol levels.[6]

Oboh[26] showed that steep water, used as a coagulant in the preparation of *Tofu*, significantly modulated its hypercholesterolemic effect. *Tofu* has been found to be a healthy meal because of its numerous nutritional components[35] which may be a promising diet for the control of the diabetic incidence in humans.[32] Findings by Mueller et al.[25] reported that unsweetened soy foods and isoflavones play a protective role by reducing the risk of type 2 diabetes.

Tofu, a low-fat and a high protein diet, may cause a significant reduction in blood LDL cholesterol and may invariably reduce the complications associated with diabetes mellitus.[37] Li et al.[23] suggested that the products and constituents of soy including soy protein and soy isoflavones may play an important role in the prevention of type 2 diabetes mellitus.

Thus, the dietary supplementation of affected individuals on a diet naturally low in carbohydrate composition and rich in vitamins and proteins may be effective in reducing the associated complications of diabetes. However, in Nigeria, only a very few studies have been conducted on the beneficial effect of *Tofu* in the management of diabetes mellitus. Most of the studies highlight the effect of soybeans generally on biochemical parameters in diabetes mellitus.

This chapter investigated the effects of *Tofu* (bean curd) on blood glucose, lipid profile, lipid peroxidation, and activity of catalase in alloxan-induced diabetic rats.

15.2 EXPERIMENTAL METHODS AND MATERIALS

15.2.1 REAGENTS AND CHEMICALS

Reagents and chemicals of analytical grade were used in the study. Alloxan monohydrate, trichloroacetic acid, and thiobarbituric acid (TBA) were procured from Sigma-Aldrich, Mumbai, India. High-density lipoprotein, triglyceride, and TC assay kits were purchased from Randox Laboratories Ltd., West Virginia, USA.

15.2.2 EXPERIMENTAL ANIMALS

Male Wistar albino rats aged 4 weeks, weighing 80–120 g, were purchased from the Animal House at the University of Ibadan, they were kept in clean, well-aerated, and well-illuminated battery cages. The albino rats were fed ad libitum with commercial rat chow, purchased from Kayus Animal Feeds, Iperu, Ogun State. Water was also made available *ad libitum*. The rats were made to acclimatize for a period of 2 weeks after which they were divided into four groups.

15.2.3 INDUCTION OF DIABETES MELLITUS

Diabetes was induced, in groups III and IV, by intraperitoneal injection of the freshly prepared aqueous solution of 140 mg/kg body weight alloxan monohydrate. Blood was extracted from the tail vein, after 48 h, for glucose analysis using the Accuchek™ Glucometer, model 35NZ, manufactured by Roche Diabetes Care, Mannheim, Germany. Rats that had fasting blood glucose ≥ 200 mg/dL were considered diabetic while rats that had blood glucose <200 mg/dL were reinjected pending the time when were found to be diabetic.

15.2.4 CONTROL AND EXPERIMENTAL DIETS

Commercial rat chow served as the control diet while the experimental diet was compounded with *Tofu.* This was prepared, using the method of Chang et al.,[11] by soaking 1.0 kg of soybean in 6 L of water for 9 h at 27–32°C. The soaked beans were drained and ground in an electric grinder (Philip, HR 2115, Eindhoven, Netherlands). A little amount of water was added to mix to homogeneity. The soymilk which was produced was then heated to 98°C, and 20 mM hydrated potassium aluminum sulfate (potassium alum) was added to the mixture as a coagulant. This was then cooled to 87°C, soymilk and mixed with a coagulant, and allowed to stand for 10 min. The resulting bean curd was then collected into a fine sieve cloth and pressed to remove any excess water. The *Tofu* formed was then stored in water at 4°C overnight before analysis commenced. The *Tofu* produced was compounded with other dietary requirements as shown in Table 15.1, to make *Tofu* diet which was fed to the animals in the experimental group.

TABLE 15.1 Preparation of the Control and Experimental Diets.

Diet	Normal (Control) diet		Tofu (Experimental) diet		Percentage composition (%)
	Source	Amount (grams)	Source	Amount (g)	
Carbohydrate	Corn flour	6700	Corn meal	6700	6.70
Protein	Fish meal	1600	*Tofu*	1600	1.60
Vitamin	Vitamin premix	110	Vitamin premix	110	0.11
Mineral	Mineral mixture	400	Mineral mixture	400	0.40
Fat	Sunola groundnut oil	400	Sunola groundnut oil	400	0.40
Dietary fiber	Oyster shell	400	Oyster Shell	400	0.40
Sweetener	Granulated sugar	390	Granulated sugar	390	0.39

NOTE: Each diet was prepared in 10,000 g (or 10 kg).

15.2.5 FEEDING REGIMEN

The animals were fed ad libitum, every morning and evening and this lasted for a period of 7 weeks. Water was made available when desired. The feeding regimen is shown as follows:

Group I: Normal rats fed on rat chow.
Group II: Normal rats fed on *Tofu.*
Group III: Diabetes-induced rats fed on rat chow.
Group IV: Diabetes-induced rats fed on *Tofu.*

15.2.6 WEIGHT GAIN OF EXPERIMENTAL ANIMALS

Animal weights were recorded at the commencement and completion of the study period, using an electronic scale. The average weight gain of the animals was obtained by subtracting the initial weight (starting weight) from the weight at the end of the study period. This represented the average weight gain of each group at the end of the study.

15.2.7 SACRIFICING OF EXPERIMENTAL ANIMALS

At the end of 7 weeks of feeding, the rats were allowed to fast overnight for 10–12 h and sacrificed following anesthesia in a box containing cotton wool soaked in diethyl ether. Blood samples of the rats were collected by cardiac puncture into plain sample bottles and the rats were then sacrificed. The brain and heart organs were excised carefully and placed in a petri dish. The organs were weighed and the result recorded. Organ excision was done in a 19.8 L Polar Bag cold box. Each organ was immediately homogenized in 0.01M phosphate buffer (1 g tissue in 5 mL phosphate buffer pH 7.4). The suspension was centrifuged for exactly 30 min at a speed of 15,000 g and temperature of 4°C, the supernatant was pipetted into fresh test tubes and kept in the refrigerator for other analyses. These were analyzed for lipid peroxidation and catalase activity while lipid profile test (TC, triglyceride, high-density lipoprotein, and low-density lipoprotein cholesterol) was carried out on serum samples.

15.2.8 LIPID PROFILE TEST

A lipid profile test was conducted for serum TC, triglycerides, high-density lipoprotein cholesterol, and low-density lipoprotein cholesterol using Randox

Kit (Randox Laboratory Limited, United Kingdom). The assay procedures were carried out based on the instructions provided on the kit.

15.2.9 LIPID PEROXIDATION ANALYSIS

The brain homogenate was assayed for lipid peroxidation using the method of Sinha[39] by measuring the production of malondialdehyde (MDA) using the TBA test. The reaction mixture contained 20% acetate buffer (pH 3.5), 8.1% sodium dodecyl sulfate, and 0.8% TBA which was mixed well with 0.2 mL of the tissue homogenate for 3 min and then incubated for 60 min at 95°C. This was then cooled and the TBA-reactive substance (MDA) was extracted using 1 mL of H_2O and 2.5 mL of *n*-butanol and pyridine mixture in the ratio 15:1, v/v, respectively. The MDA that was produced by lipid peroxidation in the upper layer was then collected and its absorbance was measured at 532 nm against a reference blank of distilled water. The blank was prepared by replacing the 0.2 mL of gastric homogenate with 0.2 mL of saline solution.

15.2.10 CATALASE ACTIVITY ANALYSIS

The activity of catalase was measured in the brain homogenate following the method of Sinha.[39] The reaction mixture contained 0.5 mL of enzyme extract and 2 mL sodium phosphate buffer (0.1 M, pH 6.8), and the reaction was started by the addition of 0.5 mL of 10 mM hydrogen peroxide. The absorbance was read at 240 nm, and the decrease in absorbance was recorded every 15 s, for 3 min. Catalase activity was expressed as nKat/min/mg of protein.

15.2.11 QUALITY CONTROL

Randox quality control serum was introduced between assays to determine the accuracy and precision of the various procedures adopted for the study.

15.2.12 STATISTICAL ANALYSIS

Data obtained from this study were presented as Mean ± SD, statistical comparison of the data was carried out with the use of one-way analysis of variance (ANOVA) to determine statistical significance among the study groups, using SPSS software, version 21.0. The post-hoc test was carried out using Duncan Multiple-Range Test. The graphs and charts were constructed with the use of Microsoft Excel © 2013. The level of statistical significance was set at 0.05.

15.3 RESULTS AND DISCUSSION

15.3.1 EFFECT OF TOFU DIET ON BLOOD GLUCOSE LEVELS OF EXPERIMENTAL ANIMALS

The blood glucose level (Table 15.2) of normal rats fed on *Tofu* diet was significantly lower ($p < 0.05$) than that of normal rats fed on rat chow. Similarly, the blood glucose levels of diabetic rats fed on *Tofu* diet showed a lower value when compared with that of diabetic rats fed on rat chow.

TABLE 15.2 Blood Glucose Levels of Experimental Animals.

Group	Fasting blood glucose levels (mg/dL)
I	108.6 ± 0.61^{b}
II	105.2 ± 0.94^{a}
III	140.3 ± 0.25^{d}
IV	122.4 ± 0.40^{c}

Values are expressed as mean ± SD. Values having different superscript alphabets have a statistically significant difference at $p < 0.05$.

Group I = Normal rats fed on rat chow;
Group II = Normal rats fed on *Tofu*;
Group III= Diabetic rats fed on rat chow;
Group IV= Diabetic rats fed on *Tofu*.

D. mellitus is characterized with elevated blood glucose levels, regular consumption of *Tofu* (a whole protein) has been shown, in the present study, to have antihyperglycemic effect on alloxan-induced diabetic rats. This finding is similar to that of Hua et al.[19] which revealed that soybean oligosaccharides helped to reduce abnormal blood glucose levels in high-fat rats.

Similarly, Ji et al.[21] in a research to examine the influence of soybean supplementation on blood glucose, serum lipid profile, and antioxidant enzyme activity of the erythrocytes in type 2 *D. mellitus* patients reported that supplementation with soybean may be helpful in controlling blood glucose in diabetic patients. In this study, lowered blood glucose levels of rats fed on *Tofu* diet may be remarkably due to the soy protein content of *Tofu* that prevents accumulation of glucose in the blood.

15.3.2 EFFECT OF TOFU ON BODY WEIGHT GAIN OF EXPERIMENTAL ANIMALS

The mean body weight gain of the rats by the end of the seventh week (Figure 15.1) was observed to be higher ($p < 0.05$) for both normal and diabetic rats fed on rat chow than those fed on *Tofu* diet. This shows that consumption of *tofu* may be effective in maintaining a healthy weight in both normal and diabetic conditions.

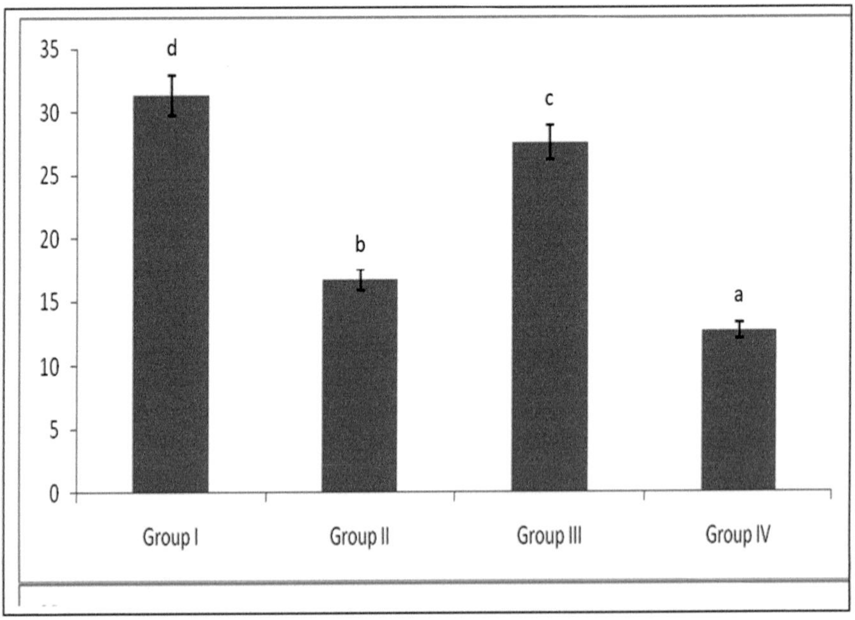

FIGURE 15.1 Average body weight gain of experimental animals. Bars with different superscript letters have statistically significant difference at $p < 0.05$.

Group I = Normal rats fed on rat chow; Group II = Normal rats fed on Tofu;

Group III = Diabetic rats fed on rat chow; Group IV = Diabetic rats fed on Tofu

Body weight increase was associated with diabetes and other cardiovascular diseases. However, findings from this study revealed that *Tofu* helped to reduce body weight gain in both diabetic and nondiabetic rats. The body weight of the experimental animals fed on *Tofu* reduced significantly, suggesting that *Tofu* consumption may be advantageous in maintaining a healthy body weight in order to prevent diseases, such as obesity, atherosclerosis, and other cardiovascular diseases associated with diabetes mellitus.

Evidence abounds in literature to show that soy protein supplement in the diet has helped to control excess in body weight gain in diabetes. A study by Sacks et al.[32] revealed that when soy protein in *Tofu* was consumed, there was a significant decrease in the weight gain of individuals who feed on *Tofu*.

In contrast to these findings, Julia (2006) reported that soy consumption led to the blockage of thyroid hormone receptors and led to hypothyroidism which consequently raised blood glucose levels and resulted in excessive weight gain. This finding was attributed to the presence of isoflavones in soy foods. However, there is a dispute in the strength of the association and on its beneficial effect.

15.3.3 EFFECT OF TOFU ON LIPID PROFILE OF EXPERIMENTAL ANIMALS

Table 15.3 shows the lipid profile of experimental animals. The levels of TC in diabetic rats which were fed on *Tofu* were significantly lower ($p < 0.05$) than that of diabetic rats fed on commercial rat chow, while no significant difference was observed ($p > 0.05$) in the TC levels of normal rats fed on commercial rat chow. There was a significant decrease ($p < 0.05$) in triglyceride (TG) levels of the groups fed on *Tofu* diet as compared with those fed on commercial rat chow. Also, a significant increase ($p < 0.05$) was observed in the high-density lipoprotein cholesterol (HDL-C) of rats in the groups fed on *Tofu* diet as compared to those fed on commercial rat chow. Low-density lipoprotein cholesterol (LDL-C) levels for both normal and diabetic rats fed on rat chow and those fed on *tofu* diet showed no significant difference ($p > 0.05$).

Studies have reported that the consumption of soy protein brought about a reduction in the level of TC. A meta-analysis conducted on the effects of soy protein on serum lipids showed that when soy protein was substituted for animal protein in a diet, it significantly lowered TC in hypercholesterolemic individuals and had no significant effect on those whose cholesterol levels were found within the normal range.[4] Anderson et al.[4] and Hodgson et al.[17]

studied soy protein effect on serum lipoproteins and some metabolizing enzymes of lipids in rats fed with cholesterol-containing diets, it was revealed that there was a significant reduction in the TC levels of rats fed with a diet containing 0% cholesterol and 20% soy protein.

TABLE 15.3 Lipid Profile of Experimental Animals.

Parameters	Group I	Group II	Group III	Group IV
TC (mg/dL)	188.16±1.08[a]	186.08±2.85[a]	220.50±3.59[c]	214.80±2.57[b]
TG (mg/dL)	339.55±2.46[b]	248.54±2.08[a]	453.70±3.19[d]	364.40±3.25[c]
HDL-C (mg/dL)	14.72±1.35[b]	28.16±1.84[d]	8.20±1.10[a]	22.40±1.24[c]
LDL-C (mg/dL)	108.09±3.56[b]	60.19±2.42[a]	126.56±2.17[c]	123.26±3.21[c]

Values are presented as mean ± SD. Data in the same row having the same superscript letters have no statistically significant difference at $p < 0.05$.

TG, triglyceride; HDL-C, high-density lipoprotein cholesterol; TC, total cholesterol level; LDL-C, low-density lipoprotein cholesterol.

The triglyceride levels of the experimental animals were also significantly lowered in rats fed on *Tofu* diet. Studies by Bowen,[8] Shepherd,[34] and Bery[6] showed that regular intake of soy protein helped to reduce triglyceride levels in both human subjects and experimental animals which fed on it for a period of 4–8 weeks. This may be attributable to the fact that the fiber in soybean delayed duodenual absorption of glucose and fatty acids and thus provided a smaller amount substrate for synthesis of triglycerides and therefore resulting in the lowered triglyceride levels of rats fed on *Tofu*.

There was no significant decrease in the serum LDL cholesterol levels of alloxan-induced rats fed on *Tofu* diet. This finding is similar to that of Jenkins et al.[21] and Simons et al.[38] who reported that soy consumption had no beneficial effect on serum lipids and lipoproteins in hyperlipidemic men and women who are at the postmenopausal stage of life. Similarly, studies by Clifton-Bligh et al.[12] showed that there was no change in serum LDL cholesterol concentrations of postmenopausal women after daily consumption of 28.5, 57, or 85.5 mg of red clover isoflavones. Furthermore, an 8-week trial by Hodgson et al.[17] with tablets containing 55 mg of subterranean clover isoflavones showed no beneficial effect on the lipid profile of men and postmenopausal women.

Contrary to this, a meta-analysis by Anderson et al.[4] showed that soy protein consumption was associated with a significant decrease in the concentrations of serum cholesterol, LDL cholesterol, and triglyceride.

Wangen et al.[42] reported that the concentrations of LDL cholesterol in the plasma samples of normocholesterolemic and mildly hypercholesterolemic postmenopausal women were reduced by 7% sequel to a daily intake of about 85 g of beverage powder gotten from isolated soy protein and consumed for 93 days. The beverage was found to have a high isoflavone content (132 mg). At a lesser dose of soy isoflavones (65 mg), there were no effects observed on plasma lipids and lipoproteins. Another trial by Gardner et al.[16] revealed that in women at the postmenopausal stage, who had moderate hypercholesterolemia, the use of 42 g of isolated soy protein powder which provided 80 mg of isoflavones for 12 weeks helped to reduce the concentration of plasma LDL cholesterol compared to 42 g of isolated soy protein which contained no isoflavones.

Findings of this study also showed that experimental animals fed on *Tofu* showed a significant increase in serum HDL concentration compared to those fed on commercial rat chow. This is related to the report of Oboh,[26] which showed that rats fed on *Tofu* meal had increased HDL cholesterol levels. Studies by Potter et al.[30] and Walsh et al.[41] reported that isolated soy protein consisting of isoflavones helped to increase HDL cholesterol concentrations in hypercholesterolemic individuals. Available evidence reveals that for every 1 mg increase in HDL cholesterol, there is a decrease to about 25% in the risk of cardiovascular diseases.[9]

Maintenance of a healthy serum lipid profile suggests that soy proteins and phytoestrogens present in soybean may exert possible hypolipidemic effect when consumed by individuals with diabetes. The possible mechanisms of the hypolipidemic effect of soy protein were described in several studies involving cell cultures, animals, or humans. These studies suggested mechanisms such as increase in the activity of LDL receptor, increase in bile acid synthesis and fecal excretion, and a reduction in the absorption of cholesterol.[23,27,28,29]

15.3.4 EFFECT OF TOFU ON OXIDATIVE STRESS IN THE EXPERIMENTAL ANIMALS

Table 15.4 presents the lipid peroxidation and catalase activity levels as markers of the antioxidant status of the rats. Diabetic animals fed on *Tofu* had significantly lower ($p < 0.05$) lipid peroxidation levels than the diabetic group fed on commercial rat chow while the normal animals fed on either commercial rat chow or *Tofu* had no significantly different ($p > 0.05$) lipid peroxidation levels. Also, catalase activity was significantly higher ($p < 0.05$) in animals fed on *tofu* as compared to those fed on commercial rat chow.

TABLE 15.4 Oxidative Stress in the Experimental Animals.

Group	Lipid peroxidation (μg/mg protein)	Catalase activity (U/L)
I	27.02±0.21[a]	0.41±0.01[c]
II	26.16±0.63[a]	0.81±0.01[d]
III	41.58±0.32[c]	0.13±0.01[a]
IV	37.19±0.48[b]	0.35±0.02[b]

Values are presented as mean ± SD. Values in the same column and having different superscript letters have statistically significant difference at p < 0.05.

Tofu has also been shown to be effective in managing oxidative stress level in alloxan-induced diabetic rats. Findings by Chang et al.[11] showed that soy protein was helpful in reducing oxidative stress levels in Wistar rats fed on high-fat diet. Studies have shown that low-lipid peroxidation levels reduce the release of free radicals which result in oxidative stress when present in high concentrations in the cells.[13,15,43] Malondialdehyde is one of the final products of lipid peroxidation which conjugates with the amino group of proteins and forms intramolecular and intermolecular cross-links that inactivate membrane-bound enzymes and receptors.[20] Hence, feeding of animals with *Tofu* may have reduced MDA concentration in diabetic rats and thus indicating its ability to prevent oxidative damage, by interfering with the free radical-induced processes that lead to intra- and inter-molecular cross-linkages.

Catalase activity was also increased significantly in both diabetic and normal rats that were fed on *Tofu*. Catalase is a hemoprotein that catalyzes reduction of hydrogen peroxide and protects body tissues from free radical damage by scavenging highly reactive hydroxyl radicals.[10] A study by Liu et al.[24] to determine the antioxidant potential of *Tofu* and an extract of soybean isoflavone showed that diets containing high isoflavone contents exhibited apparent elevated antioxidant enzyme levels in various organs. Diets containing high isoflavone contents (150 and 250 ppm) were shown to elevate catalase activity levels in various organs of rats. Also, *Tofu* prepared in the laboratory which contained approximately 50 ppm isoflavones were more effective in stimulating antioxidant enzymes in vivo than the soy extract which contained 250 ppm isoflavones. This finding indicated that molecules other than isoflavones may have a synergistic effect in increasing the activity of antioxidant enzymes. Thus, the isoflavone and other soy content present in the *Tofu* diet may be responsible for the increase in catalase activity levels of the rats which were fed on it.

Several experimental studies involving animals and humans have investigated the effect of consuming soy-containing diets on the metabolism of glucose and lipids and the hormones controlling their metabolism. Studies on the role of soy in human subjects which were healthy showed that soy polysaccharides decreased postprandial glucose and concentration of triacylglycerol, suggesting that polysaccharides present in soy may provide potential benefits in conditions where glucose tolerance is impaired and in conditions of hyperlipidemia.[1,42]

The observed activities of soy may also be attributed to the presence of proteins. It was reported in a study by Sanchez and Hubbard[33] that a lower postprandial insulin–glucagon ratio was induced by soy protein in both hypercholesterolemic and healthy subjects than did casein. Arginine and glycine are components of soy proteins and are involved in the secretion of insulin and glucagon from the pancreas. Hence, the decrease in cholesterol resulting from soy protein may be associated with the decrease in insulin–glucagon ratio triggered by arginine and glycine.[18]

A study by Taha and Wasif[40] showed the effect of soy flour-supplemented whole-durum meal in alloxan-induced diabetic hypercholesterolemic rats. The addition of soy flour with or without methionine helped to lower plasma glucose, lipids, and cholesterol levels. Another study on BioBreeding rats, which were prone to type 1 diabetes, reported that soy protein diet brought about a delay in the onset of diabetes and reduced blood glucose levels compared to a diet containing a combination of plant and animal protein.[2] This finding suggests that type 1 diabetes development is dependent on the nature of the protein contained in the diet. Similarly, in diabetic mice which were not obese, soy protein combined with or without nicotinamide helped to inhibit insulitis and thereby prevented the manifestation of diabetes.[31]

Evidence has revealed that soy proteins may potentially improve diabetes through reduction of insulin resistance and decrease in adiposity via inhibition of insulin secretion from the β cells of the pancreas or by inhibition of lipogenesis and facilitation of lipolysis within the liver and adipocytes. The activities isoflavones may also be associated with their antioxidant potential by scavenging free radicals and preventing lipid peroxidation.[5] It may, therefore, be suggested that the antidiabetic properties elicited by *Tofu* were achieved through these biochemical mechanisms.

15.4 SUMMARY

Tofu is a curd produced from soybean seeds. It has been reported to be a healthy diet with many beneficial effects. Consumption of *Tofu*-containing diets may have potential antidiabetic effect by regulating blood glucose levels, reducing excessive weight gain, maintaining a healthy lipid profile, and preventing oxidative stress. Hence, *Tofu* consumption may be a promising candidate for individuals with diabetes, to replace animal protein which may be high in fat, and to benefit from its many health-promoting properties.

KEYWORDS

- **antioxidant**
- **blood glucose**
- **diabetes**
- **isoflavone**
- **lipid profile**
- **tofu**

REFERENCES

1. Ahmed, M.; Gannon, M.; Nuttall, F. Postprandial Plasma Glucose, Insulin, Glucagon and Triglyceride Response to a Standard Diet in Normal Subjects. *Diabetologia* **1976,** *12*, 61–67.
2. Atkinson, M. A.; Winter, W. E.; Skordis, N. Dietary Protein Restriction Reduces the Frequency and Delays the Onset of Insulin Dependent Diabetes in BB Rats. *Autoimmunity* **1998,** *2*, 11–19.
3. ADA (American Diabetes Association). Classification and Diagnosis of Diabetes: Standards of Medical Care in Diabetes. *Diabetes Care*. **2019,** *42* (1), S13–S28.
4. Anderson, J. W.; Johnstone, B. M.; Cook-Newell, M. E. Meta-Analysis of the Effects of Soy Protein Intake on Serum Lipids. *N. Engl. J. Med.* **1995,** *333* (5), 276–282.
5. Bhathena, S. J.; Velasquez, M. T. Beneficial Role of Dietary Phytoestrogens in Obesity and Diabetes. *Am. J. Clin. Nutr.* **2001,** *76* (6), 1191–1201.
6. Bery, M. The Health Benefits of *Tofu* Food, 2007. http://ezinearticles.com/?The-Health-Benefits-Of-Tofu-Food&id=485422 (accessed: July 14, 2021).
7. Boyle, J.; Honeycutt, A.; Narayan, K.; Hoerger, T.; Geiss, L.; Chen, H.; Thompson, T. Projection of Diabetes Burden through 2050. *Diabetes Care*. **2001,** *24* (11), 1936–40.

8. Bowen, R. Physiologic Effects of Insulin; 2004. http://arbl.cvmbs.colostate.edu/hbooks/pathphys/endocrine/pancreas/insulin_phys.html (accessed: July 14, 2021).
9. CDC (Centers for Disease Control and Prevention). Prevalence of Overweight and Obesity among Adults with Diagnosed Diabetes—United States, 1988-1994 and 1999-2002. *Morb. Mortal. Wkly. Rep.* **2004,** *53* (45), 1066–1068.
10. Chance, B.; Greenstein, D.; Roughton, F. The Mechanism of Catalase Action 1—Steady State Analysis. *Arch. Biochem.* **1952,** *37*, 301–321.
11. Chang, J. H.; Kim, M. S.; Kim, T. W.; Lee, S. S. Effects of Soybean Supplementation on Blood Glucose, Plasma Lipid Levels, and Erythrocyte Antioxidant Enzyme Activity in Type 2 *Diabetes Mellitus* Patients. *Nutr. Res. Pract.* **2008,** *2* (3), 152–157.
12. Clifton-Bligh, P. B.; Baber, R. J.; Fulcher, G. R.; Nery, M. L.; Moreton, T. The Effect of Isoflavones Extracted from Red Clover on Lipid and Bone Metabolism. *Menopause* **2001,** *8*, 259–265.
13. Devasagayam, T.; Tilak, J.; Boloor, K.; Sane, K.; Ghaskadbi, S.; Lele, R. Free Radicals and Antioxidants in Human Health: Current Status and Future Prospects. *J. Assoc. Phys. India*. **2004,** *52*, 796–798.
14. Devlin, T. M., (Ed.) *Textbook of Biochemistry with Clinical Correlations*, 6th edn.; Wiley-Liss: Hoboken: New Jersey, 2006; p 582.
15. Drews, G.; Krippeit-Drews, P.; Dufer, M. Oxidative Stress and Beta-Cell Dysfunction. *Pflugers Arch.* **2010,** *460*, 703–718.
16. Gardner, C. D.; Newell, K. A.; Cherin, R.; Haskell, W. L. The Effect of Soy Protein with or without Isoflavones Relative to Milk Protein on Plasma Lipids in Hypercholesterolemic Postmenopausal Women. *Am. J. Clin. Nutr.* **2001,** *73*, 728–735.
17. Hodgson, J. M.; Puddey, I. B.; Beilin, L. J.; Mori, T. A.; Croft, K. D. Supplementation with Isoflavonoid Phytoestrogens does not Alter Serum Lipid Concentrations: A Randomized Controlled Trial in Humans. *J. Nutr.* **1998,** *128*, 728–732.
18. Howe, J. C. Postprandial Response of Calcium Metabolism in Postmenopausal Women to Meals Varying in Protein Level/Source. *Metabolism* **1996,** *39*, 1246–1252.
19. Hua, C.; Li-jun, L.; Jian-jun, Z.; Bo, X.; Rui, L. Effect of Soybean Oligosaccharide on Blood Lipid, Glucose Levels and Antioxidant Enzymes Activity in High Fat Rats. *Food Chem.* **2010,** *119* (4), 1633–1636.
20. Jenkins, D.; Kendall, C.; Vidgen, E.; Vuksan, V.; Jackson, C.; Augustin, L.; Lee, B.; Garsetti, M.; Agarwal, S.; Rao, A.; Cagampang, G. Fulgoni, V. Effect of Soy-Based Breakfast Cereal on Blood Lipids and Oxidized Low-Density Lipoprotein. *Metabolism* **2000,** *49*, 1496–1500.
21. Ji, H.; Min, S.; Tae, W.; Sang, S. Effects of Soybean Supplementation on Blood Glucose, Plasma Lipid Levels, and Erythrocyte Antioxidant Enzyme Activity in Type 2 Diabetes Mellitus Patients. *Nutr. Res. Pract.* **2008,** *2* (3), 152–157.
22. Julia, R. B. The Science of Soy: What Do We Really Know? *Environ. Health Perspect.* **2006,** *114* (6), 352–358.
23. Lichtenstein, A. H. Soy Protein, Isoflavones and Cardiovascular Disease Risk. *J. Nutr.* **1998,** *128*, 1589–1592.
24. Liu, J.; Chang, S.; Wiesenborn, D. Antioxidant Properties of Soybean Isoflavone Extract and *Tofu In Vitro* and *In Vivo*. *J. Agric. Food Chem.* **2005,** *53* (6), 2333–2340.
25. Mueller, N.; Odegaard, A.; Gross, M.; Yu, W.; Yuan, J.; Pereira, M. Soy Intake and Risk of Type 2 Diabetes Mellitus in Chinese Singaporeans. *Eur. J. Nutr.* **2012,** *51* (8), 1033–1040.

26. Oboh, G. Coagulants Modulate the Hypocholesterolemia Effect of *Tofu* (Coagulated Soymilk). *Afr. J. Biotechnol.* **2006,** *5* (3), 290–294.
27. Potter, S. M. Overview of Proposed Mechanisms for the Hypocholesterolemia Effect of Soy. *J. Nutr.* **1995,** *125*, 606S–611S.
28. Potter, S. M. Soy Protein and Serum Lipids. *Curr. Opin. Lipidol.* **1996,** *7*, 2604.
29. Potter, S. M. Soy protein and Cardiovascular Disease: The Impact of Bioactive Components in Soy. *Nutr. Rev.* **1998,** *56*, 231–235.
30. Potter, S.; Baum, J.; Teng, H.; Stillman, R.; Shay, N.; Erdman, J. Jr. Soy Protein and Isoflavones: Their Effects on Blood Lipids and Bone Density in Postmenopausal Women. *Am. J. Clin. Nutr.* **1998,** *68*, 1375S–1379S.
31. Reddy, S.; Bibby, N.; Wu, D.; Swinney, C.; Barrow, G.; Elliot, R. A Combined Casein-Free-Nicotinamide Diet Prevents Diabetes in the NOD Mouse with Minimum Insulitis. *Diabetes Res. Clin. Pract.* **1995,** *29*, 83–92.
32. Sacks, F.; Lichtenstein, A.; Linda, H.; Kris, E.; Penny, W. Soy Protein, Isoflavones, and Cardiovascular Health by American Heart Association Science Advisory for Professionals from the Nutrition Committee. *Circulation* **2006,** *113* (7), 1034–1044.
33. Sanchez, A.; Hubbard, R. Plasma Amino Acids and the Insulin/Glucagon Ratio as an Explanation for the Dietary Protein Modulation of Atherosclerosis. *Med. Hypotheses* **1991,** *36*, 27–32.
34. Shepherd, P. Actions of Insulin. *Acta Physiol. Scand.* **2005,** *183*, 3–12.
35. Shurtleff, W.; Aoyagi, A., Eds. *Tofu & Soymilk Production: A Craft and Technical Manual*, 3rd ed.; Soy Foods Center: Lafayette, California; 2000; http://www.soyinfocenter.com/Tofusource (accessed: July 14, 2021).
36. Shurtleff, W.; Aoyagi, A. Dou Fu Zhi Shu (The Book of Tofu). 2005; http://www.soyinfocenter.com/Tofu (accessed: July 14, 2021).
37. Shurtleff, W.; Aoyagi, A. A Comprehensive History of Soy. History of Soybeans and Soyfoods Past Present and Future, Lafayette—California; Soyinfo Center; 2005, http://www.soyinfocenter.com/HSS/history.php (accessed: July 14, 2021).
38. Simons, L.; Von Konigsmark, M.; Simons, J.; Celermajer, D. Phytoestrogens do not Influence Lipoprotein Levels or Endothelial Function in Healthy, Postmenopausal Women. *Am. J. Cardiol.* **2000,** *85*, 1297–1301.
39. Sinha A. K. Colorimetric Assay of Catalase. *Anal. Biochem.* **1972,** *47* (2), 389–394.
40. Taha, S. A.; Wasif, M. M. Hypoglycemic Effect and Protein Nutritive Quality of Soy and Methionine Supplemented Whole Durum Pasta Products. *Nahrung* **1996,** *40*, 281–287.
41. Walsh, B.; Schiff, I.; Rosner, B.; Greenberg, L.; Ravnikar, V.; Sacks, F. Effects of Postmenopausal Estrogen Replacement on the Concentrations and Metabolism of Plasma Lipoproteins. *N. Engl. J. Med.* **1991,** *325*, 1196–1204.
42. Wangen, K.; Duncan, A.; Xu, X.; Kurzer, M. Soy Isoflavones Improve Plasma Lipids in Normocholesterolemic and Mildly Hypercholesterolemic Postmenopausal Women. *Am. J. Clin. Nutr.* **2001,** *73*, 225–231.
43. West, I. C. Radicals and Oxidative Stress in Diabetes. *Diabet. Med.* **2000,** *17*, 171–180.
44. World Health Organization (WHO). Diabetes. 2018; www.who.int/news-room/fact-sheets/detail/diabetes (accessed: July 14, 2021).
45. World Health Organization (WHO). Classification of Diabetes Mellitus. 2019; Available online at https://apps.who.int/iris/rest/bitstreams/12333444/retrieve (assessed July 14, 2021).

CHAPTER 16

Antioxidant Potential of Seed and Leaf Extracts of *Adansonia digitata*

Adetoun Elizabeth Morakinyo,
Samuel Chibueze Nzekwe, Samuel Olufolarin Babarinde, Oluwatosin Elijah Olopade, Abraham Yakubu, and Ifeoluwa Dorcas Adegbite

ABSTRACT

Baobab (*Adansonia digitata*) is a tree native to coastal region of Africa and possesses biological properties, such as antimicrobial, diarrhea, anti-inflammatory, asthma, antimalarial, anemia, antiviral, and antioxidant activities. Therefore, the present study on the seeds and leaves extracts was conducted to determine the antioxidant potentials of baobab tree. The baobab seeds and leaves extracts were prepared using different solvents: ethyl acetate, butanol, and ethanol to assess total phenolics and total flavonoids present in the extracts. Also, FRAP, DPPH, ABTS, and nitrite assays were carried out to assess the antioxidant activities of *A. digitata* seeds and leaves extracts. Results obtained in this study showed that phenolic contents and FRAP level of the ethanol, butanol, and ethyl acetate leaves extracts increased significantly ($p < 0.05$) when compared to the extracts of seeds sample while the concentration of flavonoids in the seed extracts increased significantly ($p < 0.05$) when compared to the leaves in the various solvents. There was no significant difference in the nitrite concentration of ethanol extract of the seeds and the leaves. The DPPH and ABTS radical scavenging activities of the leaves extracts of all the solvents were significantly increased at ($p < 0.05$) when compared to the seed extracts. These results indicated that baobab (*Adansonia digitata*) could be a good natural antioxidant agent.

Curative and Preventive Properties of Medicinal Plants: Research on Disease Management and Animal Model Studies. Megh R. Goyal, Junaid Ahmad Malik, & Ademola Olabode Ayeleso (Eds.)

16.1 INTRODUCTION

Many natural plants, such as vegetables, fruits, and beverages, serve as sources of antioxidant. Dietary antioxidants such as flavonoids aid in reducing complications related to disease conditions such as diabetes mellitus, myocardial infarction, and cardiovascular disease.[22,41]

Presently, natural antioxidants have become the center of interest calling attention for search on possible treatment against oxidative stress. Imbalance between the antioxidants system and free-radicals components of the body system leading to accumulation of excess reactive species was the reason for oxidative stress causative damage to proteins, nucleic acids and carbohydrates macromolecules, and lipids membrane.[7,35] The human body has an antioxidant defense system that protects it from oxidative damages induced by the reactive species (free-radicals) generated during organism's cellular respiration.[7,39] However, natural antioxidants strengthen the endogenous antioxidants defense against the oxidant species by neutralizing the reactive oxygen species (ROS) thereby restoring the optimal balance.[3,7,17]

Over the years, traditional medicines or medicinal plants possess different phytochemical components that are of medicinal values that played roles in treating human or animal diseases.[20] The use of medicinal plants was due to its little or no side effects, affordable price, and efficacy compared to the synthetic drugs with several side effects and out of reach to the common man.[23] Bioactive compounds in plants (such as flavonoids, tannins, saponins, anthraquinones, quiterpenes, carotenoids, phenolic compounds, cumarin, glycosides, etc.) have shown potential of medicinal functions, such as antioxidant, anti-inflammatory, antidiabetic, antipyretic, immunomodulatory, and anticancer activities. Most medicinal plants function differently resulting from varieties in the concentrations and availability of the various bioactive compounds that have shown to possess great potential to scavenge ROS generated in the body.[13,4,11]

All parts of *Adansonia digitata* were reported to serve as traditional medicine due to active phytochemical constituents and high natural vitamin C content.[12,16] *A. digitata* has been reported to possess phytoconstituents, such as alkaloids, terpenes, flavonoids, tannins, steroids, and saponins.[7,31,21,25]

This chapter focuses on antioxidant potential of seed and leaf extracts of *A. digitata.* The antioxidant capacity of seeds and leaves of African baobab was compared after extraction using different solvent mediums. The phenolic compounds, flavonoids, FRAP, DPPH, ABTS, and nitrite concentrations of ethanol, ethyl acetate, and butanol extracts of the powdered leaves and seeds were also evaluated.

16.2 HISTORY AND HABITAT DISTRIBUTION

Baobab plant, scientifically known as *A. digitata,* is a word obtained from *bu hobab,* which is an Arabic word, which means "fruit with multiple seeds." Baobab tree is about eight to nine species existing in the world, with six of the species known to originate from Madagascar while the rest were indigenous to Australia and mainland Africa. *Adansonia perrieri*, *Aponogeton madagascariensis*, *A. za*, *Adansonia rubrostipa, Adansonia grandidieri,* and *Adansonia suarezensis* are species indigenous to Madagascar, while *Adansonia kilima* and *Adansonia digitata* species are found majorly in Arabian Peninsula as well as mainland Africa.[28] The baobab tree is recognized as significant to the African Sahara, and *A. digitata* species is mostly found in southern Sahara African countries where it is majorly located in South Africa, precisely in Mpumalanga and Limpopo provinces.[9]

Baobab is known of its distinctive silhouette, enormous-shinny-widal trunk, large-velvety fruit, and root-like branches. Baobab has special light and heat reflection adaptation, hence, it always give serene atmosphere even under heavy scotching sun.[24] *A. digitata* has special adaptation feature of tetraploid chromosome that its wide range of distribution.[9] It has a deciduous succulent stem and only native in savannah regions of tropical Africa.[38,40]

The characteristic features found on the six Madagascan baobab species were that the trunk appears gray-brown to red and possess taper shape that runs from the top to bottom in a bottle-shaped to cylindrical figure. Moreover, baobab tree is also distinctive in its shiny and slick outer bark that is scaly, which prevents water loss during drought or winter. The slippery skin helps to keep away monkeys and other climbing predators from eating its sprouting leaves and flowers. Moreso its shinny bark may serve to protect it from getting burnt in the case of wildfire outbreak.[24] Gigantic baobab can thrives on land with a limited annual range of rainfall and can survive in adverse drought conditions. The leaves of baobab are completely shed during drought periods to help conserve water, the red or yellow, or white colors of the flowers are attractive to pollination insects or large animals like bats and lemurs, and sometimes by hawk moths. The baobab tree is a legendary plant with many values and different uses, probably locally named as the "tree of life."[40]

16.3 TRADITIONAL USE OF BAOBAB TREE

In the African continent, baobab has superstitious folklore assumptions and spiritual beliefs to be tree of the gods.[14] It is very unique in the sense of its hugeness compared to other plants and it is called "monkey bread tree." It is also called Senegal Calash, upside-down tree or bottle tree with a long-life span of up to 6000 years.[27] Because of its tolerance to drought and high temperature, it can be found in Sahara and desert regions in the northern parts of Nigeria.[15,18,22] Various parts of baobab tree serve traditional medicinal purposes as a beverage if the fruit pulp is mixed with milk or water.[37 38] The pulp can also be used in local brewery production as fermenting agent or in baking as tartar cream.[30,38] The seeds, when fermented, it serves as a flavoring in soup, also, when roasted, are eaten as a peanut.[2] Baobab seed oil is also good for cooking and the shaft can be used to make seed cake.[33] Proximate analysis, assessment of phytochemical, and mineral contents of baobab have shown that its medicinal values vary with soil type, climatic condition, and geographical locations. Baobab plants indigenous to different regions differ in their species, phytochemical components, and medicinal values.[15]

16.4 PHENOLIC, FLAVONOIDS, AND FRAP LEVEL IN ETHYL ACETATE, BUTANOL, AND ETHANOL EXTRACTS OF BAOBAB SEED AND LEAF

Baobab leaves serve as a regular food in African countries for people of the region mostly in Central and Western Africa.[34] The leaves can be consumed as either fresh leaves or dried under the sun. Northern Nigeria used the leaves to prepare a popular sauce called "Miyan kuka" made by the use of baobab leaves.[1] To establish the antioxidant potentials of baobab-derived foods, there is a need to assess the antioxidant capacity of the phytochemical constituents.

Phenolic content was significantly increased ($p < 0.05$) in the ethanol and butanol leaf extract when compared to the seed extracts (Table 16.1). In ethyl acetate extracts, there was no significant difference in the phenolics content. Arowora et al.[5] expressed the high potential antioxidant capacity of baobab leaves relating to its high content of phenolic compounds. The highest phenolic compound (365.42 mg/mL) was obtained in the ethanolic-ethyl acetate equal volume fraction. Irondi et al.[23] indicated a high profile of phenolic compound in the leaf extract sample of baobab when analyzed in-vitro. They revealed that the phenolic acids content present in baobab is gallic acid (4.15 ± 0.01), chlorogenic acid (6.28 ± 0.03), ellagic acid (2.43 ± 0.01), and caffeic acid (6.31 ± 0.02) indicating high combination of phenolic content.

The flavonoids content in the leaf extracts of all the solvents was increased significantly ($p < 0.05$) when compared to the seed extracts (Table 16.1). The leaves of baobab have been shown to be rich in flavonoids content in all the solvent medium used and possess the high potent antioxidant capacity. Arowora et al.[5] revealed that ethyl acetate-ethanolic (30.05 mg/mL) extract fraction of the leaves has higher flavonoid content when assessed for the chemical composition. Braca et al.[13] also showed that baobab used for staple food and ethnomedicinal purposes contains high concentration of antioxidant phenolics and flavonoids phytochemicals, and other chelating minerals that help in scavenging reactive oxidative species (ROS).

FRAP of antioxidant defense has the property of transferring electron transfer from the reducing agent in the plant extract.[26] The ferric ion is reduced to the ferrous ion, and FRAP concentration is correspondent to the amounts of electrons donated by the antioxidant compounds present in the plant extract.[10] The assessment of Baobab leaf and seed reducing power was focused on its ability to convert ferric ion to ferrous ion (Table 16.1). FRAP was found in higher concentration in the leaves extract of the various solvents used when compared to the seed extracts.

Muhammad et al.[29] discovered that baobab leaves prepared under shade drying agents possessed significant high level of antioxidant capacity in reducing ferric iron to ferrous iron. Also, Okorie et al.[32] further study on the action of baobab leaf extract in Wistar albino rats, discovered a reduction in oxidative stress generated due to diabetes mellitus disease condition. Their study on antidiabetic properties baobab leaves extract through analysis of blood glucose level suggested that baobab possess antidiabetic effect in the experimental rats. In vitro studies on the oxidative reducing potential of baobab leaf extract by Ayele et al.[6] showed that FRAP concentration increased significantly in the leaf extract. DPPH, ABTS, FRAP, and nitrite concentration were determined to evaluate the antioxidant potentials of baobab leaves and seeds.

TABLE 16.1 Total Phenolics, Flavonoids, and FRAP Concentrations in Seeds and Leaves of Baobab.

Solvents	Leaves	Seeds
	Phenolic concentration (mg/mL)	
Ethanol	3.33 ± 0.641[b]	21.04 ± 3.996[a]
Ethyl acetate	1.521 ± 0.875[d]	12.83 ± 3.419[c]
Butanol	0.479 ± 0.240[f]	15. 50 ± 5.900[e]
	Flavonoids concentration (mg/mL)	
Ethanol	1.750 ± 0.000[b]	2.60 ± 0.029[a]
Ethyl acetate	4.100 ± 0.173[c]	2.583 ± 0.303[d]
Butanol	2.950 ± 0.000[e]	1.90 ± 0.058[f]
	FRAP (mg of AAE/g)	
Ethanol	5.175 ± 0.046[a]	1.237 ± 0.198[b]
Ethyl acetate	4.539 ± 0.091[c]	0.790 ± 0.456[d]
Butanol	4.917 ± 0.027[e]	0.342 ± 0.261[f]

NOTE: Values with the same letter superscript showed insignificant difference while different letters showed significantly difference, in the row.

16.5 NITRITE, DPPH, AND ABTS ASSAY OF ETHYL ACETATE, BUTANOL, AND ETHANOL EXTRACTS OF SEED AND LEAF OF BAOBAB

The nitrite concentration of the ethanolic extract decreased significantly ($p < 0.05$) in the leaves extract when compared to the seed. The concentration in both ethyl acetate and butanol extracts of the leaves and the seeds was insignificantly different (Table 16.2). Nitrite, as oxidative molecule, is reduced to nitric oxide which can combine with oxygen to form the nitrogen radical known as peroxinitrite. Excess of nitrite in the body can react with the proteins and nucleic acids to form modified protein and DNA structure, hence, dysfunction the cell. Ethanol leaves extract of baobab showed a significant decrease in the nitrite concentrations while it was insignificant at ($p < 0.05$) in ethyl acetate and butanol leaves and seeds extracts.

Braca et al.[13] study was on antioxidant effects of flavonol, phenolic, procyanidins, and glycosides of baobab fruits and leaves through inhibition of α-glucosidase and discovered that the leaves extract exhibited corresponding antioxidant capacity potentials in scavenging ROS.

DPPH is a molecule that donates reactive species which accept electron or hydrogen ions to form a nonreactive compound.[25] One mechanism of action of plant extract is to donate an electron or hydrogen atom to an active radical molecule, reducing to an inactive harmless molecule, hence, the free reactive oxygen scavenging characteristics of medicinal plants.[21] The results obtained in this study showed that baobab leaves possess higher significant ($p < 0.05$) antioxidant activity when compared to the seeds extracts of all the solvents (Table 16.2). The ability of baobab leaf and seed to scavenge DPPH can be linked to its antioxidant property due to its chemical structures and the interaction with biological membranes, and its ability to quench singlet oxygen with higher efficiency, which is due to its number of conjugated bonds.[19] In furtherance to establish the ability of the baobab leaves and seed extracts to scavenge reactive species, ability of baobab leaves and seeds to reduce oxidative species was determined using ABTS radical, which is a stable nitrogen-centered radical species scavenger.[36] As presented in DPPH result, baobab leaves extracts have more significant ($p < 0.05$) ABTS scavenging ability when compared to the seed extracts (Table 16.2).

TABLE 16.2 The Percentage of DPPH, Nitrite Concentration, and ABTS Radical Scavenging Ability of Seeds and Leaves of Baobab.

Solvents	Leaves	Seeds
	Percentage of DPPH	
Ethanol	52.3	47.9
Ethyl acetate	61.5	49.9
Butanol	35.0	18.0
	Nitrite concentration (mg/mL)	
Ethanol	0.821 ± 0.003^b	0.861 ± 0.008^a
Ethyl acetate	0.861 ± 0.030^c	0.828 ± 0.003^c
Butanol	0.883 ± 0.017^d	0.829 ± 0.002^d
	ABTS radical scavenging ability (mmol of TEAC/g)	
Ethanol	6.005 ± 0.123^a	2.512 ± 0.286^b
Ethyl acetate	5.163 ± 0.091^c	2.481 ± 0.203^d
Butanol	3.824 ± 0.914^e	2.624 ± 0.432^f

NOTE: Values with the same letter superscript showed insignificant difference while different letters showed significantly difference, in the row.

16.6 SUMMARY

Baobab plant has proved a potent antioxidant medicinal plant that functions in amelioration of oxidative stress caused by reactive oxidative species as a result of imbalance in oxidant/antioxidant ratio in the body. The results obtained in this study have shown the potency of the leaves extract of various solvents in reducing oxidative stress compared to the seed extracts. Polyphenols, flavonoids, and anthocyanidins phytochemicals present in the leaves play vital role in antioxidant activities compared to the seeds. The butanol, ethanol, and ethyl acetate extracts of baobab seeds and leaves have proved to contain bioactive compounds with great antioxidant activity. The nitrite radical, FRAP, ABTS, and DPPH scavenging activities of baobab seed and leaf extracts showed that it is rich in antioxidant compounds. Hence, this baobab can serve as natural source of antioxidants for therapeutic purposes to prevent or slowdown aging processes and its associated degenerative diseases. It can also be applied in the industry as natural antioxidants for food preservative additives to prevent food deterioration. The incessant and regular use of baobab leaves by the local settlers for different purposes of food and medicine has proved its benefit in both the health and beverage industries.

KEYWORDS

- **ABTS (2,2-azino-bis-3-ethylbenzthiazoline-6-sulphonic acid)**
- ***Adansonia digitata***
- **antioxidant**
- **Baobab**
- **DPPH (2,2-diphenyl-1-picrylhydrazyl)**
- **FRAP (ferric reducing antioxidant power)**
- **nitrite assay**
- **phenolic compound**
- **phytochemical**

REFERENCES

1. Abioye, V. F.; Adejuyinta, J. A.; Idowu, C. F. Effects of Different Drying Methods on the Nutritional And Quality Attributes of Baobab Leaves (*Adansonia digitata*). *Agric. Biol. J. N. Am.* **2014,** *5*, 103–108.
2. Addy, E. O.; Eteshola, E. Nutritive Value of Mixture of Tigernut Tubers (*Cyperus esculentus* L.) and Baobab Seeds (*Adansonia digitata* L.). *J. Sci. Food Agric.* **1984,** *35*, 437–440.

3. Albasha, M. O.; Azab, A. E. Effect of Cadmium on the Liver and Amelioration by Aqueous Extracts of Fenugreek Seeds, Rosemary, and Cinnamon in Guinea Pigs: Histological and Biochemical Study. *J. Cell Biol.* **2014,** *2*, 34–44.
4. Al-Mamary, M; Al-Meeri, A; Al-Habori, M. Antioxidant Activities and Total Phenolics of Different Types of Honey. *Nutr. Res.* **2002,** *22*, 1041–1047.
5. Arowora, K. A.; Yakubu, O. E.; Shaibu, C.; Iornenge, T. J.; Ugwuoke, K. C. Chemical Composition of Baobab Leaves and Fractionation of its Ethanolic Extract Using Column Chromatography. *Int. J. Sci. Res.* (IJSR) **2018,** *8*, 812–821.
6. Ayele, Y.; Kim, J.; Park, E; Kim, Y.; Retta, N.; Dessie, G.; Rhee, S.; Koh, K.; Nam,K.; Kim, H. S. (2013). A Methanol Extract of *Adansonia digitata* Leaves Inhibits Pro-Inflammatory iNOS Possibly via Inhibition of NF-κB Activation. *Biomol. Ther.* **2013,** *21*, 146–152.
7. Azab, A. E.; Albasha, M. O. Hepatoprotective Effect of Some Medicinal Plants and Herbs Against Hepatic Disorders Induced by Hepatotoxic Agents. *J. Biotechnol. Bioeng.* **2018,** *2*, 8–23.
8. Azab, A. E.; Albasha, M. O.; Elsayed, A. S. I. Prevention of Nephropathy by Some Natural Sources of Antioxidants. *Yangtze Med.* **2017,** *1*, 235–266.
9. Baum, D.; Oginuma, K. A review of Chromosome Numbers in Bombacaceae with New Counts for Adansonia. *Taxon* **1994,** *43*, 11–20.
10. Bente, H.; Kari, H., Mari, M.; Rune, B. A Systematic Screening of Total Antioxidants in Dietary Plants. *J. Nutr.* **2002,** *132*, 461–471.
11. Besco, E; Braccioli, E; Vertuani, S; Ziosi, P; Brazzo, F; Bruni, R; Sacchetti, G; Manfredini, S. The Use of Photochemiluminescence for the Measurement of the Integral Antioxidant Capacity of Baobab Products. *Food Chem.* **2007,** *102*, 1352–1356.
12. Blomhoff, R.; Carlsen, M.; Halvorsen, B.; Holte, K.; Bohn, S.; Dragland, S.; Sampson, L.; Willey, C.; Senoo, H.; Umezono, Y.; Sanada, C.; Barikmo, I.; Berhe, N.; Willett, W.; Phillips, K.; Jacobs, D. The Total Antioxidant Content of More Than 3100 Foods, Beverages, Spices, Herbs and Supplements Used Worldwide. *Nutr. J.* **2010,** *9*, 3–6.
13. Braca, A.; Sinisgall, C.; De Leo, M.; Muscatello Pier, B.; Cioni, L.; Milella, L.; Ostuni, Giani, S.; Sanogo, R. Phytochemical Profile, Antioxidant and Antidiabetic Activities of Adansonia digitata L. (Baobab) from Mali, as a Source of Health-Promoting Compounds. *Molecules* **2018,** *23*(12), 3104–3110.
14. Bremer, B.; Bremer, K.; Chase, M. W.; Reveal, J. L.; Soltis, D. E.; Soltis, P. S.; Stevens, P. F.; Anderberg, A. A.; Fay, M. F.; Goldblatt, P.; Judd, W. S.; Kallersjo, M.; Karehed, J.; Kron, K. A.; Lundberg, J.; Nickrent, D. L.; Olmstead, R. G.; Oxelman, B.; Pires, J. C.; Rodman, J. E.; Rudall, P. J.; Savolainen, V.; Sytsma, K. J.; Van der Bank, M.; Wurdack, K.; Xiang, J. Q. Y.; Zmarzty, S. An update of the Angiosperm Phylogeny Group Classification for the Orders and Families of Flowering Plants: APG II. *Bot. J. Linn. Soc.* **2003,** *141*, 399–436.
15. Danbature, W. L.; Fai, F. Y.; Usman, A.; Ayim, P. Nutritional Evaluation of Babobab Seed. *Int. J. Res. Agric. Food Sci.* **2013,** *2*, 2311–2476.
16. Diederich, M.; Muller, F.; Cerella, C., Cardiac Glycosides: From Molecular Targets to Immunogenic Cell Death. *Biochem. Pharmacol.* **2017,** *125*, 1–11.
17. Francis, G.; Kerem, Z.; Makkar, H. P.; Becker, K. The Biological Action of Saponins in Animal Systems: A Review. *Br. J. Nutr*. **2002,** *88*, 587–605.
18. Gebauer, J.; El-Siddig, K.; Ebert, G. Baobab (*Adansonia digitata* L.): A Review on a Multipurpose Tree With Promising Future in the Sudan. *Garten Bauwissenschaft* **2002,** *67*, 155–160.

19. Gulcin, I. Antioxidant Activity of Food Constituents: An Overview. *Arch. Toxicol.* **2011,** *86*, 345–391.
20. Gyamfil, M. A.; Yonamine, M.; Aniya Y. Free Radical Scavenging Action of Medicinal Herbs From Ghana: *Thomningia sanguinea* on Experimentally-Induced Liver Injuries. *Gen. Pharmacol. Vascul. Syst.* **1999,** *32*, 661–667.
21. Hu, C.; Zhang, Y.; Kitts, D. D. Evaluation of Antioxidant and Prooxidant Activities of Bamboo *Phyllostachys niger* Leaf Extract *In Vitro*. *J. Agric. Food Chem*. **2000,** *48*, 3170–3176.
22. Igboeli, L. C.; Addy, E. O. H.; Salami, L. I. Effects of Some Processing Techniques on the Antinutrient Contents of Baobab Seeds (*Adansonia digitata*). *Bioresour. Technol.* **1997,** *59*, 29–31.
23. Irondi, E. A.; Akintunde, J. K.; Agboola, S. O.; Boligon, A. A.; Athayde, M. L. Blanching Influences the Phenolics Composition, Antioxidant Activity, and Inhibitory Effect of *Adansonia Digitata* Leaves Extract on *A*-Amylase, *A*-Glucosidase, and Aldose Reductase. *Food Sci. Nutr*. **2016,** *5*, 233–242.
24. Jackson, S. **Baobab: The Tree of Life: An Ethnopharmacological Review.** *J. Am. Botanical Council* **2015,** *108*, 42–53.
25. Je, J. Y.; Park, P. J.; Kim, E. K.; Ahn, C. B. Antioxidant and Angiotensin I Converting Enzyme Inhibitory Activity of *Bambusae caulis* in Liquamen. *Food. Chem.* **2009,** *113*, 932–935.
26. Keyvan, D.; Damien, D.; Muberra, K.; Raimo, H. Chemical Composition and *In Vitro* Antioxidant Evaluation of a Water-Soluble Moldavian Balm (*Dracocephalum moldavica* L.) Extract. *LWT- Food Sci. Technol.* **2007,** *40*, 239–248.
27. Magaji, B. M.; Danbature, W. L.; Muzakir, M. M.; Abubakar, B. A. Production of Biodiesel From Baobab Seed Oil. *Greener J. Agric. Sci.* **2014**, *4*, 22–26.
28. Maranz, S.; Niang, A.; Kalinganire, A.; Konate, D., Kaya, B. Potential to Harness Superior Nutritional Qualities of Exotic Baobabs if Local Adaptation can be Conferred Through Grafting. *Agrofor. Syst.* **2007,** *72*, 231–239.
29. Muhammad, I. A.; Musa, A.; Ajingi, Y. S.; Mika'il, T. A.; Haruna, M.; Yahaya, S. M.; Bello, A. M.; Dalhatu, M. M.; Mahmud, T.; Bichi, S. A.; Yunusa, A.; Muhammad, M.; Abubakar, S. A. Phytochemical Screening and Antioxidant Potential of Chloroform Extract of *Adansonia digitata* Leaves. *Dutse J. Pure Appl. Sci.* (DUJOPAS) **2019,** *5*, 55–64.
30. Obizoba, T. C.; Amaechi, N. A. The effects of Processing Methods on the Chemical Composition of Baobab (*Adansonia digitata*) Pulp and Seed. *Ecol. Food Nutr.* **1983,** *229*, 199–205.
31. Oboh, G.; Akinyemi, A. J.; Ademuluyi, A. O. Antioxidant Properties and Inhibitory Effect of Ethanolic Extract of *Struchium Sparaganophora* Leaves on α–Amylase and α- Glucosidase Activities. *Afr. J. Tradit. Complement. Altern. Med.* **2009,** *9*, 342–349.
32. Okorie, P.; Agu, F.; Ani, C.; Alozie, I.; Nworgu, C.; Anyaeji, P.;Ugwu,P.; Uzoigwe J.; Igwe, U.; Ejim, N.; Nwachukwu, D. The Effect of Aqueous Leaf Extract of *Adansonia Digitata* (Baobab) on *Diabetes Mellitus* and the Anterior Pituitary of Adult Male Wistar Rats. *J. Diabetes Endocrinol.* **2019,** *10*, 18–29.
33. Osman, M. A. Chemical and Nutrient Analysis of Baobab (*Adansonia digitata*) Fruit and Seed Protein Solubility. *J. Plant Foods Human Nutr.* **2004,** *59*, 29–33.
34. Rahul, J. *Adansonia digitata* (baobab): A Review of Traditional Information and Taxonomic Description. *Asian Pac. J. Trop. Biomed.* **2015,** *5*, 79–84.

35. Ramadan, A.; Harraz, F. M.; El-Mougy, S. A. Anti-inflammatory, Analgesic and Antipyretic Effects of the Fruit Pulp of *Adansonia digitata*. *Fitoterapia* **1993,** *65*, 418–422.
36. Re, R.; Pellegrini, N.; Proteggenta, A.; Pannala, A.; Yang, M.; Rice-Evans, C. Antioxidant Activity Applying an Improved ABTS Radical Cation Decolorization Assay. *Free Radic. Biol. Med.* **1999,** *26*, 1231–1237.
37. Shukla, Y. N.; Dubey, S.; Jain, S. P.; Kumar, S. Chemistry, Biology and Uses of *Adansonia digitata* – A Review. *J. Med. Aromat. Plant Sci.* **2001,** *23*, 429–434.
38. Sidibe, M.; Williams, J. T. Ed.; Baobab-Adansonia digitata. International Centre for Under Utilised Crops: Southampton, UK, 2002; pp 14–36.
39. Vertuani, S.; Braccioli, E.; Buzzoni, V.; Manfredini, S. Antioxidant Capacity of *Adansonia digitata* Fruit Pulp and Leaves. *Acta Phytotherapeutica* **2002,** *86*, 2–7.
40. Wickens, G. E. *The Baobabs: Pachycauls of Africa, Madagascar and Australia*. Springer Science & Business Media: Cham, Switzerland, 2008; pp 1–30.
41. Yao, Y. W.; Sang, M. Z.; Ren, G. Phenolic Composition and Antioxidant Activities of Eleven Celery Cultivars. *J. Food Sci.* **2010,** *75*, C9–C13.

CHAPTER 17

Antioxidant, Anti-Inflammatory, and Thrombolytic Activities of *Monodora tenuifolia*, Benth Seed Extract

Ademola Olabode Ayeleso, Olagunju Doris Imuetiyanosa, Samuel Chibueze Nzekwe, and Oluokun Oluboade Oyedapo

ABSTRACT

In this study, in vitro antioxidant, anti-inflammatory, and thrombolytic activities of *Monodora tenuifolia* Benth seed extract were explored. The procedure involved identification, collection, authentication, and extraction of *M. tenuifolia* seed extract with 80% (v/v) ethanol. The evaluation of antioxidant capacity (DPPH, FRAP, H_2O_2, and ABTS) of the extract was done using established protocols. The anti-inflammatory as well as thrombolytic activities of the extract were also investigated with standard methods. Results obtained showed that alkaloids, flavonoids, tannins, and proteins were present in the extract while saponin was absent. Quantitatively, the extract contained total phenolic 21.43 ± 0.03 mg/g extract (TAE), flavonoid 190.40 ± 2.52 QE/g, Vitamin C content was 40.60 ± 0.60 mg/g, and vitamin E content was 14.20 ± 0.04 mg/g QE and protein 6.80 ± 0.01 mg/g. Moreover, extract exhibited anti-inflammatory potential by inhibiting albumin denaturation in all the different concentrations used in the study. The extract also showed appreciable thrombolytic activities in concentration-dependent manner but not as effective as the standard drug. Even though the extract showed antioxidant potentials, these activities were significantly lower than the standard drugs used. Summarily, the study indicated that *M. tenuifolia* seeds contain bioactive compounds with anti-inflammatory and thrombolytic effects.

Curative and Preventive Properties of Medicinal Plants: Research on Disease Management and Animal Model Studies. Megh R. Goyal, Junaid Ahmad Malik, & Ademola Olabode Ayeleso (Eds.)

17.1 INTRODUCTION

Inflammation can be defined as the mechanism of immune response to injury or invasion of pathogenic organisms and initiate the process of healing. It is, therefore, a vital mechanism of wound healing, pathogen, or toxic compound eradication healing of damaged cells or tissues.[33] Inflammatory response brings efficient interaction of the cells and cellular molecules to minimize injuries and infections, hence, contributing to tissue restoration and inflammatory resolution commonly known as homeostasis. However, prolonged and uncontrolled inflammation is a disease condition of chronic inflammation characterized with heat, redness of skin, pain, swelling, and loss of tissue.[2,34] Embolism is the process of formation of blood clotting taking place within the blood vessel or arterial endothelium that may cause full or partial blockage of blood flow in the form of plaque that results to a severe disease known as arteriosclerosis or thrombosis. In the case of pulmonary embolism or blood clot, it causes difficulty in breathing, chest pain in the case of cardiovascular blockage, and oedema when the big vein of the limb is blocked.[20] These are linked to the generation of ROS, such as $O_2^{\cdot -}$, H_2O_2 pro-oxidant, peroxi-nitrite, singlet oxygen that cause damage to cells or tissues and triggers inflammatory responses. Antioxidant remains the class of biochemical molecules responsible for neutralization of the free radicals to prevent oxidative stress from cells and tissue damage.[32] Oxidative stress has been implicated in several diseases associated with inflammation such as cancer, cellular necrosis, muscular dystrophy, neurological disorder, Parkinson's dementia, and liver disorder.[6]

Research and clinical trials have proved the inefficiency of nonsteroidal anti-inflammatory synthetic drugs in the treatment of thromboembolic disease as well as in oxidative and inflammatory diseases due to several side effects.[12] Moreso, the thrombolytic agents readily available possess certain deficiencies that involve that necessitate the usage of large doses before it becomes effective. The drug is nonspecific to fibrin degradation and is also associated with bleeding, hence there is a need to develop improved recombinant therapy or use of less toxic and readily available natural products of medicinal plant origin in order to prevent the antagonistic response of the thrombolytic drugs.[21] Different plant phytochemicals, such as ajoene, thiosulfate, diallyl disulfide, allicin, and diallyl trisulfide, are found in *Allium sativum* and sesamol, sesamolin, and sesamin from

Sesamum indicum have shown to be effective inhibitors in the formation of thrombosis.[23]

This study focuses on the in vitro investigation of antioxidant, anti-inflammatory, and thrombolytic potentials of *Monodora tenuifolia* seed extract.

17.2 HISTORY AND DISTRIBUTION OF THE *MONODORA TENUIFOLIA*

Medicinal plants or herbal medicines have formed the basis of the primary health care system in both developed and developing countries for ages because these are less toxic, efficient, and cost-effective in comparative to synthetic drugs.[16] Monodora plant genus contains about 35 species belonging to the family of annonaceae and is widely distributed in Africa with two majorly known species found in Western Africa, Nigeria. The two species of *M. myristica* and *M. tenuifolia* grow mostly in Eastern and Western parts of Nigeria and are known as "African Nutmeg" in English and "Ehu" in Igbo or "Ariwo" in Yoruba.

M. tenuifolia is a rich ethno-botanical traditional medicinal plant used in toothache relief, diarrhea, skin rashes, dysentery, dermatitis, and vermifuge.[13,14] *M. tenuifolia* is a spice that has been shown to possess antioxidant activity, hence, it is assumed to be used for effective management and treatment of diseases related to oxidative stress.[26] The curative or preventive potentials of *M. tenuifolia* plants have been attributed to their phytochemical constituents such as phenolic acids, tannins, flavonoids, terpenoids, alkaloids, saponins, and stilbenes.[4] *M. tenuifolia,* Benth (Figure 17.1) is a wild growing orchid flowering tree that consists of rarely woody vines, deciduous, or evergreen, with aromatic bark, fruits, leaves, and flower.[14]

The fruits are in orange sizes that appear greenish with pale spots when unripen but turn orange or sunshine-yellow when ripe. The fruit contains multiple pea-sized seeds that are brown with aromatic scent when broken to remove the endocarp. *M. tenuifolia* is a seasonal plant, fruiting between May and August during the rainy season and when ripped, the fruits are plucked, then cut open to harvest the seeds. The growth of the seedlings of *M. tenuifolia* is best done under shade as its growth is slow and can take up to 12 weeks after germination to adapt to full sunny climatic conditions. The seeds have a good aroma, hence can serve as spice and condiments in almost all foods in Nigeria.[18,26]

FIGURE 17.1 The fruit and seeds of *M. tenuifolia* Benth plant (own camera).

17.3 MEDICINAL USE OF *MONODORA TENUIFOLIA* BENTH

M. tenuifolia has ethnobotanical archive in traditional treatment of different diseases including tooth ache, skin rashes, dysentery, vermifuge, and diarrhea.[14] It has been shown to possess good antioxidant potentials, which is a vital property in the treatment and management of stress-induced diseases.[26] In food, it gives flavor and enhances appetite, hence it is used as spices and condiments.[14] Spices are well-known plant products used for food spicing and appetite enhancement to give maximum satisfaction during consumption. *M. tenuifolia* is an important member of spices that have been shown to possess antipyretic, antioxidant, antidiarrhea, and antilipidaemia properties.[14,26]

17.4 MATERIALS AND METHODS

17.4.1 COLLECTION OF MONODORA TENUIFOLIA SEEDS

Matured ripped fruits of *M. tenuifolia* were harvested from Obafemi Awolowo University Ile-Ife, Nigeria, and identified by a botanist. A copy of the specimen was deposited at the University Herbarium with a voucher number-17979.

17.4.2 PREPARATION OF BLOOD SAMPLES

Ethical clearance was obtained from the Ethical Review Committee, Adeleke University, Ede, Nigeria (Reference number: AUERC/FOS/BCH/02). Fresh bovine blood sample was collected from an abattoir at Halleluya Estate, Osogbo, Nigeria into EDTA bottles. The blood was spinned at 3000 rpm with a centrifuge for 10 min and rinsed with normal saline (0.85% NaCl) until supernatant becomes colorless. Two milliliters of erythrocytes were added up with normal saline to make 100 mL of 2% RBC suspension. Also, blood sample (3 mL) from healthy human volunteers, without a history of anticoagulant therapy or oral contraceptive were collected into dry sterile Eppendorf tubes and stand for 45 min at room temperature to coagulate, then, the samples were centrifuged for 5 min at 10,000 rpm to obtain the serum.

17.4.3 PROCESSING OF PLANT MATERIALS

The fruits were sliced and opened to expose the seeds which were collected. The fresh seeds of *M. tenuifolia* were removed from the opened fruits, washed clean with water, and sun-dried over a period of 2 weeks. The seed coat was removed by cracking, the cotyledon was removed, and followed by grinding to powder using both hand operator grinder and electric blender.

17.4.4 PREPARATION OF HYDRO-ETHANOL EXTRACT OF M. TENUIFOLIA SEEDS

Hydro-ethanol solution (80%) was prepared by mixing 800 mL ethanol with 200 mL of distilled water according to the method of Oyedapo and Amos.[30] Typically, 200 g of powdered seed was soaked in 100 mL of the solvent at the ratio of 1:5 (w/v) for 72 h and stirred occasionally, then filtered through doubled-layered cotton cloth. The residue was washed severally and refiltered until a colorless solution is obtained. The filtrates were concentrated using rotary evaporator at 40°C. The slurry was carefully dispensed into clean-dry petri dish, exposed to hot air to yield dry ethanol seed extract of *M. tenuifolia*. The residue was kept in the desiccator for further use.

17.4.5 QUALITATIVE PHYTOCHEMICAL ANALYSIS OF THE EXTRACT OF *M. TENUIFOLIA* SEEDS

Different phytochemicals such as cardiac glycoside, alkaloids, flavonoid, saponin, tannin, protein, and phenolics in the extract were identified using different standard methods of Oyedapo et al.[31] and Al-Tannak et al.[5]

17.4.5.1 TEST FOR ALKALOIDS

Extract of *M. tenuifolia* (0.5 g) was dissolved in 10 mL of HCl (10% v/v) and gently warmed, then filtered. In four different test tubes, 1 mL of the filtrate was added into each, followed by the addition of a few drops of Draggendorff's, Wagner's, and Mayer's reagents separately. Equal volume of 10% (v/v) HCl was used as parallel control. Cloudiness or suspension indicated that alkaloid was present.

17.4.5.2 TEST FOR FLAVONOIDS

Powdered seed (0.25 g) was soaked with 10 mL ethyl acetate for 30 min, then filtered. Ethanolic potassium hydroxide solution (0.5 M) was prepared and added in a few drops to 1 mL of filtrate, then observed for cloudy yellowish suspension which indicated the presence of flavonoid.

17.4.5.3 TEST FOR TANNINS

Powdered seed (10 mg) was added 10 mL distilled water, shaken, heated in the warm water bath, and then filtered. Then, 0.5 M $FeCl_3$ (few drops) dissolved in glacial acetic acid was added to the filtrate (1 mL). The appearance of blue-black precipitate was an indication of the presence of tannin.

17.4.5.4 TEST FOR PROTEIN

The powdered seeds (0.1 g) were soaked in 5 mL distilled water, shaken with a vortex mixer, and filtered. Biuret reagent (1 mL) was added to the filtrate (1 mL), gently shaken, and allowed to stand for 5 min. A violet color appearance showed the presence of protein.

17.4.5.5 TEST FOR SAPONIN

The presence of saponin was tested by soaking 50 mg of the sample in distilled water (2 mL) and shaken extensively to observe froth formation. It was then warmed at 55°C for 5 min and vigorously shaken again. The absence of frothing before or after warming indicated the absence of saponin.

17.4.6 QUANTITATIVE PHYTOCHEMICALS ANALYSIS OF THE SEED EXTRACT OF *M. TENUIFOLIA*

17.4.6.1 DETERMINATION OF PHENOLIC CONTENT

This was carried out by the procedure of Singleton et al.[37] Typically, standard sample stock of 1 mg/mL extract was prepared using 10 mg extract in 200 μL of DMSO and made up to 10 mL. Tannic acid (20 μg/mL) was prepared as standard stock and divided into five dilutions (0.0–1.00 μg/mL). Then, Na_2CO_3 (1.5 mL, 7.5% w/v) and Folin–Ciocalteu reagent (1:10 v/v) were added into the different test tubes containing 500 μL of both the sample and standard, respectively, and incubated for 1.5 h. The solutions were read at 725 nm to obtain the absorbance values and the result was presented in milligram per gram tannic acid equivalent (μg/g TAE).

17.4.6.2 DETERMINATION OF FLAVONOID CONTENT

This was done using Sun et al.[38] method. Briefly, $NaNO_2$ (3 mL, 5% w/v) with $AlCl_3$ (0.3 mL, 10% w/v) and NaOH (2.0 mL, 4% w/v) were added to 200 μL of 1 mg/mL standard and sample solutions. The resulting solution was measured at 500 nm to obtain absorbance values after 10 min incubation at 25°C against reagent blank and the result was expressed in mg/g QE.

17.4.6.3 EVALUATION OF VITAMIN C CONTENT

This was estimated according to Omaye et al.[27] method and as explained by Fajobi et al.[15] The mixture contained 0.5 mL extract (1 mg/mL) and 2,4-dinitrophenyl hydrazine (DNPH) reagent (0.5 mL), then incubated at 37°C for

10 min and cooled in ice cold. Thereafter, 8.5% (v/v) H_2SO_4 (2.5 mL) was added and incubate at 25°C for 30 min. Ascorbic acid (50 μg/mL) was used as standard stock divided into 0–50 μg/mL, and blank containing 0.5 mL of distilled water. The absorbance was obtained at 530 nm and the result expressed in mg/mL after plotting a calibration curve.

17.4.6.4 EVALUATION OF VITAMIN E CONTENT

This was evaluated according to the method of Baker and Frank[8] and as described by Fajobi et al.[15] Briefly, 0.2 mL extract in 1 mg/mL was added with 2′,2′,α-bipyridyl reagent (0.5 mL, 0.12% w/v) and ferric chloride (0.5 mL, 0.12% w/v). Trolox standard sample (25 μg/mL) in ethanol was divided into six dilutions of 0–0.25 μg/mL and treated in the similar way as the sample. Absorbance at 492 nm was read against the reaction blank that contains 0.2 mL distilled water and expressed in mg TE/g of sample.

17.4.7 ANTIOXIDANT POTENTIAL OF THE SEED EXTRACT OF M. TENUIFOLIA

17.4.7.1 DPPH SCAVENGING ABILITY

This was determined using Blois (1985) method as modified and reported in the work of Aina and Oyedapo.[3] Extract stock (1 mg/mL) was divided into six dilutions ranging from 0 to 250 μg/mL and standard ascorbic acid (0.25 μg/mL) divided in six varying volumes 0–0.6 mL. In total, 200 μL of the extract and standard ascorbic acid was mixed with DPPH (1 mL, 0.3 mM) in methanol, then incubated inside cupboard for 30 min. The absorbance at wavelength 517 nm was taken against the blank containing distilled water. DPPH radical percentage scavenging ability was calculated as stated below:

$$\%\ \text{DPPH Scavenging Ability} = \frac{(\text{Absorbance of control} - \text{Absorbance of test})}{\text{Absorbance of control}} \times 100$$

17.4.7.2 FERRIC REDUCING ANTIOXIDANT POWER

The method of Benzie and Strain[9] was employed for this assay. It consisted of different concentrations ranging from 0 to 300 μg/mL of extract. Then, 500 μL of the sample and different concentrations of ascorbic acid as the standard was added into a solution of potassium ferricyanide ([$K_3Fe(CN)_6$] (2.5 mL, 1% w/v) and phosphate buffer (2.5 mL, 0.2 M, pH 6.6). The resulting solution was heated in the warm water bath at 50°C in 20 min and added with tricholoroacetic acid (2.5 mL, 10% (w/v), centrifuge with 3000 rpm for 10 min and decanted. Thereafter, 2.5 mL of supernatant was added to dH_2O (2.5 mL) and $FeCl_3$ (0.5 mL, 0.1% w/v) and absorbance was read at 700 nm. The result was expressed in μg/g ascorbic acid equivalent.

17.4.7.3 ABTS SCAVENGING ABILITY

This was carried out with the method of Rezzagui et al.[1] The cation radical ($ABTS^+$) of ABTS was prepared by reacting the stock solution (7 mM) with potassium persulphate (2.45 mM) then made up to 1 L with distilled water and incubated in the dark cupboard overnight. The assay mixture was extract of varying volumes (0–0.05 mL) adjusted with dH_2O (2 mL) and $ABTS^+$ solution (1.0 mL) was added to 500 μL from each test tube and incubated at 25°C for 30 min. The absorbance values were obtained at 734 nm against distilled water as blank. Trolox served as the standard and percentage ABTS scavenging potential was determined as stated below:

$$\%\ \text{ABTS Scavenging potential} = \frac{(\text{Absorbance of control} - \text{Absorbance of test})}{\text{Absorbance of control}} \times 100$$

17.4.7.4 H_2O_2 SCAVENGING ASSAY

This was done as described by Modi et al.[25] and as reported by Archana and Vijayalakshmi.[7] The reaction mixture contained extract (0–250 μg/ mL) in varying volumes from 0 to 0.25 mL and added up to 1 mL dH_2O. Thereafter, 200 μL of the extract was added with phosphate buffer (0.5 mL, 20 mM, pH 7.4), hydrogen peroxide (0.4 mL, 2 mM), then incubate at 25°C for 5 min. Then, a 2 mL dichromate-acetic acid reagent (5 %, w/v) was added and read at 570 nm. Ascorbic acid was used as the standard.

$$\%\ H_2O_2\ \text{inhibition} = \frac{(\text{Absorbance of test} - \text{Absorbance of control})}{\text{Absorbance of test}} \times 100$$

17.4.8 ANTI-INFLAMMATORY POTENTIAL OF THE SEED EXTRACT OF *M. TENUIFOLIA*

17.4.8.1 ASSAY OF MEMBRANE STABILITY ACTIVITY

The membrane stability potential of *M. tenuifolia* seed extract was determined by the method of Aina and Oyedapo[3] using a freshly collected bovine blood into an EDTA bottle. The blood was spinned at 3000 rpm for 10 min and washed in normal saline (0.85% NaCl) until supernatant becomes colorless. The erythrocyte residue was prepared in 2% suspension made up of normal saline (0.85%). The assay mixture contained different concentrations (0–350 µg/mL) of the extract measured in varying volumes (0.00, 0.05, 0.10, 0.15, 0.20 mL) with (0.85% (w/v) NaCl normal saline (1 mL) followed by addition of hypo saline NaCl (1.0 mL, 0.42% w/v), phosphate buffer (0.5 mL, 0.15M, pH 7.4), then erythrocyte suspension (0.5 mL, 2% (v/v) was added and incubated at 56°C for 30 min, cooled and spun at 3000 rpm for 10 min. Ibuprofen served as the standard drug (1 mg/mL). The supernatant collected was read at 560 nm and result presented as percentage stability.

$$\text{Percentage stability} = 100 - \left[\frac{\text{Abs}_{\text{cnt}} - \text{Abs sample}}{\text{Abs}_{\text{control}}}\right] \times 100$$

17.4.8.2 INHIBITORY ASSAY OF ALBUMIN DENATURATION

The inhibitory effect of *M. tenuifolia* in albumin denaturation using the method described by Mizushima and Kobayashi[24] and as reported by Aina and Oyedapo.[3] Different concentrations of extracts ranging between 0 and 350 µg/mL was prepared in normal saline (0.85%) and phosphate buffered saline pH 6.3 (1.5 mL) and bovine serum albumin (0.5 mL) was added. The mixed solution was incubated at 37°C for 20 min and absorbance was read at 660 nm against phosphate buffered saline as blank. Ibuprofen (1 mg/mL) served as the standard drug and inhibition were calculated as described below:

$$\%\ \text{inhibition} = \frac{100\ (1\ \text{Absorbance of samples})}{\text{Absorbance of blank}}$$

Then graph of % inhibition plotted against extract concentration.

17.4.8.3 ASSAY OF INHIBITION OF HYPOTONIC-INDUCED RBC HEMOLYSIS

The inhibitory effect of the hypotonic-induced RBC hemolysis was done by the method of Shinde[36] and slightly modified as reported by Oyedapo et al.[29] Briefly, extract of different concentrations (0–250 µg/mL) was prepared at varying volumes (0–0.125 mL) made up to 3.0 mL with phosphate-buffered saline (2.5 mL, 0.15 M, pH 7.4) and red blood cell (0.5 mL, 2% v/v) was added, then mixed slowly by turning the test tubes upside-down. The solution was then incubated for 5 min under cupboard and then at 37°C for 15 min in a water bath. After cooling, the resulting solution was spun for 10 min at 3000 rpm and absorbance values were obtained at 540 nm. Ibuprofen served as a standard drug and percentage inhibition was calculated as described below:

$$\% \text{ inhibition} = \frac{100 \times (1 - \text{Absorbance of samples})}{\text{Absorbance of blank}}$$

Then the graph of % inhibition plotted against extract concentration.

17.4.8.4 ASSAY OF INHIBITION OF HEAT-INDUCED RBC HEMOLYSIS

The inhibitory effect *M. tenuifolia* seed extract on heat-induced RBC hemolysis was assayed by the method of Mizushima and Kobayashi[24] with modifications by Aina and Oyedapo.[3] Different concentrations (0–250 µg/mL) of the extract were pipetted varying volumes (0–0.125 mL), then adjusted to 1.0 mL with normal saline and phosphate buffered saline (1.5 mL, 0.15 M, pH 7.4) and RBC (0.5mL, 2%) suspension were added. The solution was slowly mixed by turning upside down severally and incubated for 5 min at 37°C and then for 15 min at 54°C, cooled, and spinned for 10 min at 3000 rpm. The absorbance values were obtained at 540 nm while ibuprofen was used as a standard drug and % inhibition was calculated as:

$$\text{Percentage inhibition} = \frac{100\ (1 - \text{Absorbance of samples})}{\text{Absorbance of blank}}$$

17.4.9 THROMBOLYTIC ACTIVITY OF THE SEED EXTRACT OF *M. TENUIFOLIA*

Thrombolytic activity of *M. tenuifolia* seed extract was done using the method described by reyad-ul-Ferdous et al.[35] Briefly, 500 μL of blood sample was collected and transferred to sterile preweighed eppendorf tubes (M_1) and stood to coagulate for 45 min at 25°C and centrifuge at 3000 rpm for 5 min. Blood clots residue obtained was weighed (M_2) and added with 100 μL of extract (1 mg/mL) and 100 μL of Aspirin (1 mg/mL) as positive control. They were all incubated in water bath for 1.5 h at 37°C. The released fluid was removed after incubation and tubes containing the blood clot were reweighed (M_3). The percentage of clot lysis is calculated as:

$$\text{Percentage clot lysis} = \frac{\text{Weight of released clot}(M3 - M1)}{\text{Clot Weight}(M2 - M1)} \times 100$$

17.5 STATISTICAL ANALYSIS OF DATA

One-way ANOVA was used to determine mean ± standard error of mean. Significant differences in the mean values were considered when $p < 0.05$ using GraphPad Prism Software.

17.6 RESULTS AND DISCUSSION

17.6.1 PHYTOCHEMICAL CONSTITUENTS OF THE SEED EXTRACT *M. TENUIFOLIA* SEEDS

The phytochemical constituent screened showed that alkaloids, protein, flavonoids, tannins, and saponin were present (Table 17.1). This implies that the *M. tenuifolia* is rich in vital biomolecules with known biological activities. The extract of *M. tenuifolia* seeds contained total phenolic (190.40 ± 2.52 mg/g TAE), total flavonoids (21.43 ± 0.03 mg/g QE), total protein (26.80 ± 0.01 mg/g), vitamin C (40.60 ± 0.06 mg/g), and vitamin E (14.21 ± 0.60 mg/g) (Table 17.2). These bioactive components with beneficial effects have been demonstrated to possess and exhibit different health-promoting properties ranging from antioxidants, anti-inflammatory, antiproliferative, anticancer, antispasmodic, antidepressant, antiulcer, and protection against degenerative and chronic diseases, that is, cardiovascular diseases.[44]

TABLE 17.1 Phytochemical Constituents of the Seed Extract of *M. tenuifolia.*

Phytoconstituents	Test	Observation	Remark
Saponin	Extract + shake + warm	Frothing	Present
Flavonoid	0.5 M Ethanolic KOH + Extract	Yellow cloudy Precipitate	Present
Protein	Extract + Biuret reagent	Pink/violet precipitate	Present
Tannin	Extract + $FeCl_3$ in acetic acid	Blue/black precipitate	Present
Alkaloids			
Mayer's reagent	Extract + Mayer's reagent	Cream precipitate	Present
Wagner's reagent	Extract + Wagner's reagent	Brownish/yellow precipitate	Present
Dragendorff's reagent	Extract + Dragendorff's reagent	Orange-red precipitate	Present

TABLE 17.2 Quantitative Levels of Phytoconstituents in the Seed Extract of *M. tenuifolia.*

Phytochemical constituent	Concentration (mg/g)
Total phenol	190.40 ± 2.52 (mg/g QE)
Total flavonoid	21.43 ± 0.03(mg/g TAE)
Total protein	26.80 ± 0.01 (mg/g)
Vitamin C	40.60 ± 0.60 (mg/g)
Vitamin E	14.21 ± 0.60 (mg/g)

17.6.2 ANTIOXIDANT POTENTIALS OF M. TENUIFOLIA SEED EXTRACT

17.6.2.1 DPPH SCAVENGING ACTIVITY OF THE SEED EXTRACT OF M. TENUIFOLIA

The study showed that *M. tenuifolia* seed extract exhibited an appreciable amount of DPPH scavenging ability at 62.86% compared to that ascorbic acid standard at 94.83%. This is an indication that the extract exhibited an appreciable scavenging activity in a concentration-dependent manner but was not as effective as the standard drug. This effect might be a result of the presence of phenolic compounds especially tannins in the extract. Hagerman et al.[17] have shown that high molecular weight phenolic compounds such as tannins have the capacity to scavenge free radicals. This study agreed with the previous reports of Usunobun and Igwe[42] and Yembeau et al.[45] that studied the concentration-dependent DPPH scavenging activities for extracts of *P. mildbraedii* and *Theobroma cacao*, respectively. This study revealed that the seed extract of *M. tenuifolia* scavenged DPPH radical maximally at a percentage of 62.86% while that of the standard drug scavenged DPPH radical maximally at 94.83%. This showed that there was an appreciable antioxidant activity by the extract though not as high as that of the standard drug (Table 17.3).

TABLE 17.3 DPPH Scavenging Activity of the Seed Extract of *M. tenuifolia*.

Concentration (μg/mL)	DRSA of extract (%)	DRSA of ascorbic acid(%)
50	17.07 ± 0.48	19.38 ± 0.78
100	30.18 ± 0.48	58.05 ± 0.26[a]
150	38.53 ± 1.17	75.50 ± 0.43[a]
200	48.72 ± 0.64	88.78 ± 2.26[a]
250	54.87 ± 1.08	93.97 ± 0.00[a]
300	62.86 ± 2.20	94.83 ± 0.43

Each value represents the mean ± SEM, (Standard Error of Mean), n = 3.
[a]*Indicates significant difference between the extract and standard (ascorbic acid) at $p < 0.05$*

17.6.2.2 H_2O_2 SCAVENGING ACTIVITY OF THE SEED EXTRACT OF *M. TENUIFOLIA*

Hydrogen peroxide is a weak oxidizing agent that renders some enzymes inactive by the oxidation of essential thio (–SH) group and can sometimes be toxic to the cell when it crosses the cell membrane giving rise to hydroxyl radicals through the Fenton reaction.[7] This study revealed that the extract scavenged hydrogen peroxide maximally at 17.09% while the standard drug scavenged hydrogen peroxide at 45.16%. It implies that though the extract showed a level of scavenging activity that was concentration dependent, the extract was not as effective as the standard drug. This is in agreement with a previous study that reported a dose-dependent inhibition of hydrogen peroxide by phloroglucinol, a polyphenol component of Laminariaceae family.[7] The results revealed that the extract of *M. tenuifolia seed* scavenged hydrogen peroxide radical maximally at 17.09% while the standard drug scavenged maximally at 45.16%. This implies that the hydrogen peroxide scavenging action of the seed extract is significantly lower compared to the standard drug (Table 17.4).

TABLE 17.4 H_2O_2 Scavenging Activity of the Seed Extract of *M. tenuifolia*.

Concentration (µg/mL)	HPRSA of extract (%)	HPRSA of ascorbic acid
0	0	0
50	1.83 ± 1.05	24.19 ± 1.88[a]
100	8.06 ± 0.91	37.30 ± 1.82[a]
150	9.37 ± 5.85	39.26 ± 5.18[a]
200	17.09 ± 0.91	45.16 ± 1.82[a]

Each value represents the mean ± SEM, (Standard Error of Mean), n = 3.

[a]*Indicates significant difference between the extract and standard (ascorbic acid) at* $p < 0.05$

17.6.2.3 FRAP OF THE SEED EXTRACT OF M. TENUIFOLIA

FRAP assay evaluates the reduction potential of the extract of *M. tenuifolia* seeds when it reacted with potassium ferrocyanide, which then reacts with ferric chloride to form a Perl's Prussian blue ferric ferrous complex when reduced, thereby showing reducing ability of the agent or compound.[28] The reducing ability of the extract gave a maximum FRAP value of 49.37 mg/g while the standard drug had a maximum FRAP value of 272.90 mg/g. This finding reveals that though the extract has a reducing ability that was concentration dependent, this activity is far less than that of the standard drug (Table 17.5).

TABLE 17.5 Ferric Reducing Antioxidant Power (FRAP) of the Seed Extract of *M. tenuifolia*.

Concentration (µg/mL)	HPRSA of extract (%)	HPRSA of ascorbic acid
0	0	0
50	43.86 ± 0.41	53.80 ± 0.41[a]
100	45.01 ± 0.45	94.74 ± 0.99[a]
150	47.82 ± 1.93	149.55 ± 1.63[a]
200	47.99 ± 1.12	161.22 ± 13.22[a]
250	49.37 ± 0.13	272.90 ± 1.19[a]

Each value represents the mean ± SEM, (Standard Error of Mean), n = 3.

[a]*Indicates significant difference between the extract and standard (ascorbic acid) at* $p < 0.05$

17.6.2.4 ABTS RADICAL SCAVENGING ABILITY OF SEED EXTRACT OF M. TENUIFOLIA

ABTS assay is a measure of the capacity of antioxidants to scavenge the produced ABTS in the aqueous phase when compared with the Trolox standard.[11] ABTS radical maximally at 11.77% while the standard drug scavenged maximally at 90.88%. This further implies that the antioxidant activity of *M. tenuifolia* seed extract by inhibiting ABTS radical was significantly lower than the standard drug in a dose-dependent manner (Table 17.6).

TABLE 17.6 ABTS Radical Scavenging Ability of the Seed Extract of *M. tenuifolia*.

Concentration (μg/mL)	ARSA of extract (%)	ARSA of trolox
0	0	0
10	6.56 ± 0.48	55.02 ± 1.47[a]
20	7.40 ± 1.68	60.89 ± 0.30[a]
30	8.75 ± 1.25	76.04 ± 3.55[a]
40	11.77 ± 1.03	90.88 ± 2.13[a]

Each value represents the mean ± SEM, (Standard Error of Mean), n = 3.
[a]Indicates significant difference between the extract and standard (ascorbic acid) at p < 0.05

17.6.3 ANTI-INFLAMMATORY ACTIVITIES OF SEED EXTRACT M. TENUIFOLIA

17.6.3.1 INHIBITORY EFFECT OF M. TENUIFOLIA SEED EXTRACT ON HEAT-INDUCED RBC HEMOLYSIS

The exposure of cell membranes and RBCs to injuries such as heat and hypotonic solution leads to the lysis of membrane associated with hemolysis.[3,39] This study revealed that hydro-ethanolic extract of *M. tenuifolia* seeds inhibited the hemolysis of RBC maximally at 50.52% while that of the standard drug was 81.23%. The result showed that the inhibitory activity of the extract was significantly lower in comparison to the standard drug at $p < 0.05$ in concentration-dependent manner (Table 17.7).

17.6.3.2 INHIBITORY EFFECT OF M. TENUIFOLIA SEED EXTRACT ON HYPOTONIC-INDUCED RBC HEMOLYSIS

The inhibitory effect of *M. tenuifolia* seed extract on hypotonic RBC hemolysis is shown in Table 17.8. The result revealed that the extract of seeds of *M. tenuifolia* inhibited hemolysis of hypotonic induced red blood cell maximally at 55.27% and that of the standard drug with a percentage inhibition of 74.35%. It was shown in this study that the percentage inhibition by the extract was significantly lower when compared with the standard drug at $p < 0.05$ in concentration-dependent manner.

TABLE 17.7 Inhibitory Effect of *M. tenuifolia* Seed Extract on Heat-Induced RBC Hemolysis.

Concentration (µg/mL)	IE of extract (%)	IE of ibuprofen (%)
0	0	0
50	19.82 ± 0.29	22.90 ± 0.38[a]
100	22.65 ± 0.14	29.37 ± 0.97[a]
150	33.75 ± 0.08	45.93 ± 0.87[a]
200	47.85 ± 0.42	62.95 ± 0.38[a]
250	50.52 ± 0.93	81.23 ± 0.43[a]

Each value represents the mean ± SEM, (Standard Error of Mean), n = 3.
[a]*Indicates significant difference between the extract and standard (ascorbic acid) at p < 0.05.*

TABLE 17.8 Inhibitory of *M. tenuifolia* Seed Extract on Hypotonic Induced RBC Hemolysis.

Concentration (µg/mL)	IE of extract (%)	IE of ibuprofen (%)
0	0	0
50	10.38 ± 0.67	24.56 ± 1.19[a]
100	22.19 ± 0.37	24.55 ± 1.25
150	25.40 ± 0.74	50.38 ± 1.41[a]
200	46.84 ± 0.73	55.11 ± 0.30[a]
250	55.27 ± 0.37	74.35 ± 0.47[a]

Each value represents the mean ± SEM, (Standard Error of Mean), n = 3.
[a]*Indicates significant difference between the extract and standard (ascorbic acid) at p < 0.05.*

17.6.3.3 INHIBITORY EFFECT OF M. TENUIFOLIA SEED EXTRACT ON ALBUMIN DENATURATION

The mechanism of denaturation probably involves the alteration of the bonds associated with the tertiary structure of proteins. It has been well established that denaturation of tissue proteins leads to inflammatory disorders, such as arthritis.[43] The inhibition of albumin denaturation by the extract is shown in Table 17.9. The result revealed that the percentage inhibitory activity of the extract on albumin denaturation was maximal at 87.84% compared to the standard drug with 84.23%. The study showed that the percentage inhibitory activity of the extract was not significantly different when compared with the standard drug at $p < 0.05$. This further implies that the *M. tenuifolia* seeds could be used as an agent against diseases like rheumatoid arthritis and other inflammatory diseases.

TABLE 17.9 Inhibitory Effect of *M. tenuifolia* Seed Extract on Albumin Denaturation.

Concentration (μg/mL)	IE of extract (%)	IE of ibuprofen (%)
0	0	0
50	66.67 ± 1.19	20.72 ± 2.51[a]
100	71.17 ± 1.91	33.78 ± 5.40[a]
150	77.48 ± 1.62	50.00 ± 3.40[a]
200	81.53 ± 1.19	61.26 ± 3.52[a]
250	85.14 ± 0.78	70.27 ± 2.34[a]
300	87.84 ± 0.78	84.23 ± 2.25

Each value represents the mean ± SEM, (Standard Error of Mean), n = 3.
[a]Indicates significant difference between the extract and standard (ascorbic acid) at p < 0.05

17.6.3.4 MEMBRANE STABILITY EFFECT OF M. TENUIFOLIA SEED EXTRACT

This study showed a significant increase in membrane stabilization in a dose-dependent manner similar to that standard drug. The membrane stability activity of *M. tenuifolia* seed extract is shown in Table 17.10. The percentage membrane stability of the extract has a maximum activity at 96.59% while that of the standard drug has a percentage membrane stability activity of 91.28%. The potency of the extract could be as a result of synergistic activities of its many constituents and the high content of its polyphenol constituents.[29,41]

TABLE 17.10 Membrane Stability Effect of *M. tenuifolia* Seed Extract.

Concentration (μg/mL)	MSA of extract (%)	MSA of ibuprofen (%)
0	0	0
50	67.92 ± 5.41	85.80 ± 3.41[a]
100	84.73 ± 11.12	88.77 ± 5.39
150	87.51 ± 0.87	88.86 ± 2.92
200	96.59 ± 12.73	91.28 ± 3.60

Each value represents the mean ± SEM, (Standard Error of Mean), n = 3.
[a]Indicates significant difference between the extract and standard (ascorbic acid) at p < 0.05

17.6.4 THROMBOLYTIC ACTIVITY OF M. TENUIFOLIA SEED EXTRACT ON HUMAN RBC CLOT

Thrombosis refers to the formation of a typical mass (blood clot) in the artery of a living vascular system linked with myocardial infarction and stroke, while thrombolysis is regarded as the breakdown of blood clots through pharmacological means involving thrombolytic drugs which dissolve clots.[40,19] Thrombolytic activity of the extract of *M. tenuifolia* seeds on human RBC clot is shown in Table 17.11. The result revealed that the extract has an appreciable thrombolytic activity of 47.54% when compared with that of the standard drug at 69.24%.

TABLE 17.11 Thrombolytic Activity of the Extract of *M. tenuifolia* Seeds.

Concentration (µg/mL)	MSA of extract (%)	MSA of ibuprofen (%)
1	83.87	50.00
2	64.00	46.15
3	75.00	50.01
4	73.91	40.60
5	87.27	66.67
6	57.14	40.62
7	58.62	45.45
8	75.00	41.94
9	61.11	48.48
10	56.52	45.50
Mean	**47.54 ± 4.37**	**69.24 ± 6.50**[a]

Each value represents the mean ± SEM, (Standard Error of Mean), n = 3.
[a]*Indicates significant difference between the extract and standard (ascorbic acid) at $p < 0.05$*

17.7 SUMMARY

M. tenuifolia is known as African nutmeg commonly used as spices in most African dishes. The study concludes that the ethanol seed extract of *M. tenuifolia* could act as anti-inflammatory and thrombotic agent. Though the seed extract showed antioxidant activities but not significantly in comparison to standard drug. These activities may be a function of certain phytochemical present in the seeds. Further studies are recommended for an in vivo investigation of the extract of *M. tenuifolia* seeds in animal models to confirm the findings of this study.

ACKNOWLEDGMENTS

Profound appreciation to the Department of Biochemistry and Molecular Biology, Obafemi Awolowo University, Ile-Ife, Nigeria, and the Department of Biochemistry, Adeleke University, Ede, Nigeria, for making available their laboratories and equipment toward the success of this research.

KEYWORDS

- **anti-inflammatory**
- **antioxidant**
- ***Monodora tenuifolia***
- **phytochemicals**
- **thrombolysis**

REFERENCES

1. Abir, R.; Abderrahmane, S.; Soumia, B.; Hamama, B. Free Radical Scavenging Activity, Reducing Power and Anti-Hemolytic Capacity of Algerian *Drimia maritima* Baker Flower Extracts. *J. Drug Deliv. Ther.* **2020,** *10*(4), 70–78.
2. Aggarwal, B. B. Inflammation, a Silent Killer in Cancer is not so Silent. *Curr. Opin. Pharmacol.* **2009,** *9*(4), 347–350.
3. Aina, O. I.; Oyedapo, O. O. *In vitro* Investigations into the Anti-Oxidant and Anti-Inflammatory Potential of the Fractions and Ethanolic Extract of *Cyclosorus afer*. *Ife J. Sci.* **2013,** *15*(2), 235–249.
4. Akinwunmi, K. F.; Oyedapo, O. O. *In Vitro* Anti-Inflammatory Evaluation of African Nutmeg (*Monodora myristica*) Seeds. *Eur. J. Med. Plants* **2015,** *8*, 167–174.
5. Al-Tannak, N. F.; Ibrahim, K., Igoli, N. P.; Igoli, J. O. LC-MS Analysis of Oils of Monodora Myristica and Monodora Tenuifolia and Isolation of a Nove Cyclopropane Fatty Acid. *Nat. Prod. Res.* **2019,** *34* (5), 1–6.
6. Amit, K.; Priyadarsini, K. I. Free Radicals, Oxidative Stress and Importance of Antioxidants in Human Health. *J. Med. Allied Sci.* **2011,** *1*(2), 53–60.
7. Archana, I.; Vijayalakshmi, K. Antioxidant Potential of Phloroglucinol: An *In-Vitro* Approach. *Int. J. Pharm. Sci. Res.* **2017,** *8*(7), 2947–2951.
8. Baker, H.; Frank, O. Tocophero. In *Clinical Vitaminology: Methods and Interpretation*; InterScience Publisher: New York, NY, pp 1968; 172–173.
9. Benzie, I.; Strain, J. Ferric Reducing/Antioxidant Power Assay: Direct Measure of Total Antioxident Activity of Biological Fluids and Modified Version for Simultaneous

Measurement of Total Antioxidant Power and Ascorbic Acid Concentration. *Meth. Enzymol.* **1999,** *299*, 15–27.
10. Blois, M. S. Antioxidant Determination by use of Stable Free Radicals. *Nature* **1985,** *29*, 1199–1200.
11. Cano, A.; Alcaraz, O.; Manuel Acosta, M.; Arnao, M. B. On-Line Antioxidant Activity Determination: Comparison of Hydrophilic and Lipophilic Antioxidant Activity Using the ABTS•+Assay. *Redox Rep.* **2002,** *7*(2), 103–109.
12. Dorfmüller, P.; Günther, S.; Ghigna, M.; de Montpréville, V. T.; Boulate, D.; Paul, J. F.; Jaïs, X.; Decante, B.; Simonneau, G.; Dartevelle, P.; Humbert, M.; Fadel, E.; Mercier, O. Microvascular Disease in Chronic Thromboembolic Pulmonary Hypertension: A Role for Pulmonary Veins and Systemic Vasculature. *Eur. Respir. J.* **2014,** *44*, 1275–1288.
13. Ekeanyanwu, R. C.; Njoku, O. U. Acute and Subacute Oral Toxicity Study on the Flavonoid Rich Fraction of *Monodora tenuifolia* Seed in Albino Rats. *Asian Pac. J. Trop. Biomed.* **2014,** *4*(3), 194–202.
14. Ezenwali, M. O.; Njoku, O. U.; Okoli, C. O. Studies on the anti-diarrheal properties of the seed extract of *Monodora tenuifolia. Int. J. Appl. Res. Nat. Prod.* **2010,** *2*(4), 20–26.
15. Fajobi, O. A.; Emma-Okon, B. O.; Oyedapo, O. O. *In vitro* Antioxidant, Anti-Inflammatory and Thrombolytic Properties of Leaf Extract. *Sci. World* **2020,** *4*(1), 23–26.
16. Fennell, C. W.; Lindsey, K. L.; McGaw, L. J.; Sparg, S. G.; Stafford, G. I.; Elgorashi, E. E.; Grace, O. M.; van Staden, J. Assessing African Medicinal Plants for Efficacy and Safety: Pharmacological Screening and Toxicology. *J. Ethnopharmacol.* **2004,** *94*(2–3), 205–217.
17. Hagerman, A. E.; Ried, K. M.; Jones, G. A.; Sovik, K. N.; Ritchard, N. T.; Hartzfeld, P. W.; Riechel, T. L. High Molecular Weight Plant Polyphenolics (Tannins) as Biological Antioxidants. *J. Agric. Food Chem.* **1998,** *46*, 1887–1892.
18. Irvine, F. R. *Woody Plants of Ghana With Special Reference to Their Uses*; Oxford University Press: London, 1961; p 351.
19. Jakaria, M.; Mukimu, I.; Sarifu, I.; Mohammad, B.T.; Chayan, D. C.; Mohammed, I. Thrombolysis Potential of Methanol Extracts From the Five Medicinal Plants Leaf Available in Bangladesh. *Pharmacologia* **2017,** *8*, 78–82.
20. Kasper D. L.; Braunwald E.; Fauci A. S. Eds.; In *Harrison's Principles of Internal Medicine*, 16th ed.; McGraw-Hill: New York, USA, 2005; vol *6*, pp 1561–1565.
21. Kinugasa, C.; Naemura, A.; Hyodo, K.; Nakai, Y.; Katsuta, M.; Yamamoto, J. Experimental Antithrombotic Effects of Sesame Seed Whole Grains and Extracts. *Blood Coagul. Fibrinol.* **2011,** *22*(6), 526–531.
22. Bhoyar, M.; Mishra, G. P.; Naik, P. K.; Srivastava, R. B. Estimation of Antioxidant Activity and Total Phenolics Among Natural Populations of Caper (*Capparis spinosa*) Leaves Collected From Cold Arid Desert of Trans-Himalayas. *Aust. J. Crop Sci.* **2011,** *5*(7), 912–919.
23. Mikaili, P.; Maadirad, S.; Moloudizargari, M.; Aghajanshakeri, S.; Sarahroodi, S. Therapeutic Uses and Pharmacological Properties of Garlic, Shallot, and Their Biologically Active Compounds. *Iran J. Basic Med. Sci.* **2013,** *16*(10), 1031–1048.
24. Mizushima, Y.; Kobayashi, M. Interaction of Anti-Inflammatory Drugs With Serum Proteins, Especially With Some Biologically Active Proteins. *J. Pharm. Pharmacol.* **1968,** *20*, 169–173.
25. Modi, S.; Behere, D. V.; Mitra, S. Horseradish Peroxidase Catalyzed Oxidation of Thiocyanate by Hydrogen Peroxide: Comparison With Lactoperoxidase-Catalysed Oxidation and Role of Distal Histidine. *Biochim. Biophys. Acta* **1991,** *1080*(1), 45–50.

26. Njoku, U. O., Akah, P. A., Okonkwo, C. C. Antioxidant Activity of Seed Extracts of *Monodora tenuifolia* (annonaceae). *Int. J. Basic Appl. Sci.* **2012,** *12*(2), 80–87.
27. Omaye, S. T.; Turnbull, J. D.; Sauberlich, H. E. Selected Methods for the Determination of Ascorbic Acid in Animal Cells, Tissues and Fluids. In *Methods of Enzymology*; McCormick, D. B., Wright, D. L., Eds.; Academic Press: MA, USA, 1979; vol *62*, pp 3–11.
28. Oyaizu, M. Studies on Product of Browning Reaction Prepared From Glucose Amine. *Jpn. J. Nutr.* **1986,** *44*, 307–315.
29. Oyedapo, O. O.; Akinpelu, B. A.; Akinwunmi, K. F.; Adeyinka, M. O.; Sipeolu, F. O. Red Blood Cell Membrane Stabilizing Potentials of Extracts of *Lantana camara* and its Fractions. *Int. J. Plant Physiol. Biochem.* **2010,** *2*(4), 46–51.
30. Oyedapo, O. O.; Amos, S. Further Investigations into the Bioactivity of Root Extract of *Plumbago zeylanica*. *Phytother. Res.* **1997,** *11*, 62–63.
31. Oyedapo, O. O.; Sab, F. C.; Olagunju, J. A. Bio-Activity of Fresh Leaves of *Lantana camara*. *Biomed. Lett.* **1999,** *59*, 175–183.
32. Parween, T.; Jan Mahmooduzzafar, S.; Fatma, T. Evaluation of Oxidative Stress in *Vigna radiata* L. in Response to Chlorpyrifos. *Int. J. Environ. Sci. Technol.* **2012,** *9*, 605–612.
33. Prasad, S.; Sung, B.; Aggarwal, B. B. Age-Associated Chronic Diseases Require Age-Old Medicine: Role of Chronic Inflammation. *Prev. Med.* **2012,** *54*(l), 29–37.
34. Reuter, S.; Gupta, S. C.; Chaturvedi, M. M.; Aggarwal, B. B. Oxidative Stress, Inflammation, and Cancer: How are They Linked? *Free Radic. Biol. Med.* **2010,** *49*(11), 1603–1616.
35. Reyad-ul-ferdous, M.; Jisha, H. R.; Azam, M. Evaluation of In-Vitro Anti-Inflammatory, Thrombolytic and Cytotoxic Properties of *Lablab niger* Leaves. *Int. J. Pharm. Sci. Res.* **2014,** *5*(2), 350–353.
36. Shinde, U. A.; Phadke, A. S.; Nari, A. M; Mungantiwar, A. A; Dikshit, V. J.; Saraf, M. N. Membrane Stabilization Activity-A Possible Mechanism of Action for the Anti-Inflammatory Activity of Cedrus Deodara Wood Oil. *Fitoterapia* **1999,** *70,* 251–257.
37. Singleton, V. L.; Orthofera, R.; Lamuela, R. R. M. Analysis of Total Phenols and Other Oxidation Substrates and Antioxidants by Means of Folin-Ciocalteu's Reagent. *Methods Enzymol.* **1999,** *299,* 152–179.
38. Sun, P. X.; Yie, L. K.; Zhang, Z. L.; Hu, M.; Lu, L. Colometric Determination of the Total Content of the Flavonoids in Epimedium Capsules. *J. Shenyang Pharm. Univ.* **1999,** *16,* 68–70.
39. Tantary, S.; Masood, A.; Bhat, A. H.; Dar, K. B.; Zargar, M. A.; Ganle, S. A. *In Vitro* Antioxidant and RBC Membrane Stabilization Activity of *Euphorbia wallichii*. *Free Radic. Antioxid.* **2017,** *7*(1), 13–22.
40. Tanvir, A. C.; Kamal, A. T. M. M.; Kazi, A. A. C.; Afrina, J.; Shakhawat, H.; Arafatul, M.; Mahmudul, H.; Javed, H. Cytotoxic and Thrombolytic Activity of Methanolic Extract of *Macaranga denticulate* Bark. *Parma Innov. J.* **2015,** *4*(5), 36–39.
41. Umapathy, E.; Ndebia, E. J.; Meeme, A.; Adam, B.; Menziwa, P.; Nkeh-Chungaga, N. B.; Iputo, J. E. An Experimental Evaluation of *Albuca setosa* Aqueous Extract on Membrane Stabilization, Protein Denaturation and White Blood Cell Migration During Acute Inflammation. *J. Med. Plants Res.* **2010,** *4*, 789–795.
42. Usunobun, U.; Igwe, V. C. Phytochemical Screening, Mineral Composition and *In Vitro* Antioxidant Activities of *Pterocarpus mildbraedii* Leaves. *Int. J. Sci. World* **2016,** *4*(1), 23–26.

43. Williams, H.; Walker, D.; Orchard, T. R. Extra-Intestinal Manifestations of Inflammatory Bowel Disease. *Curr. Gastroenterol. Rep.* **2008,** *10*, 597–605.
44. Yadav, A.; Kumari, R.; Yadav, A.; Prabha, S. Antioxidant and its Functions in Human Body–A Review. *Res. Environ. Life Sci.* **2016,** *9*(11), 1328–1331.
45. Yembeau, N. L.; Biapa, P. C. N.; Chetcha, B.; Nguelewou, F. L.; Kengne, F.; Nkwikeu, P. J. N.; Telefo, P. B.; Pieme, C. A. *In Vitro* Evaluation of Antioxidant and Antisickling Properties of *Theobroma cacao* (Sterculiaceae) Extract From East and South Regions in Cameroon. *Invest. Med. Chem. Pharmacol.* **2018,** *1*(1), 1–9.

CHAPTER 18

Effects of *Taraxacum officinale* Extract on Hematological Parameters in Female Wistar Rats With Ethanol-Induced Toxicity

Ademola Olabode Ayeleso, Samuel Chibueze Nzekwe, Favour Oghenefejiro, Samuel Olufolarin Babarinde, and Ayodeji Emmanuel Adepoju

ABSTRACT

The ability of *Taraxacum officinale* (TO) (dandelion) aqueous leaf extract to ameliorate the effect of ethanol-induced toxicity on hematological parameters in female Wistar Rats was evaluated. Twenty-four female Wistar rats were grouped into four, each group had six (6) rats each: normal, TO extract, ethanol (EtOH), and TO extract + EtOH, where each group included six animals. TO leaf extract was administered orally for 7 days before orally administering a single dose of 50% ethanol. Results obtained showed that the plant extract caused WBC count, hematocrit, mean corpuscular volume, hemoglobin, and platelet in the TO extract group to increase significantly compared with the normal group at ($p < 0.05$). TO extract + EtOH group increased significantly in platelet and lymphocyte levels when compared with the EtOH group while the levels of hemoglobin and hematocrit in the TO extract + EtOH group significantly decreased compared to EtOH group. From this study, alcohol intake has a significant negative effect on some blood parameters; however, when TO leaf extract was prophylactically administered to the rats, it prevented the ethanol-induced toxicity effect in the blood parameters, such as platelet, lymphocyte, hemoglobin, and

Curative and Preventive Properties of Medicinal Plants: Research on Disease Management and Animal Model Studies. Megh R. Goyal, Junaid Ahmad Malik, & Ademola Olabode Ayeleso (Eds.)

hematocrit. It is concluded that TO leaf extract improved some hematological abnormalities in ethanol toxicity.

18.1 INTRODUCTION

Alcoholic beverages contain ethanol, which is an organic compound with a hydroxyl functional group.[15] Researches have established reasons behind excessive alcohol intake, these include to enhance relationship, increase energy, lighten up difficult moments, cope with stress, entertainment or celebration, and social influences.[25] Research has proved that poor dietary habits practiced by alcoholics cause nutritional deficiencies, which may be compounded by the burden of the alcohol, and the absorption and utilization of vitamins.[31] Alcoholics are predisposed to signs of vitamin deficiency, especially thiamine, riboflavin, pyridoxine, ascorbic acid, and folic acid. The frequency of these deficiencies is related to the amount of alcohol consumed, with a corresponding decrease in vitamin absorption.[16] Alcohol is metabolized in the liver by cytoplasm enzyme alcohol dehydrogenase (ADH). The metabolism of alcohol occurs through the oxidative pathway that is controlled by two different enzymes, ADH which is firstly involved in the oxidative breakdown of alcohol to aldehyde, while aldehyde dehydrogenase then converts aldehyde to acetate.[22,8]

Alcohol intake affects nutrient absorption by damaging the stomach and intestine endothelial cell linings, thereby preventing the absorption and transport of nutrients through the blood causing organ injury as a result of alcohol interactions with nutrients.[28] Alcohol abuse is associated with deterioration in behavior and cognitive processes such as learning and memory[7], therefore, impairing some of the brain function. The indirect consequences of this excessive alcohol intake include impaired yellow marrow of the bone, decreased precursors production of the blood cells, and immature erythrocytes, lymphocytes, and platelets production that cause immunodeficiency.[5,27] Therefore, hematological abnormalities are dominant in chronic alcoholics, leading to anemia as a general symptom in alcohol consumers where RBC count, MCH concentration, PCV, mean corpuscular volume (MCV), platelet, and WBC count, also cell morphology are affected.[25,32]

This chapter examined the ability of leaf extract of *Taraxacum officinale* (TO), through evaluation of hematological parameters in ethanol-induced toxicity in female Wistar rats, to reduce the damaging effects on the cell-wall linings of the gastrointestinal tract, and impaired nutrients transport into the bloodstream due to interactions of the cells with alcohol, which may be as a result of digestive organ damage in the presence of alcohol.

18.2 HISTORY AND HABITAT DISTRIBUTION

TO (L.) Wigg is a medicinal plant also known as "Dandelion" majorly found growing naturally in different parts of the world such as Europe, America, Asia, and Africa.[30] The distribution of dandelion in the disturbed rainforest of south west Nigeria (where it is called "efo-yanrin" or referred to as forest vegetable) has been recorded.[2,29] The seeds of TO are planted through dispersion, and when germinated, the young shoot forms a rosette of blade-like leaves.[1] Thick root develop later on and remain in the ground during the dry season. New rosettes spring out along the roots in the rainy season, which are good for consumption as a vegetable as it is not yet very bitter in this early stage. Flower shoot grows from the rosette, the stem leaves are very bitter mostly when they are matured while fruiting, and seeds maturity occurs within 3 to 4 months after planting. Seed dispersion is majorly by wind enhanced through the pappus which is attached to the fruits. The rate of germination of TO seeds is low, hence, this is compensated through the production of multiple seeds from each plant.[1]

TO which has a high regenerative potential is able to survive different environmental conditions, among diverse plant communities. It can also survive in different weather conditions and grow in large geographical areas.[26] Additionally, TO is able to successfully spread to new areas through dispersion actions from several farm practices, wind, water, animal, and bird vectors (Figure 18.1). Majority of the plant species are dominant in non-native areas, where the plants are being exposed to different environmental conditions.[24] This suggests that invasive species of TO show vegetative and reproductive evolution for it to maximize their performance and establish in those new environments.[11,23]

18.3 MEDICINAL USE OF *TARAXACUM OFFICINALE*

TO is known to contain several biologically active constituents (such as taraxasterol, sesquiterpene lactones, taraxerol, chicoric acid, and chlorogenic acid), which are reported to possess potent anti-inflammatory, antirheumatic, antioxidative, and chloretic activities.[4] The medicinal use of dandelion plants lies in the bioactive substances, most notably: alkaloids, tannins, steroids, flavonoids, and phenolic compounds. For this reason, dandelions have been employed to treat hepatic disorders.[3]

FIGURE 18.1 *Taraxacum officinale* plant in the tropical rain forest.

TO is utilized in many traditional and contemporary herbal medicinal systems to enhance the response of the immune system to infections of the upper respiratory tract, pneumonia, or bronchitis.[20] Almost all the plant parts are been used as a herbal remedy for many ages. The roots are regarded as digestion enhancers. In addition, the root and leaf extracts are generally used in rehydration and to correct electrolyte imbalance. The high potassium, sodium, and calcium contents of leaves posited for the plant's use as diuretics and also as stimulants for digestion.[12,18] The roots of TO are reported as a treatment for gastrointestinal conditions, thereby enhancing digestion and the function of liver.[21] The flowering part of TO revealed different bioactive phytochemicals that have antioxidant activities against free radicals, preventing the cells and macromolecules from radical-induced damage.[19] Furthermore, it is used in treating some viral infections like hepatitis B and indigestion, and also enhances androgenic hormones.[2,6]

18.4 MATERIALS AND METHODS

18.4.1 PLANT MATERIALS

The leaves of the plant TO were collected from a local farm in Ede, Osun State, Nigeria. The plants were authenticated at the Plant Biology Programme of Adeleke University, Osun State, Nigeria.

18.4.2 CHEMICALS

Reagents and chemicals purchased and utilized for the study were of analytical grade.

18.4.3 PREPARATION OF PLANT EXTRACT

The collected fresh leaves of TO were carefully rinsed on a running tap and air dried at room temperature for 7 days. This was followed by pulverizing the dried leaf sample into powder. In total, 250 g of plant sample was soaked in 2500 mL of distilled water in a ratio 1:100 (w/v) of the sample for 24 h, filtered with a cellophane material, and then with 0.1 mm Whatman filter paper. The filtrate was concentrated by freeze-drying to obtain the crude plant extract, kept in a tight container, and stored in the refrigerator for further use.

18.4.4 ANIMAL CARE

Twenty-four female Wistar rats of weight ranging from 130 to 280 g were procured from the Department of Anatomy, Adeleke University. The animals were kept in the animal facility of the Biochemistry Department, Adeleke University, where the animal experiment was conducted. The rats were initially acclimatized for 2 weeks and fed throughout the study with standard rat pellets and clean water ad libitum. The animals were subjected to natural photoperiod of 12 h daylight and night throughout the study period and other necessary care was provided for all the animals.

18.4.5 EXPERIMENTAL DESIGN

The experiment was designed to consist of four groups, with six rats per group. TO leaf extract was administered orally for 7 days before a single oral dose of ethanol was administered as below:

Group 1 (Normal): Normal group rats, taken 1 mL of distilled water only.

Group 2 (TO extract): Rats treated with leaf extract of TO (200 mg/kg body weight).

Group 3 (Ethanol): Ethanol group rats were treated with a single dose of 1 mL of 50% ethanol.

Group 4 (TO extract + EtOH): Rats pretreated with leaf extract of TO (200 mg/kg body weight) followed by 1 mL of 50% ethanol.

The weights of all experimental animals were determined before and after the experiment. Following the treatment, the Wistar rats were fasted for 12 h and then sacrificed. The blood samples were collected under light diethyl ether anesthesia for hematological analysis.

18.4.6 HEMATOLOGICAL PARAMETERS

A fully automated Beckman Coulter (LH 700 Series System) was used to analyze hematological indices.

18.5 RESULTS AND DISCUSSION

18.5.1 TARAXACUM OFFICINALE PROPHYLAXIS ON WHITE BLOOD CELL, RED BLOOD CELL COUNTS, AND PLATELET CONCENTRATION

Earlier studies have shown that chronic alcohol consumption through direct or indirect effects, suppresses hematopoiesis in individuals with alcohol dependence. This results in a decrease in the counts of red blood cells, white blood cells, and platelets.[10,12,13] It is observed that excessive consumption of alcohol makes the host susceptible to pulmonary infections in particular as well as several other complications of infections. Chandini and Mathai[9] reported that alcohol consumption results in a decreased level of WBC. In this study, the ethanol (EtOH) group (9.0 ± 1.42 μL^{-1}) recorded a significant decrease in the WBC when compared with the TO extract group. However, prophylactic 200 mg/kg TO extract had WBC (12.73 ± 2.86 μL^{-1}) significantly ($p < 0.05$) higher than the WBC in control group (5.97 ± 1.33 μL^{-1}). Upon exposure to ethanol, there was a significant ($p < 0.05$) decrease in the WBC of TO extract + EtOH group (8.62 ± 1.38 μL^{-1}) when compared with the unexposed TO extract group (Figure 18.2A).

The RBC count in the EtOH group (7.06 ± 0.34μL^{-1}) significantly ($p < 0.05$) increased than the level in the control group (5.96 ± 0.31 μL^{-1}). The groups treated with TO and TO + EtOH (6.23 ± 0.44μL^{-1} and 6.64 ± 0.26

μL^{-1}, respectively) had no a significant difference in RBC concentration as compared to control group(Figure 18.2B).

Reduced platelet count levels are noted with increased duration of alcohol use and period of alcohol dependence. Earlier studies have found an increased prevalence of thrombocytopenia in individuals with alcohol dependence syndrome. Platelet levels showed a significant decrease in EtOH group when compared with the TO + EtOH group (Figure 18.2C). The groups of rats administered with 200 mg/kg TO extract and TO + EtOH, platelet concentration ($414.83 \pm 82.83\ \mu L^{-1}$ and $420.33 \pm 39.05\ \mu L^{-1}$, respectively) was significantly ($p < 0.05$) higher than the platelet concentration of control ($327.83 \pm 72.03\ \mu L^{-1}$).

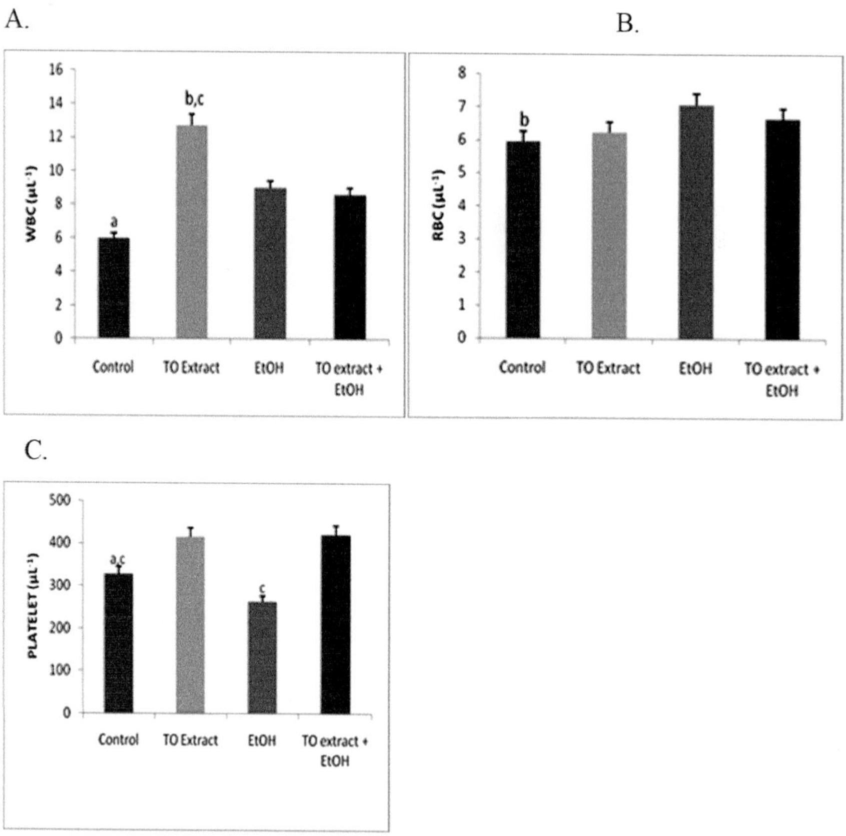

FIGURE 18.2 Effect of 200 mg/kg extract pretreatment on white blood cell count (A); red blood cell count (B); and platelet volume (C) in alcoholic rats. a,b,c above bars imply significance when compared with TO extract, EtOH, and TO extract + EtOH groups, respectively.

18.5.1.1 TARAXACUM OFFICINALE PROPHYLAXIS ON HEMOGLOBIN AND MEAN CORPUSCULAR HEMOGLOBIN CONCENTRATIONS

Hisham et al. reported that *Moringa oleifera* improves the constituents of the blood with the erythrocytes, platelets, and other hematological indices in rats and rabbits.[17] This study reveals that 200 mg/kg plant extract and EtOH alone caused hemoglobin (12.47 ± 1.01 g/dL and 13.77 ± 0.40 g/dL, respectively) concentration significantly higher than the control. The hemoglobin concentration increased significantly in the ethanol group when compared to TO extract + EtOH group (Figure 18.3A). There were insignificant ($p > 0.05$) differences in MCHC of all TO extract exposed and unexposed groups compared to the normal group (Figure 18.3B).

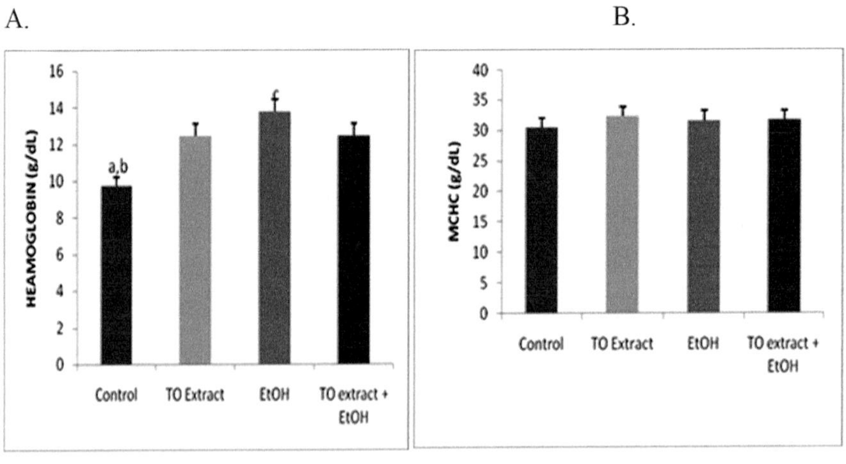

FIGURE 18.3 Effects of 200 mg/kg extract pretreatment on hemoglobin concentration (A); and mean corpuscular hemoglobin concentration (MCHC) in alcoholic rats (B). a,b,c above bars imply significance when compared with TO extract, EtOH, and TO extract + EtOH groups, respectively.

18.5.2 TARAXACUM OFFICINALE PROPHYLAXIS ON HEMATOCRIT AND LYMPHOCYTE PERCENTAGES, MEAN CORPUSCULAR VOLUME (MCV), AND MEAN CORPUSCULAR HEMOGLOBIN (MCH)

Rubina et al.[27] reported a reduced MCV in social and chronic alcoholics. In this study, percentage hematocrit count significantly ($p < 0.05$) increased in TO extract and ethanol groups (38.58 ± 2.99% and 42.83 ± 1.04%, respectively) when compared to the normal group (33.18 ± 1.73%). The TO and TO extract + EtOH were insignificantly raised than the control group. The EtOH group revealed a significant increase in hematocrit percentage compared with TO extract + EtOH (Figure 18.4A). In the TO extract + EtOH group, there was a significant increase in the lymphocyte concentration when compared with both TO extract and EtOH groups. There was an insignificant difference between the 200 mg/kg TE and the control (Figure 18.4B). Administration of 200 mg/kg TO extract caused MCV (61.77 ± 0.93) to increased significantly ($p < 0.05$) when compared with the normal control group. Also, EtOH group has MCV (63.2333 ± 1.38) level significantly ($p < 0.05$) higher than the control (Figure 18.4C). An insignificant difference was observed in MCH of all the TO extract exposed and unexposed groups compared to the normal group (Figure 18.4D).

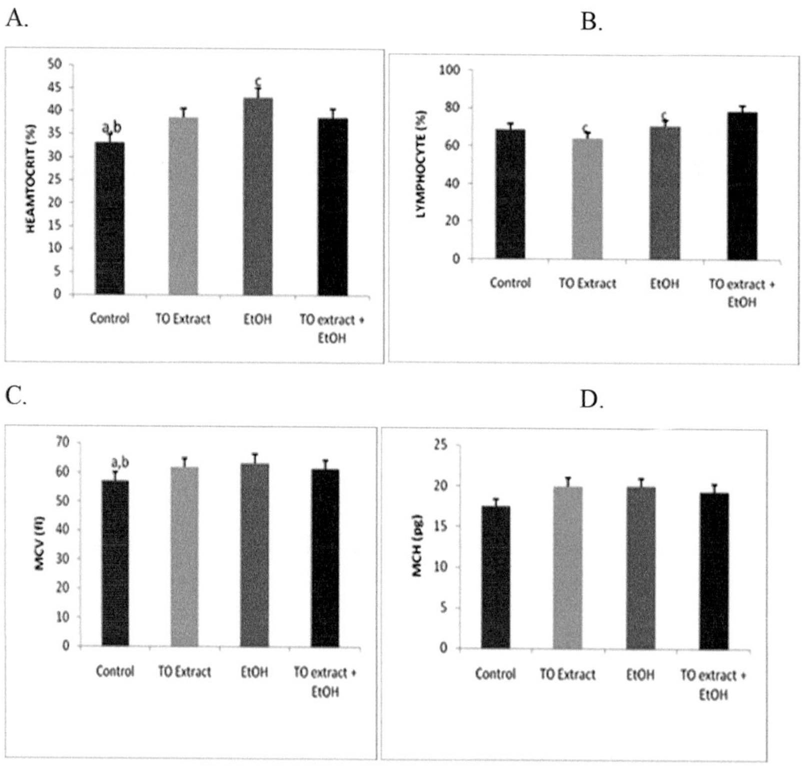

FIGURE 18.4 Effect of 200 mg/kg extract pretreatment on percentage % hematocrit (A), % lymphocyte (B), mean corpuscular volume (MCV) (C), and mean corpuscular hemoglobin (MCH) (D) in alcoholic rats. a,b,c above bars imply significance when compared with TO extract, EtOH, and TO extract + EtOH groups, respectively.

18.6 SUMMARY

Chronic alcohol intake results in changes in full blood count and different adverse effects on the functionality of blood cells, which can result in moderate anemia due to reduced platelet count or thrombocytopenia, abnormal RBC, reduced WBC (basically the neutrophils), and complex aberrations of hematopoiesis may develop which can result into death. This chapter revealed the phytochemicals found in aqueous leaf extract of TO, such as cardiac glycosides, phenols, flavonoids, steroids, tannins, and alkaloids. There were increasing levels of blood parameters in ethanol-induced groups except for the value of platelet which was reduced significantly when compared to the control. Prophylactic administration of the plant extract normalized the majority of the assayed hematological parameters. This, therefore, suggests a clear indication that TO leaf extract protects against blood parameters abnormalities as a result of the acute toxicity of ethanol.

KEYWORDS

- **ethanol**
- **hematological parameters**
- ***Taraxacum officinale***
- **Wistar rats**

REFERENCES

1. Adebisi, A. A. *Launaea taraxacifolia* (Wild.) Amin ex C. Jeffrey. Grubben, G. J. H., Denton, O. A., Eds.; PROTA: Wageningen, Netherlands, 2004; pp 47–59.
2. Ahmed, D.; Gulfraz, M.; Ahmad, M. S.; Tahir, R. M.; Anwar, P. Cytoprotective Potential of Methanolic Leaves Extract of *Taraxacum Officinale* on Ccl4 Induced Rats. *Pensee J.* **2013,** *75*, 220–227.
3. Amin Mir, M.; Sawhney, S. S.; Jassal, M. M. S. Qualitative and Quantitative Analysis of Phytochemicals of *Taraxacum officinale*. *Wudpecker J. Pharm. Pharmocol.* **2013,** *2*(1), 1–5.
4. Arpadjan, S.; Celik, G.; Taskesen, S.; Gücer, S. Arsenic, Cadmium and Lead in Medicinal Herbs and Their Fractionation. *Food Chem.Toxicol.* **2008,** *46*(8), 2871–2875.
5. Ballard, H. S. The Hematological Complications of Alcoholism. *Alcohol Health Res. World* **1997,** *21*(1), 42–52.

6. Blumental, M.; Cladbery, A.; Brinkman, J. In *Herbal Medicine: Expanded Commission E Monographs*; Integrative Medicine Communications: Newton, MA, USA, 2000; p 105.
7. Brust, J. C. Ethanol and Cognition: Indirect Effects, Neurotoxicity and Neuroprotection: A Review. *Int. J. Environ. Res. Public Health* **2010,** *7*(4), 1540–1557.
8. Cederbaum, A. I. Alcohol Metabolism. *Clin. Liver Dis.* **2012,** *16*, 667–685.
9. Chandini, D.; Mathai, P. J. Haematological Parameters in Patients With Alcohol Dependence Syndrome. *IOSR-JDMS* **2017,** *16*(5), 11–16.
10. Charles, A. T.; Everton, E.; Osaro, E. Evaluation of Hematological Parameters Among Alcoholics in Port Harcourt Nigeria. *Brit. J. Med. Heal. Sci.* **2015,** *5*, 3244–3250.
11. Chun, Y. J.; Collyer, M. L.; Moloney, K. A.; Nason, J. D. Phenotypic Plasticity of Native vs. Invasive Purple Loosestrife: A Two-State Multivariate Approach. *Ecology* **2007,** *88*(6), 1499–512.
12. Das, S. K.; Vasudevan, D. M. Biochemical Diagnosis of Alcoholism. *Indian J. Clin. Biochem.* **2005,** *20*(1), 35–42.
13. Das, S. K.; Mukherjee, S.; Vasudevan, D. M.; Balakrishnan, V. Comparison of Haematological Parameters in Patients With Non-Alcoholic Fatty Liver Disease and Alcoholic Liver Disease. *Singapore Med. J.* **2011,** *52*(3), 175–181.
14. Elanchezhian; Yoganandh, T.; Mayilsamy, S.; Shankar, R. Comparison of Haematological Parameters Between Alcoholics and Nonalcoholics. *Int. J. Res. Med. Sci.* **2017,** *5*(11), 5041–5047.
15. Giesbrecht, N.; Ashley, W.; Walker, E. Beer, Wine and Distilled Spirits in Ontario: A Comparison of Recent Policies, Regulations and Practices. *Nord. Stud. Alcohol Drugs* **2012,** *29*(1), 79–102.
16. Giorgia, S.; Borrás-Novell, C. The Effects of Alcohol and Drugs of Abuse on Maternal Nutritional Profile during Pregnancy. *Nutrients* ***2018,*** *10*(8), 38–52.
17. Osman, H. M.; Mohamed-Shayoub, E.; Elsiddig-Babiker, M. The Effect of *Moringa* oleifera Leaves on Blood Parameters and Body Weights of Albino Rats and Rabbits. *Jordan J. Biol. Sci.* **2012,** *5*(3), 147–150.
18. Hook, I.; McGee, A.; Henman, M. Evaluation of Dandelion for Diuretic Activity and Variation in Potassium Content. *Int. J. Pharmacogn.* **1993,** *31*(1), 29–34.
19. Hu, C.; Kitts, D. D. Dandelion (*Taraxacum officinale*) Flower Extract Suppresses Both Reactive Oxygen Species and Nitric Oxide and Prevents Lipid Oxidation *In Vitro*. *Phytomedicine* **2005,** *12*, 588–597.
20. Ivanov, I. G. Polyphenols Content and Antioxidant Activities of *Taraxacum officinale* (Dandelion) Leaves. *Int. J. Pharmacogn. Phytochem. Res.* **2014,** *6*(4), 889–893.
21. Koh, Y. J.; Cha, D. S.; Ko, J. S.; Park, H. J.; Choi, H. E. Anti-inflammatory Effects of *Taraxacum officinale* Leaves on Lipopolysaccharide-Induced Inflammatory Reslonses in RAW 264.7 Cells. *J. Med. Food* **2010,** *13*(4), 870–878.
22. Lieber, C. S. Metabolism of Alcohol. *Clin. Liver Dis.* **2005,** *9*, 1–35.
23. Molina-Montenegro, M. A.; Atala, C.; Gianoli, E. Phenotypic Plasticity and Performance of *Taraxacum Officinale* (Dandelion) in Habitats of Contrasting Environmental Heterogeneity. *Biol. Invasions* ***2010,*** *12*, 2277–2284.
24. Molina-Montenegro, M. A.; Palma-Rojas, C.; Alcayaga-Olivares, Y.; Rómulo, O. Ecophysiological Plasticity and Local Differentiation Help Explain the Invasion Success of *Taraxacum officinale* (dandelion) in South America. *Ecography* ***2013,*** *36*, 718–730.
25. Moses-Lugos, D.; Bitrus-Lekshak, N.; Changjul-Singnap, L.; Umanka-Polit, Y.; Benedict -Mangs, H. Investigating the Impact of Alcohol Consumption on Some

Haematological Variables among Alcohol Consumers in Jos Metropolis, Nigeria. *Int. J. Hematol. Blo. Dis.* **2018,** *3*(3), 1–6.
26. Rejmánek, M.; Richardson, D. M.; Higgins, S. I.; Pitcairn, M. J.; Grotkopp, E. Ecology of Invasive Plants, State of the Art. In *Invasive Alien Species a New Synthesis*; Island Press: Washington, DC, 2005; pp 104–161.
27. Rubina, M. P.; Basheer, R. K. B.; Safeer, M. T. K.; Soumya, V.; Ali, M. K. T.; Ramachandran, C.C. A Comparative Study on Hematological Parameters Among the Social and Problem Drinkers Admitted in a Tertiary Care Rehabilitation Centre. *Int. J. Res. Med. Sci.* **2019,** *7*, 2981–2984.
28. Shirish, B.; Shao-Yu, C.; Irina, K. Development, Prevention, and Treatment of Alcohol-Induced Organ Injury: The Role of Nutrition. *Alcohol Res.* **2017,** *38*(2), 289–302.
29. Sofowora, A. In *Medicinal Plants and Traditional Medicine in Africa*, 1st ed.; John Wiley & Sons: Chichester, New York, 1982; p 256.
30. Sweeney, B.; Vora, M.; Ulbricht, C.; Basch, E. Evidence-Based Systematic Review of Dandelion (*Taraxacum officinale*) by Natural Standard Research Collaboration. *J. Herb. Pharmacother*. **2005,** *5*, 79–93.
31. Tardelli, V. S.; Lago, M. P. P. D.; Silveira, D. X. D.; Fidalgo, T. M. Vitamin D and Alcohol: A Review of the Current Literature. *Psychiatry Res.* **2017,** *248*, 83–86.
32. Thoma, E. L.; Bitri, S.; Mucaj, K.; Tahiri, A.; Pano, I. Changes of Some Blood Count Variables in Correlation with the Time of Alcohol Abuse. *J. Addict. Res. Ther.* **2015,** *6*, 2–9.

CHAPTER 19

Antimutagenic Activity of *Moringa oleifera* Seeds

Temitope Abiola, Olakunle Akinyode, Bisola Jimoh,
Babawale Emmanuel Abiola, Ibukun Temitope Sossou,
Adedayo Olukanni, and Olumide Olukanni

ABSTRACT

Moringa oleifera has medicinal and nutritional values. This study assessed the antimutagenic potential of seeds of *M. Oleifera*. Ethanol was used for the extraction. Chemical analysis was done by following standard methods. Antimutagenicity was determined against nitrate-treated 1-amino pyrene in acid solution using the *Salmonella assay.* Preliminary phytochemical screening of seed extract *M. Oleifera* revealed tannin, alkaloid, flavonoid, terpenoid, saponin, cardiac glycoside, anthraquinone, and steroid; phlobatanin was not detected. Tannin was found to be in the highest amount (28.2 ± 0.63) mg/100 g, flavonoid (23.21 ± 0.44) mg/100 g, steroid (20.50 ± 2.15) mg/100 g, alkaloid (18.68 ± 2.15) mg/100 g, and saponin (15.22 ± 0.63) mg/100 g. The extract showed a strong percentage inhibition in both strains at higher doses. The observed antimutagenic effect of the seed extract could be adduced to some phytochemical constituents therein.

19.1 INTRODUCTION

The second principal cause of death after cardiovascular diseases is cancer which is a group of over 100 diverse diseases that results in uninhibited growth of cells, confined tissue incursion, and spreading to other remote parts of the body, with genetic and epigenetic levels changes.[26,43] Mutagens are

Curative and Preventive Properties of Medicinal Plants: Research on Disease Management and Animal Model Studies. Megh R. Goyal, Junaid Ahmad Malik, & Ademola Olabode Ayeleso (Eds.)

substances capable of altering the hereditary material in living things, thereby leading to build-up of genetic changes that exceed the normal fundamental rate.[7] Mutagens cause alterations in the DNA which upon accumulation may lead to cancer as detrimental mutation can result in anomalous, damaged, or nonfunctional gene.[41] An example of a mutagen is 1-amino pyrene (1-AP).[22]

Cancer can be linked to mutagens that induce oxidative stress in the cells as a result of overproduction of free radicals.[21] These free radicals are generated in living systems as electrons escape from the electron transport chain to oxygen in the mitochondria.[31] They can damage biomolecules as they are implicated in mutagenesis, genotoxicity, and carcinogenesis.[35] A widely employed test for identifying mutagens at the early stage is the Ames test. It involves the use of *Salmonella typhimurium* bacterial strains which lack the ability to produce histidine.[29]

Antimutagenic agents are natural or synthetic compounds that can reduce or eliminate the genotoxic properties of mutagenic and carcinogenic factors.[14] This can be by inactivating the mutagen–DNA reaction or by preventing the conversion of a mutagenic compound into mutagen.[9]

Plants are rich sources of nutrients and in addition also contain various phytochemicals employed in the treatment of a wide range of diseases. Medicinal plants are now being utilized more than before as a source of potential chemotherapeutic drugs. Compounds from plants could behave as defensive substances against cancer-formation processes in humans[16] or, perhaps, by destroying or inhibiting the mutagens that destroy DNA outside the cells, thereby forestalling cell mutations.[38]

This research study evaluated the medicinal and nutritional values; and antimutagenic potential of seeds of *Moringa oleifera.*

19.2 ORIGIN AND DISTRIBUTION OF *MORINGA OLEIFERA*

Moringa, popularly tagged "miracle tree," belongs to *Moringaceae* family, grown for a long time especially, in bush thickets, and flourishes more in the tropics. Every of its component is utilized in different ways in earth regions at the middle of the globe.[5,17] *M. oleifera* has fruits that resemble drumstick, tiny white flowers, and leaves with round shape (Figure 19.1).

19.3 PHARMACOLOGICAL PROPERTIES OF *M. OLEIFERA*

Diverse components of *M. Oleifera* plant possess pharmacological properties. It has been demonstrated that its leaves have potent hypotensive,[18] useful against diabetes,[28] antimicroorganisms, spasms, and diuresis,[46] and also reported to be helpful against inflammation in anaphylaxis, helminthesis, and scurvy.[11] The pits are reported to be useful against microbes,[10] demonstrate antifungal,[12] antifertility,[34] antihepatotoxic,[39] analgesic,[36] antiarthritic,[27] anti-tumor,[20] antipyretic, and anti-inflammatory effects.[48] Dried *M. oleifera* seeds have been utilized as purgative, tonic, eye ointments, and in the treatment of venereal infections.[5]

FIGURE 19.1 *Moringa* seeds.

19.4 METHODS AND MATERIALS

19.4.1 ASSORTMENT AND IDENTIFICATION

The seeds of *Moringa* were obtained in a local market. It was then identified at the Biological Science Department, University of Lagos, Akoka, Lagos State, and allotted the specimen numeral APR/2018/MO-2564.

19.4.2 CHEMICALS

Standard chemicals used in this study were bought from Sigma Aldrich, Munich, Germany. The Lagos University, Teaching Hospital Central Research Laboratory, supplied the technical grade chemicals, and TA98 and TA100 variants of *S. typhimurium*. The remaining chemicals utilized conformed to standard specifications.

19.4.3 PREPARATION OF PLANT EXTRACT

The method of Ref. [47] was used in preparing the extract. The seeds were washed and allowed to dry in the sun for 7 days before being blended into finer particles using a milling machine. In total, 1000 mL of ethanol was added to 100 g of the powdered *M. oleifera* seeds in a flask which resulted in a mixture. This was stirred gently and covered for 72 h followed by evaporation using a vacuum drier at 40°C. A yield of 10.5% was obtained from the extract. It was then stored in the dark at +4°C until when needed.

19.4.4 DETERMINATION OF THE CHEMICAL CONSTITUENTS

An initial qualitative investigation of the chemicals in the *M. oleifera* seed was done to identify the constituents using a method described by Ref. [45], followed by a quantitative analysis of some of the chemicals present.

19.4.5 EVALUATION OF ANTIMUTAGENICITY

The ability of the seed extract at different concentrations to counteract the actions of a known mutagen was determined by employing TA 98 and TA 100 variants of *S. typhimurium* using the method of Ref. [29] with slight modifications. The inhibitory effect (antimutagenic efficacy) of seed extract of *M. oleifera* was calculated and expressed in percentage as previously reported by Ref. [32] using the following equation:

$$\text{Inhibition (\%)} = [(A\ B)/(A - C)] \times 100$$

where A represents the number of colonies observed on the plate after being exposed to the mutagen; B represents the number of colonies observed on the plate after being exposed to the mutagen and *M. oleifera* seed extract; C represents the number of spontaneous colonies observed per plate.

Antimutagenic effect was concluded to be efficacious if percentage inhibition is greater than 60, modest (40–60), small (20–40), and insignificant if less than 20.

19.4.6 STATISTICAL TEST

Results were expressed as average ± standard deviation. The results were subjected to a statistical test by first employing one-way analysis of variance (ANOVA) and afterward, Dunnett's test (Graphpad software, San Diego, CA, USA). Probability (p) values at <0.05 were considered statistically significant when compared to the concentration of the ethanol extract (3.5, 7.5, 15, 30 mg) on the number of revertants/plate (1000/Plate) and inhibition rate (%).

19.5 RESULTS

19.5.1 CHEMICAL ANALYSIS

Initial investigation revealed the presence of alkaloid, tannin, flavonoid, steroid, saponin, cardiac glycoside, anthraquinone, terpenoid in *M. oleifera* seed, while phlobatanin was not detected as shown in Table 19.1. The amount of some of the chemicals contained in the seed is also shown in Table 19.1. As seen from the table, tannins were most abundant, followed by flavonoid, steroid, alkaloid, and saponin.

TABLE 19.1 Qualitative Phytochemical Analysis of Seed Extract of *Moringa oleifera*.

Phytochemical	Moringa oleifera seed extract	
	Present	**mg/100 g**
Tannin	+	28.2 ± 0.63
Alkaloid	+	18.7 ± 2.15
Flavonoids	+	23.2 ± 0.44
Phlobatanin	–	**n/a**
Terpenoid	+	**n/a**
Saponin	+	15.2 ± 0.63
Cardiac glycoside	+	**n/a**
Anthraquinone	+	**n/a**
Steroid	+	20.5 ± 2.15

(+) is present; (–) is not detected.

Values = mean ± SD of triplicate experiments.

19.5.2 ANTIMUTAGENIC ACTIVITY

As seen in Table 19.2, a higher percentage of inhibition and a lower number of revertants per plate were recorded at higher concentrations for both TA 98 and TA 100 variants. A decrease ($p < 0.05$) in the number of TA 98 and TA 100 revertants was recorded at *M. Oleifera* seed extract concentrations of 15 and 30 mg/mL as compared to those obtained at lower doses of the extract percentage, which resulted in a higher percentage of inhibition at those doses.

TABLE 19.2 Antimutagenicity Assay of Seed Extract *Moringa Oleifera* Against Nitrosated Products.

Moringa oleifera (mG/mL)	TA98 number of revertants/plate)	TA98 inhibition (%)	TA100 number of revertants/plates	TA100 inhibition (%)
3.5	$388.20 \pm 19.40^{**}$	$11.92 \pm 4.50^{*}$	$858.60 \pm 37.85^{*}$	$12.36 \pm 4.40^{*}$
7.5	$307.60 \pm 18.74^{**}$	$30.61 \pm 4.35^{*}$	$729.00 \pm 65.94^{*}$	$27.43 \pm 7.67^{*}$
15	$180.80 \pm 17.53^{***}$	$60.02 \pm 4.16^{**}$	$560.40 \pm 20.77^{**}$	$47.06 \pm 2.42^{*}$
30	$101.00 \pm 11.25^{***}$	$78.53 \pm 2.61^{***}$	$421.80 \pm 25.78^{*}$	$63.18 \pm 3.00^{**}$

**Figures within the same column differ ($p < 0.05$); **, ***Figures differ greatly ($p < 0.05$).*

19.6 DISCUSSION

In this research on *M. oleifera* seed extract, results revealed the presence of medicinally active constituents. These results agree with that of Refs. [15,35]. The explicit physiological action of diverse medicinal plants against an extensive array of diseases in the human body is due to the various phytochemicals in those plants.[44]

The quantitative phytochemical analysis showed that seed extract of *M. Oleifera* is richer in tannins than other phytochemicals. Tannins are employed as caustic medicine for treating ailments that affect the intestine, for example, dysentery and diarrhea.[37] They also prevent carcinogenesis and the treatment of swollen tissues.[1]

The seed extract of *M. Oleifera* also contains considerable amounts of flavonoids which possess antioxidant properties that are also able to counteract allergies, inflammation, viruses, proliferation, and carcinogenesis, thereby influencing some aspects of mammalian metabolism.[3]

Saponins are used medically for treating high blood cholesterol and are valuable in managing arteriosclerosis, hypertension, and the control of postmenopausal syndrome. Saponins impede the assimilation of surplus cholesterol in the intestine, thereby reducing the likelihood of cardio dysfunction of which hypertension is an example.[4,17] Alkaloids are used to treat malaria, cold, cough, hypertension, diabetes, and cancer.[4] *Moringa* seeds contain saponins and terpenoids; hence, it is employed in managing hyperglycemia, which is a characteristic symptom of diabetes.[24] Steroids are significant signaling molecules and membrane components.[25]

AMES test is a simple, cheap, and convenient preliminary screening test for mutagens and it involves the use of bacteria that rely on amino acid and is able to revert genetically upon exposure to mutagens a.[30] The cells cannot form colonies when histidine is not available. However, on the reversion of mutation, the cells regain their ability to synthesize histidine and can produce colonies.

Table 19.2 shows the antimutagenic effect of ethanol extracts of *M. oleifera* seeds at various concentrations on *S. typhimurium* variants TA98 and TA100 exposed to 1-aminopyrene. As it pertains to TA98, an efficacious antimutagenic property of the extract was observed at a concentration of 30 mg/mL plate with 78.53% as the inhibition percentage. Regarding TA100, a relatively high inhibition percentage of 63.18% was also obtained at 30 mg/plate. This might imply that mutagenic substances that work directly are generated through contact with acidified nitrosated products such as 1-AP to suppress some components in the *M. oleifera* extract. The method involved in the antimutagenic property of the extract remains vague. An antimutagenic substance is able to hamper the conversion of pro-mutagen to mutagenic substances, incapacitate the DNA changing agent, or thwart the association of the agent with the genetic material. Enzymes involved in the pathways of the repair, recombination, and replication of DNA may be stimulated by some antimutagens by direct or indirect mechanisms.[9]

Moreover, the ability of the seed extract to counteract the effect of the mutagenic agent might be due to its ability to form a complex attributed to its constituents which nullifies the mutagen. One other theory for the anti-mutagenic property of the seed extract could be the activation of the trans-membrane export system in bacteria which allows mutagens to be quickly eradicated from the bacteria thereby preventing contact between the DNA and the mutagen.[2] Specific chemicals in plants like flavonoids group[8,16,19] and tannins[13] have been linked to antimutagenic activities of certain plants.

Flavonoids do contribute hydrogens and behave as antioxidants[6] by clearing free radicals, thereby preventing oxidative cell damage. The relatively high amounts of tannins in seeds of *M. oleifera* could also be linked to the considerable amounts of tannins present in the plant, as tannins are known to demonstrate antimutagenic activity.[23,40,42]

19.7 SUMMARY

Mutagenesis plays a significant role in carcinogenesis. *M. Oleifera* plant with antimutagenic properties can protect organisms against the initiation stage of carcinogenesis.

KEYWORDS

- **1-amino pyrene (1-AP)**
- **antimutagenic**
- **cancer**
- ***Moringa oleifera***
- **mutagen**
- ***S. typhimurium***

REFERENCES

1. Adegboye, M. F.; Akinpelu, D. A.; Okoh, A. The Bioactive and Phytochemical Properties of *Garcinia kola* (Heckel) Seed Extract on Some Pathogens. *Afr. J. Biotechnol.* **2008,** *7*(21), 3934–3938.
2. Ajith, T. A.; Janardhanan, K. K. Antimutagenic Effect of *Phellinus rimosus* (Bark) Pilat Against Chemical Induced Mutations of Histidine Dependent *Salmonella Typhimurium* Strains. *Food Chem. Toxicol.* **2011,** *49*, 2676–2680.
3. Akaneme, F. Identification and Preliminary Phytochemical Analysis of Herbs that Can Arrest Threatened Miscarriage in Orba and Nsukka towns of Enugu State. *Afr. J. Biotechnol.* **2008,** *7*(1), 006–011.
4. Akpuaka, M. U. Essentials of Natural Product Chemistry; Mason Publishers: Enugu, Nigeria, 2009; p 34.
5. Anwar, F.; Ashraf, M.; Bhanger, M. I. Inter Provenance Variation in the Composition of *Moringa Oleifera* Oil Seeds from Pakistan. *J. Am. Oil Chem. Soc.* **2005,** *82*, 45–51.
6. Bamforth, C. W.; Muller, R. E.; Walker, M. D. Oxygen and Oxygen Radicals in Malting and Brewing: a Review. *J. Am. Soc. Brew. Chem.* **1993,** *53*, 79–88.

7. Bennett, R. N., Kiddle, G.; Wallsgrove, R. M. Biosynthesis of Benzylglucosinolate, Cyanogenic Glucosides and Phenylpropanoids in *Carica papaya*. *Phytochemistry* **1997,** *45*(1), 5966–5971.
8. Beudot, C.; De Méo, M. P.; Dauzonne, D.; Elias, R.; Laget, M.; Guiraud, H.; Balasard, G.; Duménil, G. Evaluation of the Mutagenicity and Antimutagenicity of Forty-Two 3-Substituted Flavones in the Ames Test. *Mutat. Res.* **1998,** *417*, 141–153.
9. Sanjib, B. Natural Antimutagens: A Review. *Res. J. Med. Plant* **2011,** *5*, 116–126.
10. Caceres, A. B.; Cabrera, O.; Mirals, O.; Mollinedo, O.; Imendia, A. Preliminary Screening for Antimicrobial Activity of *Moringa oleifera*. *J Ethnopharmacol.* **1991,** *33*, 213–216.
11. Caceres, A. A.; Saravia, A.; Rizzo, S.; Zabala, L.; Leon, D. E.; Nave, F. Pharmacological Properties of *Moringa oleifera*: Screening for Antispasmodic, Anti-inflammatory and Diuretic Activity. *J Ethnopharmacol.* **1992,** *36*, 233–236.
12. Chuang, P. H.; Lee, C. W.; Chou, J. Y.; Murugan, M.; Shieh, B. J.; Chen, H. M. Anti-Fungal Activity of Crude Extracts and Essential Oil of *Moringa oleifera Lam*. *Bioresour. Technol.* **2007,** *98*, 232–236.
13. Dauer, A.; Hensel, A.; Lhoste, E.; Knasmüller, S.; Mersch-Sundermann, V. Genotoxic and Antigenotoxic Effects of Catechin and Tannins from the Bark of *Hamamelis Virginiana* in Metabolically Competent, Human Hepatoma Cells (HEP G2) Using Single Cell Gel Electrophoresis. *Phytochemistry* **2003,** *63*, 199–207.
14. De Flora, S. Mechanisms of Inhibitors of Mutagenesis and Carcinogenesis. *Mutat. Res.* **1998,** *402*(1–2), 151–8.
15. Doughari, J. H.; Pukuma, M. S.; De, N. Antibacterial Effects of *Balanites aegyptiaca L. Drel.* and *Moringa oleifera* L. on *Salmonella typhi*. *Afr. J. Biotechnol.* **2007,** *6*(19), 2212–2215.
16. Edenharder, R.; Van Petersdorf, I.; Rauscher, V. Antimutagenic Effects of Flavonoids, Chalcones and Structurally Related Compounds on the Activity of 2-Amino-3-Methylimidazol (4,5 -*f*) Quinoline (IQ) and other Heterocyclic Amine Mutagens From Cooked Food. *Mutat. Res.* **1993,** *287*, 261–274.
17. Fahey, J. W. *Moringa oleifera*: A Review of the Medical Evidence for its Nutritional, Therapeutic and Prophylactic Properties, Part 1. *Trees Life J.* **2005,** *1*(5), 1–15.
18. Faizi, S.; Siddiqui, B. S.; Saleem, R.; Siddiqui, S.; Aftab, K.; Gilani, A. H. Fully Acetylated Carbamate and Hypotensive Thiocarbamate Glycosides From *Moringa oleifera*. *Phytochemistry (Oxford)* **1995,** *38*, 957–963.
19. Ferreira, I. C. F. S.; Vargas, V. M. F. Mutagenicity of Medicinal Plant Extracts in Salmonella/Microsome Assay. *Phytother. Res.* **1999,** *13*, 397–400.
20. Guevara, A. P.; Vargas, C.; Sakurai, H.; Fujiwara, Y.; Hashimoto, K.; Maoka, T.; Kozuka, M.; Ito, Y.; Tokuda, H.; Nishino, H. An Anti-tumor Promoter From *Moringa oleifera* Lam. *Mutat. Res.* **1999,** *440*, 181–188.
21. Halliwell, B.; Gutteridge, J. M. Oxygen Toxicity, Oxygen Radicals, Transition Metals and Disease. *Biochem. J.* **1984,** *219*(1), 1–14.
22. Howard, P. C.; Beland, F. A.; Cerniglia, C. E. Reduction of the Carcinogen 1-nitropyrene to 1-aminopyrene by Rat Intestinal Bacteria. *Carcinogenesis* **1983,** *4*(8), 985–990.
23. Imanishi, H.; Sasaki, Y. F.; Ohta, T.; Watanabe, M.; Kato, T.; Shirasu, Y. Tea Tannin Components Modify the Induction of Sister-Chromatid Exchanges and Chromosome Aberrations in Mutagen-Treated Cultured Mammalian Cells and Mice. *Mutat. Res.* **1991,** *259*, 79–88.

24. Krishnaiah, D. T.; Bono, A.; Sarbatly, R. Studies on Phytochemical Constituents of Six Malaysian Medical Plants. *J. Med. Plant Res.* **2009,** *3*(2), 67–72.
25. Lednicer, D. Ed.; In *Steroid Chemistry at a Glance*; John Wiley & Sons: Hoboken, NJ, 2011; p 152.
26. Madhusudan, S.; Middleton, M. R. The Emerging Role of DNA Repair Proteins as Predictive, Prognostic and Therapeutic Targets in Cancer. *Cancer Treat. Rev.* **2005,** *31*, 603–617.
27. Mahajan, S. G.; Mali, R. G.; Mehta, A. A. Effect of *Moringa oleifera Lam.* Seed Extract on Toluene Diisocyanate-Induced Immune-Mediated Inflammatory Responses in Rats. *J. Immunotoxicol.* **2007,** *4*(2), 85–96.
28. Marles, R. J.; Farnsworth, N. R. Anti-Diabetic Plants and Their Active Constituents. *Phytomedicine* **1995,** *2*, 137–189.
29. Maron, D.; Ames, B. Revised Methods for the Salmonella Mutagenicity Test. *Mutat. Res. Environ. Mutagen. Relat. Subj.* **1983,** *113*(34), 173–215.
30. McCann, J.; Gold, L. S.; Horn, L.; McGill, R.; Graedel, T. E.; Kaldor, J. Statistical Analysis of Salmonella Test Data and Comparison to Results of Animal Cancer Tests. *Mutat. Res.* **1998,** *205*(1), 183–195.
31. Moller, I. M. Plant Mitochondria and Oxidative Stress: Electron Transport, NADPH Turnover, and Metabolism of Reactive Oxygen Species. *Annu. Rev. Plant Physiol. Plant Mol. Biol.* **2001,** *52*, 561–591.
32. Negi, P. S.; Jayaprakasha, G. K.; Jena, B. S. Antioxidant and Antimutagenic Activities of Pomegranate Peel Extracts. *Food Chem.* **2003,** *80*, 393–397.
33. Nweze, N. O.; Nwafor, F. I. Phytochemical, Proximate and Mineral Composition of Leaf extracts of *Moringa oleifera* Lam. From Nsukka, South-Eastern Nigeria. *IOSR J. Pharm. Biol. Sci.* **2014,** *9*(1), 99–103.
34. Prakash, A. O.; Pathak, S.; Shukla, S.; Mathur, R. Pre- and post -Implantation Changes in the Uterus of Rats: Response to *Moringa oleifera* Extract. *Ancient Sci. Life* **1988,** *8*, 49–54.
35. Ramarathnam, N.; Osawa, T.; Namiki, M.; Tashiro, T. Studies on the Relationship Between Antioxidative Activity of Rice Hull and Germination Ability of Rice Seeds. *J. Sci. Food Agric.* **1986,** *37*, 719–726.
36. Rao, C. V.; Ojha, S. K.; Mehrotra, S. Analgesic Effect of *Moringa Oleifera* Leaf Extract on Rats. In *Proceedings of the Second World Congress on Biotechnological Developments of Herbal Medicine*, Lucknow, India, 2003; pp 1–42.
37. Riganti, C.; Campia, I.; Kopecka, J.; Gazzano, E.; Doublier, S.; Aldieri, E.; Bosia, A.; Ghigo, D. Pleiotropic Effects of Cardioactive Glycosides. *Curr. Med. Chem.* **2011,** *18*(6), 872–885.
38. Ruan, C. Antimutagenic Effect of Foods and Chemoprevention of Cancer. *Acta Guangxi Med. Coll.* **1989,** *1*, 68–71.
39. Ruckmani, K.; Kavimani, S.; Anandan, R.; Jaykar, B. Effect of *Moringa oleifera* on Paracetamol Induced Hepatotoxicity. *Ind. J. Pharm. Sci.* **1998,** *60*, 33–35.
40. Sakai, Y.; Nagase, H.; Ose, Y.; Kito, U.; Sato, T.; Kamai, M; Mizuno, M. Inhibitory Action of *Paeony* Root Extract on the Mutagenicity of Benzo(a)pyrene. *Mutat. Res.* **1990,** *244*, 129–134.
41. Stadler, L. J. Mutations in Barley Induced by X-Rays and Radium. *Science* **2005,** *68*(1756), 186–187.
42. Tanaka, K.; Hayatsu, T.; Negishi, T; Hayatsu, H. Inhibition of N-nitrosation of Secondary Amines *In Vitro* by Tea Extracts and Catechins. *Mutat. Res.* **1998,** *412*, 91–98.

43. Tomasetti, C.; Li, L.; Vogelstein, B. Stem Cell Divisions, Somatic Mutations, and Cancer Etiology and Cancer Prevention. *Science* **2017,** *355*(6331), 1330–1334.
44. Toshihiro, A.; Harukuni, T.; Hiroyuki, A. Curcubitane-type Triterpenoids From the Fruits of *Momordica charantia* and Their Cancer Chemopreventive Effects. *J. Nat. Prod.* **2007,** *70*, 1233–1239.
45. Trease, G.E.; Evans, W.C. In *Trease and Evans Pharmacognosy*, 11th ed.; W.B. Saunders Elsevier Science Ltd: New York, 1983; pp 122–134.
46. Udupa, S. L.; Udupa, A. L.; Kulkarni, D. R. Studies on the Anti-inflammatory and Wound Healing Properties of *Moringa oleifera* and *Aegle marmelos*. *Fitoterapia* **1994,** *65*, 119–23.
47. Usunobun, U.; Okolie, N. P; Anyanwu, O. G.; Adegbegi, A. J.; Egharevba, M. E. Phytochemical Screening and Proximate Composition of *Annona Muricata* Leaves. *Eur. J. Botany Plant Sci. Phytol.* **2015,** *2*(1), 18–28.
48. Varier, V. P. Moringa oleifera. In *Indian Medicinal Plants Compendium of 500 species*, Chapter 15; Warrier, P. K., Nambiar, V. P., Ramankutty C., Eds.; Orient Longman Ltd: Madras, India, 1997; vol *4*, pp 58–62.

CHAPTER 20

Scope of Medicinal Plants for Uterotonic, Tocolytic, and Wellness Effects in Pregnant Women: A Cultural Perspective

Hertha Kasiku Haikera, Larai Aku-Akai, and Yapo Guillaume Aboua

ABSTRACT

Before hospitals were established, childbirth took place in homes assisted by relatives or elders in the community. Newborn babies were breastfed and lived into adulthood to experience life. Cultural practices were applied to ensure safety, augment labor, and save lives. Cultural practices include the use of plants/herbs in caring for a pregnant woman and newborn baby. Despite the existence of hospitals, there are still women that deliver their babies at home and there are those that take labor cultural/traditional remedies and yet come to deliver at hospitals. A study conducted explored in-depth the cultural practices regarding pregnancy, childbirth, and postpartum care in the north-eastern part of Namibia in Southern Africa. A qualitative case study design was used, and data were collected through in-depth one-on-one interviews with key participants (such as elders), who are experts in cultural practices/traditional birth attendants (TBAs), pregnant women, mothers with a history of home deliveries, and health-care practitioners. The data obtained were analyzed using content analysis. Findings of this study revealed that till today some women in the rural and urban areas use traditional remedies during pregnancy, labor, and postnatal period, and practice treatment specific for newly born babies. The purpose for taking the concoctions ranges from safe childbirth, speeding up labor progress (uterotonic effect), and wellness of

Curative and Preventive Properties of Medicinal Plants: Research on Disease Management and Animal Model Studies. Megh R. Goyal, Junaid Ahmad Malik, & Ademola Olabode Ayeleso (Eds.)

babies. The administration of these remedies is facilitated by TBAs. The effects are positively described by givers and users while health care practitioners do not favor such practices. Some of the negative effects highlighted by health care practitioners are placenta abruption, uterine rupture, postpartum hemorrhage, fetal distress, and infertility highlighted as a late complication. The study also unearthed some practices during home delivery that is not safe. Evidence-based findings of this research can enhance understanding of the practices and this shared with relevant stakeholders to improve maternal and child health care in the regions. Based on the findings in this chapter, recommendations include, first, developing strategies to educate TBAs and pregnant women on the effects of using traditional remedies in labor; and second, to advise them on safe practices. Further research on these remedies is still needed to ascertain the strength and the actual effect it may possess.

20.1 INTRODUCTION

Midwifery as an art relating to childbirth is the oldest health care activity. When the first human being was born, someone severed the cord, removed the placenta, and placed the infant in the mother's arms for warmth and put it to her breast for nourishment.[39] In Namibia, the first hospital was established around 1893 in Windhoek.[39] Before modern health care practices, traditional healers and traditional midwives played a very important role for their communities in healing the sick and helping women to give birth. According to World Health Organization (WHO), a traditional birth attendant (TBA), or traditional midwife is "a person who assists a mother during childbirth and who initially acquired her skills by delivering babies herself or through apprenticeship to other traditional attendants."[3]

In some rural communities in Namibia, there were and still, are practices carried out when a woman is pregnant, when contractions start, during childbirth as well as during the postpartum period the first 6 weeks after delivery. TBAs played a very important role in the implementation of these practices as they are regarded as midwives in the community.[36] Cultural midwifery practices can still be observed in rural communities.[36] The cultural practices involve administration of traditional remedies for labor; however, the traditional remedies used are not being tested in any laboratories to analyze the medicinal properties although there is subjective and objective evidence of the effect that it has either in augmenting labor or suppressing contraction. Evidence in literature reveal that administration of traditional concoctions for labor can result in an overdose to cause harm to the laboring mother and/or baby. A study done in Turkey stated that the traditional remedies have an adverse effect on maternal or baby's condition.[37]

In the Namibian Region where this study was conducted, speculations are that the faster labor progress observed among some women in labor are revelations of them taking traditional concoctions to speed up/augment labor progress though they might not reveal it to the midwife. Other observations made in the clinical area are that after visiting hour by family members in the maternity ward, a woman who had slow progress suddenly experiences strong contractions followed by delivery of her baby. Although there is no physical proof of family members caught giving these remedies in the labor ward. There are assumptions that during visiting hour, the elders bring these remedies to the admitted woman. There are known TBAs that prepare women culturally for the birth of the newborn in different ways. Some of the maternity records for patients observed in the labor ward do show partographs/labor care guides that have faster progress than normal.

This chapter focuses on the sociocultural roles of TBAs and some of the traditional remedies used for purpose of augmenting or suppressing labor progress/initiation.

20.2 THE SOCIOCULTURAL ROLES OF TRADITIONAL BIRTH ATTENDANTS IN RURAL NAMIBIA

In rural Namibia, TBAs play a very important role in their communities. The traditional scope of practice for TBAs ranges from treating infertility, caring for pregnant women, assisting during childbirth/conducting delivery, care in the postpartum period, as well as caring for the newborn. Pregnant women approach TBAs for various services although they still book and attend antenatal care at health facilities or even when they plan to deliver at the hospital, visiting a TBA is still vital. TBAs educate pregnant women on how to behave during pregnancy, what to eat, and what is prohibited. They also give traditional treatment from locally available plants believed to help the baby to be active in utero as well as for safe delivery. Certain pregnancy-related ailments are also solved culturally either by giving traditional treatment or advice. Literature explored on sociocultural role for TBAs in rural Namibia did not yield any conclusions. However, these roles in the rural Namibian context do not differ from those explored in other countries. Research on the sociocultural roles of TBAs in Ghana shows that pregnant women do seek support ranging from nonorthodox therapy and psychological support to relieve fear and anxiety as well as herbal medications to assist keep fetus strong and healthy.[10] Evidence from the study shows care-seeking behavior of pregnant women is largely mediated by sociocultural influences that shape individual perceptions of threats to pregnancy and these beliefs are often associated with early disclosure, sorcery, and witchcraft.[10]

20.3 THE IMPACT OF CULTURAL PRACTICES ON PREGNANCY, CHILDBIRTH, AND POSTPARTUM ROLES IN RURAL NAMIBIA

A study that was done in one of the rural areas of Oshikoto, Oshana, and Ohangwena regions in Namibia indicated that some or most of the cultural practices relating to pregnancy have a greater influence on pregnancy and childbirth.[28] Some of the health advice given is beneficial to the pregnant women, for instance, resting, being faithful, and eating well. However, the study revealed that there are some cultural restrictions that could be harmful or deprive the pregnant woman of something beneficial, for example, telling the woman not to eat a certain type of food as it could deprive her of valuable nutrients. WHO recommends that all pregnant women should be attended by skilled health care providers at health facilities from antenatal period and also at delivery. However, because some pregnant women prefer to be attended to by TBAs, there are records of home deliveries. The NDHS data show that about 87% of births occurred in health facilities.[27] More studies need to be done in rural Namibia concerning the impact of traditional treatment given to pregnant women on pregnancy, childbirth, and postpartum. The research that was done in this area in other countries revealed that the use of herbal remedies could cause harm to the mother and newborn or may cause complication.

20.4 HERBAL/TRADITIONAL REMEDIES CONSUMED BY PREGNANT WOMEN

There are various herbal remedies consumed by some pregnant women influenced by cultural beliefs with the aim of keeping a pregnant woman healthy throughout the semesters of pregnancy, facilitating childbirth, and for postpartum wellbeing. Some of the vegetation commonly used are summarized in Table 20.1 in relation to stages in or after pregnancy, where they are relevant.

TABLE 20.1 Herbal and/or Traditional Remedies Used by Some Pregnant Women in Namibia.

Plant/traditional remedy	Scientific name	Stage used	Purpose
Eromborora	*Sesanum triphyllum*	During pregnancy	For perineal massage. If used, the perineal muscles will stretch and the woman will not sustain tears during the vaginal delivery nor require episiotomy
Kawesa-Masini	Unknown	Postpartum period	For milk production and works as a family planning method
Mbuumbu (local name)	*Vangueria esculenta*	During pregnancy	Wellness of pregnant mother and unborn baby
Mpindu	*Ancylantus bainesii*	During pregnancy and at onset of labor	Keeps baby active in uterus, and when mixed with other roots it augments labor
Mukekete	*Ziziphus Mucronata*		For wellness of the pregnant mother and for speedy delivery when given as a concoction with other herbs
Mukwevo	Unknown	In forth stage of labor	Aid in expelling clots if a woman bleeds excessively after delivery
Munandwa	*Pavonia senegalensis*	During pregnancy	For perineal massage. If used, the perineal muscles will stretch and the woman will not sustain tears during the vaginal delivery nor require episiotomy
Muroro	*Helinus integrifolis*	During pregnancy and at onset of labor	For wellness of the pregnant mother and for speedy delivery when given as a concoction with other herbs *(sivatu)*
Mutundungu	*Burkea afrikana hook*	During pregnancy	For treatment of preterm labor as a tocolytic
Other remedies			
Chicken droppings		In a complicated third stage of labor	Loosens/separates a stuck/retained placenta from the uterine wall
Elephant Dung		At onset of labor	Tea made from dry elephant dung is believed to speed up/shorten labor duration

Participants, who have used the traditional remedies during pregnancy, describe it as vital in the cultural context. Traditional care includes abdominal and perineal massages and oral treatments based on certain herbal concoctions ranging from roots, leaves, bark/phloem of medicinal plants to the elephant dung. It is alluded to that cultural treatment is necessary for every pregnant woman even if she plans to deliver at hospital. The persons involved in this practice believe that the concoctions given are powerful and efficient to prevent maternal or fetus/neonatal death. The participants also feel that the care at home prepares the pregnant women for safe delivery. The expectation is that when the perineal massage is done at home; on the delivery time, the vaginal opening will stretch to allow the baby to pass through without sustaining any perineal tears. The participants claim that this kind of information or care is not being given at health facilities even though they attend antenatal care. These sentiments are echoed in a study conducted in Nigeria, where participants described the services of TBAs as loving, caring, more considerate, and always available when needed.[1]

Another study done in Ethiopia also revealed that culturally pregnant women feel that the care they receive from TBAs assures them of less complications from pregnancy.[26] The use of elephant dung to speed up labor is common in the northern region of Ethiopia. A study done in Zimbabwe showed a high prevalence rate in women using elephant dung, anthill soil, *Fadogia ancylantha*, and muddy nest of mud daubers.[25] *F. ancylantha* (mentioned in a study for Zimbabwe) and *Vangueria esculenta* (mentioned in this research) seems to be the same plants, which means this particular plant is used for the same purpose in Namibia as well as in Zimbabwe.

Preterm labor in the cultural context of the *vaKavango* people is treated using the phloem or bark of *mutundungu* tree scientifically known as *Burkea afrikana* mixed with mud from the nest of mud dauber wasp. It is not yet known what medical properties are found in this concoction but users are certain that it is highly effective for that purpose. Some of the ethno-pharmacological studies done in South Africa found evidence of *Burkea Africana* to have antioxidant, anti-inflammatory, and has the highest 15 antilipoxygenase activities with 86% inhibition at 100 μg/mL.[13]

A study that was done in Kigoma region in Tanzania also found that local herbs (no specific names mentioned) are very useful in preventing miscarriage or preterm labor.[15] However in modern medicine more is done not just focusing on the mother but the baby as well. WHO recommends that preterm labor is managed holistically to make sure fetal and maternal wellbeing is not compromised. Treatment, therefore, includes giving tocolytics to suppress contractions, if possible, corticosteroids to mature the fetal lungs, and last but not least prophylactic antibiotics to prevent infection.[38]

20.5 THE EFFECT OF TRADITIONAL REMEDIES ON LABOR OUTCOME

The health-care professionals other than skilled birth attendants believe that these traditional remedies do more harm than good based on the experience they have with the clients. Midwives, whose primary role is to execute the monitoring of labor, shared that the contractions are hypertonic (very strong) and more frequent than normal in pregnant women that have taken concoction to speed up the labor. This results in maternal exhaustion and once the mother is tired whilst still in labor, she may end up with fetal distress (a decrease or increase in fetal heart rate) whereby the woman can end up being operated to extract the baby because regardless of the strong contractions, the cervix fails to dilate sometimes. There are also those pregnant mothers that progress abnormally fast and deliver very fast because of taking the traditional oxytocin. The midwives always discourage this practice because of the complications.

Literature supports the views stated here "the uterine contractions compress the placenta in the upper uterine segment, and this can retard the amount of oxygen to the fetus to a greater or lesser degree especially at the peak of a contraction."[35] This means in the case of continuous contractions secondary to having consumed traditional Pitocin, the fetus is subjected to a certain amount of hypoxia (not enough oxygen), hence the decrease in fetal heart rate (fetal bradycardia). In such cases, even the delivery outcome might be affected because the baby might be born floppy and need resuscitation.

Doctors have the same opinion as midwives but also added that some of the complications do present immediately (like in the case of fetal distress) but some may present later in life (e.g., infertility). The unfortunate part is both mother and baby can be affected by this and suffer complications. In worst-case scenarios, some women end up with uterine rupture, which may lead to permanent infertility if uterus is removed (hysterectomy), DIC, HELLP syndrome, sepsis, necrotic bowels, and some even succumb to this phenomenon due to multiple organ failure. In babies, possible complications may include fetal distress, IUFD, hypoxic ischemic encephalopathy, which may affect the child's performance as he/she grows. A study in Rundu Hospital in Kavango east region revealed that in a period of 3 months (January–March 2017), 242 caesarean sections were done of which 25% were attributed to fetal distress and 6.7% were due to prolonged labor.[19]

Several studies also found that the use of herbs in pregnancy and in labor may have negative effects on both mother and baby causing heart defects, liver damage, cardiorespiratory depression, and poor Apgar score in babies.[12] Another study done in Malawi also revealed that the use of herbs leads to obstetric complications. There is also very little evidence with regard to safety when using traditional medicine in labor.

A plant known as *Cissampelos mucronata* was researched using rat models. The result of this study indicated that this particular plant induced labor with an increasingly high level of maternal stress, which led to premature labor and cannibalism in rodents; however, no uterine rupture was observed.[21] Several other studies also concurred with the views of skilled birth attendants that herbal medicine leads to continuous contractions not corresponding to slow cervix dilatation.[21] Postnatal complications revealed by other studies include vaginal discharge, pelvic pains, and fever which may lead to puerperal sepsis.[15]

20.6 SUMMARY

This chapter explored the local plants used by some women in some regions in the northern part of Namibia. Participants that provided the necessary data included the pregnant women, the elders, and/or TBAs, mothers with a history of home deliveries, and skilled birth attendants. Various plants/herbs locally available and a few only found in neighboring Angola are used for either tocolytic or uterotonic effects. Participants had different views regarding utilization of herbs/roots or other traditional remedies while pregnant, in labor or postpartum. Participants speak greatly about the use of herbs that it has many advantages ranging from helping prepare for the birth, keeping mother and baby safe, keeping baby active in uterus, preventing maternal or neonatal death, preparing perineum to prevent tears during childbirth, speeding up labor process, and many more. Once again, this is their understanding and beliefs according to culture. These are not unique to the northern region but it is evident from the literature that these are common practices in Africa and even beyond. In America, women use a plant known as blue cohosh for the same purpose. Postnatal-care culturally includes staying indoors, sitz bath, giving some sort of vaginal enema to expel clots, and giving *kawesa-masini*. On the contrary, health professionals have an opposite view and there are many reports also supporting their views. From the health professionals' experience, they observed more harm or negative effects secondary to herbal use and in worst-case scenarios the woman may suffer uterine rupture due to hypertonic uterus, permanent coagulation disorders, infertility, multiple organ failure, and even death.

Regardless of the different views, culture is what defines people or communities and they do certain practices according to their beliefs because they feel if it is not done something might go wrong. And yet, there is no evidence of toxicity on these plants commonly used locally. In conclusion, the evidence presented in this report can serve as a stepping stone for further studies to analyze traditional herbs utilized in rural area (for the purpose of inducing or augmenting labor) for chemical properties, safety, and physiological effects.

ACKNOWLEDGMENTS

Special thanks to my participants, who voluntarily took part in this study and played a very important role of providing this very useful and relevant data used in this chapter. Appreciation goes to Dr. Yapo Aboua and Dr. Aku-Akai for the mentorship all the way. Thanks to my husband and children for supporting me all the time.

KEYWORDS

- **augment labor**
- **cultural practices**
- **labor onset**
- **medicinal plants**
- **postpartum**
- **pregnant woman**
- **tocolytic**
- **traditional birth attendants**
- **uterotonic**

REFERENCES

1. Afoyalan, J. A.; George, B. I. Care of a New-Born by Traditional Birth Attendants in Oguobiri Community, Bayelsa State, Nigeria. *FWU J. Social Sci.* **2010,** *4*(2), 1–19.
2. Aikins, D. A. Food Beliefs and Practices During Pregnancy in Ghana: Implications for Maternal Health Interventions. *J. Health Care Women Int.* **2014,** *2014*, 954–972.

3. Aziato, L.; Omenyo, C. N. Initiation of Traditional Birth Attendants and Their Traditional and Spiritual Practices During Pregnancy and Childbirth in Ghana. *BMC Pregnancy Childbirth* [Online] **2018,** *18*(64). https://doi.org/10.1186/s12884-018-1691-7
4. Bailey, P. E.; Szaszdi, J. A.; Glover, L. Obstetric Complications: Does Training Traditional Birth Attendants Make a Difference? *Rev. Panam. Salud. Publica.* **2002,** *11*(1), 15–23.
5. Bakhta, A.; Mani, S. Cultural, Social, Religious Beliefs and Practices During Post-Partum Period Among Post-Natal Women in Selected Rural & Urban Health Care Settings. *Indonesian J. Glob. Health Res.* **2019,** *1*(1), 1–12.
6. Bell, J.; Waters, S. Eds.; In *Doing Your Research Project. A Guide for First Time Researchers*, 7th ed.; Open University Press: Glasgow, Great Britain, 2018; E-book.
7. Brink, H. I. Ed.; In *Fundamentals of Research Methodology for Health-Care Workers*; Juta & Co, Ltd.: Cape Town, South Africa, 2012; p 239.
8. Cheikhyoussef, A.; Shapi, M.; Matengu, K.; Muashekele, H. Ethnobotanical Study of Indigenous Knowledge on Medicinal Plant Use by Traditional Healers in Oshikoto Region, Namibia. *J. Ethnobiol. Ethnomed.* **2011,** *7*, 10–18.
9. Coalter, W, S.; Patterson, S, L. Sociocultural Factors Affecting Uptake of Homebased Neonatal Thermal Care Practices in Africa: A Qualitative Review. *J. Child Health Care* **2017,** *21*(2), 132–141.
10. Dako-Gyeke, P.; Aikins, M.; Aryeetey, R., McCough, L.; Andongo, P, B. The Influence of Socio-cultural Interpretations of Pregnancy Threats on Health Seeking Behavior Among Pregnant Women in Urban Accra, Ghana. *BMC Pregnancy Childbirth* **2013,** *13*, 111–116.
11. De Vos, A. S.; Strydom, H.; Fouche, C, B.; Delport, C. S. L. Eds.; Research at Grass Roots for the Social Sciences and Human Service Professions, 4th ed.; Van Schaik Publishers: Pretoria, SA, 2011. *SA J. Ind. Psychol.* **2004,** *30*(1), 132–134.
12. Dika, H.; Dismas, M.; Iddi, S.; Rumanyika, R. Prevalent Use of Herbs for Reduction of Labour Duration in Mwanza, Tanzania: Are Obstetrician Aware? *Tanzan. J. Health Res.* [Online] **2017,** *19*(2). DOI: http://dx.doi.org/10.4314/thrb.v19i2.5
13. Farouk Z. L.; Slusher, T. M.; Danzomo, A. A.; Slusher, J. L. Factors Influencing Neonatal Practice in a Rural Community in Kano (Northern), Nigeria. *J. Trop. Pediatr.* **2019,** *65*(6), 569–575.
14. Fraser, D. M.; Cooper, M. A. Eds.; In *Myles Textbook for Midwives*, 15th ed.; Elsevier Limited: London: United Kingdom, 2009.
15. Fukunaga, R.; Morof, D.; Blanton, C., Maro, G.; Serbanescu, F. Factors Associated With Local Herb use During Pregnancy and Labor Among Women in Kigoma Region, Tanzania 2014–2016. *BMC Pregnancy Childbirth* [online] **2020**, *20*(1), 122–128. <https:doi.org/10.1186/s12884-020-2735-3
16. Government of Republic of Namibia. *Namibia 2011 Population and Housing Census Main Report*; GRN Printers: Windhoek, Namibia, 2011; p 214.
17. Grady, K.; Van den Broek, N.; Kongnyuy, E.; Ameh, C. Eds.; *Life-Saving Skills: Essential Obstetric Care and New-born Care*. Royal College of Obstetricians & Gynecologists (ROCG) Press: London, 2007; p 136.
18. Haruna, U.; Kansanga, M. M.; Galaa, S. Examining the Unresolved Conundrum of Traditional Birth Attendants in Maternal Child Health Care Delivery in Ghana. *Health Care Women Int.* **2019,** *40*(12), 1336–1354.
19. Hatupopi, S. K.; Nghamukamo, M.; Nghitanwa, E. M.; Tuhadeleni, O. N. Indications for Caesarean Section in Rundu State Hospital in Kavango East Region, Namibia. *Glob.*

J. Health Sci. [Online] **2019,** *11*(11). https://www.ccsenet.org/journal/index.php/gjhs/article/view/0/40751 (accessed Sept 30, 2021).
20. Irakunda, F. Food Taboos During Pregnancy. *Health Care Women Int.* **2019,** *41*(2), 159–168.
21. Lampiao, F.; Maliwichi-Nyirenda, C.; Mponda, J.; Tembo, L.; Clements, C. Preliminary Investigation of the Effect of Labor Inducing Plants, *Cissampelos mucronata*, on the Outcomes of Pregnancy Using Rat Models. *Malawi Med. J.* **2018,** *30*(3), 159–161.
22. Majid, U. Research Fundamentals: Study Design, Population and Sample Size. *URNCST J. Res. Gate* [Online] **2018,** *2*. https://doi.org/10.26685/urncst.16 (accessed Sept 30, 2021).
23. Maliwichi-Nyirenda, C. P.; Maliwichi, L. L. Medicinal Plants Used to Induce Labor and Traditional Techniques Used in Determination of Onset of Labor in Pregnant Women: A Case Study of Mulanje District. *J. Med. Plant Res.* **2010,** *4*(24), 2609–2614.
24. Mathole, T.; Lindmark, G. Competing Knowledge Claims in the Provision of Antenatal Care: A Qualitative Study of Traditional Birth Attendants in Rural Zimbabwe. *Health Care Women Int.* **2005,** *26*, 937–956. DOI: 101080/07399330300301796
25. Mawozo, T.; Nhachi, C.; Magwali, T. Prevalence of Traditional Medicine use During Pregnancy, at Labor and for Postpartum Care in a Rural Area, Zimbabwe. *Clin. Mother Child Health* **2002,** *16*(2), 1–7.
26. Masele, H. A. Traditional Maternal Health Beliefs and Practices in Southern Tigray: The Case of Raya Alamata District. *Anat. Physiol. Curr. Res.* **2018,** *8*(2), 298–305. DOI: 10.4172/2161-0940.1000298
27. MoHSS. *Namibia Demographic Health Survey*; Government Printers: Windhoek, Namibia, 2013; p 315.
28. Mulenga, E.; David, A. S.; Penihas, L. N. Taboos, Traditional Practices and Beliefs Affecting Pregnancies in Ohangwena, Oshana & Oshikoto Regions: University of Namibia's Fourth Year Students' Rural Placement Experience of 2016, Oshakati Campus, Namibia. *Int. J. Adv. Nurs. Stud.* **2018,** *7*(1), 68–71.
29. Muntenda, B. M.; Stern, R.; Nuuyoma, V. The Perception of Women on Obstetric Care in a Public Health Facility in a Peri-Urbarn Area in Kavango East Region. *Obstet. Gynecol. Int. J.* **2019,** *10*(2), 98–104.
30. NSA. *Namibia Inter-censal Demographic Survey 2016 Report;* Namibia Statistics Agency (NSA): Windhoek, 2017; p 119.
31. Olusanya, B.; Inem, V.; Abosede, O. Infants Delivered in Maternity Homes Run by Traditional Birth Attendants in Urban Nigeria: A Community-Based Study. *Health Care Women Int.* **2011,** *32*(6), 474–491.
32. *Oxford Advanced Learner's Dictionary of Current English*, 6th ed.; Oxford University Press: Oxford, UK, 2000; p 1539.
33. Rundrum, S. Traditional Birth Attendants in Rural Northern Uganda: Policy, Practice and Ethics. *Health Care Women Int.* **2016,** *37*(2), 250–269.
34. Scotland, J. Exploring the Philosophical Underpinnings of Research: Relating Ontology and Epistemology to the Methodology and Methods of the Scientific, Interpretive, and Critical Research Paradigms. *English Lang. Teach.* **2012,** *5*(9), 9–18.
35. Sellers, P. M. Ed.; In *Midwifery,* 2nd ed.; Juta & Co Ltd.: Cape Town, South Africa, 2013; p 317.
36. Shirungu, M. In *Preserving "Mpo vyetu" (Our Culture), Contestation and Negotiation of Likuki/ Shipumuna Ritual in Kavango Northeast of Namibia*, Proceedings of the

Conference on the Intersection of Indigenous and Traditional Knowledge and Technology Design, Informing Science Press: Santa Rosa, California, 2015; [Online]. https://indiknowtech.nust.na/?p=1445 (accessed Sept 30, 2021).
37. Sultan, A; Sengul, Y. E. Potential Harmful Traditional Practice During Pregnancy and Postpartum. *Eur. J. Contracept. Reprod. Health Care* **2008,** *13*, 282–288.
38. WHO. WHO recommendations on interventions to improve preterm birth outcomes. WHO Press: Geneva, Switzerland, 2015; p 97.
39. Van Dyk, A. Ed.; In *The History of Nursing in Namibia*. Gamsberg Macmillan Publishers: Windhoek, Namibia, 1997; p 132.
40. Yajima, S.; Asano, H.; Fukano, H. Necrosis of the Small Intestine Leading to a Diagnosis of Polyarteritis Nodosa: A Case Report. *J. Med. Case Rep.* **2019,** *13*, 55–63.
41. Zubaida, L. F.; Slusher, T. M.; Danzomo, A, A.; Slusher, J. L. Factors Influencing Neonatal Practice in a Rural Community in Kano (Northern) Nigeria. *J. Trop. Pediatr.* **2019,** *65*(6), 569–575.

CHAPTER 21

Role of Phytonutrients in Chemoprevention of Cancer: A Review

Oluwaseyi Adegoke Adetunji, Modupe Olusola Adetayo, Babafemi Tosin Ogunbiyi, and Gogonte Hezekiah Amah

ABSTRACT

The evidence from in vivo and in vitro research studies have proven that plant dietary source reduces the incidence of diseases due to the presence of biologically active plant compounds or phytochemicals. Cancer is a progressive disease that develops in distinct stages (initiation, promotion, and progression) that are separated by a long period of latency. The bioactive compound's anticancer activity depends on multitarget mechanisms that blocking mutagenic activities of other compounds, reduce macromolecular damage induced by oxidative stress, suppress abnormal cell proliferation, and modulate immune response, inflammation, cell metabolism, and genome instability. The bioactive compounds used in targeting cancer include curcumin that inhibits nuclear factor-kappa B, induces anti-inflammatory, antioxidant, and inhibits cancer cell growth, quercetin modulation of apoptotic ligand-induced cytotoxicity by tumor necrosis factor, epigallocatechin-3-gallate which has antiangiogenic effect. Other bioactive compounds in cancer chemoprevention and treatment are camptothecin, luteolin, lycopene, resveratrol, β-sitosterol, genistein, sulforaphane, indole-3-carbinol, 3,3′-diindolylmethane, vinblastine, vincristine, and piperin. Based on the understanding of the several molecular signaling pathways in cancer chemoprevention by bioactive compounds, a synergistic and additive approach combined with other cancer therapeutics including radiotherapy, immunotherapy, and chemotherapy could limit toxicity in the management and prevention of cancer.

Curative and Preventive Properties of Medicinal Plants: Research on Disease Management and Animal Model Studies. Megh R. Goyal, Junaid Ahmad Malik, & Ademola Olabode Ayeleso (Eds.)

21.1 INTRODUCTION

The incidence of cancer is increasing despite the intense study on carcinogenesis for the past six decades.[9] Cancer is defined as a population of cells that have lost their normal control of cell growth and differentiation and are proliferating without check with the ability to spread to other parts of the body.[13,22] Intentional moves can be taken to minimize the risk of cancer initiation, recurrence, or progression via primary and secondary cancer chemoprevention.

Bioactive compounds found in natural products have been recorded to have chemopreventive functions that are highly effective at preventing or reversing the carcinogenic process.[9] Bioactive compounds are referred to as secondary metabolites with eliciting pharmacological potentials.[1] Most food plants are capable of producing bioactive compounds, but average medicinal plants contain higher concentration of active compounds.[1]

Chemoprevention uses natural or synthetic substances to reduce or hinder the onset of cancer progression. Oncogenesis involves the change in the genetic and regulation system of the cell induced by carcinogens.[3] The use of chemotherapy involves the development of biological, nutritional, and pharmacological interventions to prevent, reverse, or delay carcinogenesis.[4]

This chapter focuses on the role of plant-based phytochemicals in the chemoprevention of cancer.

21.2 PATHOLOGY OF CANCER

Hippocrate coined "carcinoma" from *karkinos* (the word crab).[10] Cancer as a group of diseases has been in existence for all human history. Francois de la BoeSylvius believed that all diseases were the outcome of chemical processes, and that cancer is caused by acidic lymph fluid. Campbell De Morgan (an English surgeon) formulated between 1871 and 1874 that cancer is a poison that proliferate from primary tumor through the lymph nodes to other organs.[27]

Environmental factors are major causative agents in the development of cancer in which genes that encode important regulatory proteins are mutated that results to cells with uncontrollable growth and destroying surrounding normal tissue, spreading to vital organs, and leads to disseminated disease.[13] Effects of exogenous and endogenous agents on DNA damage during carcinogenesis are shown in Figure 21.1.

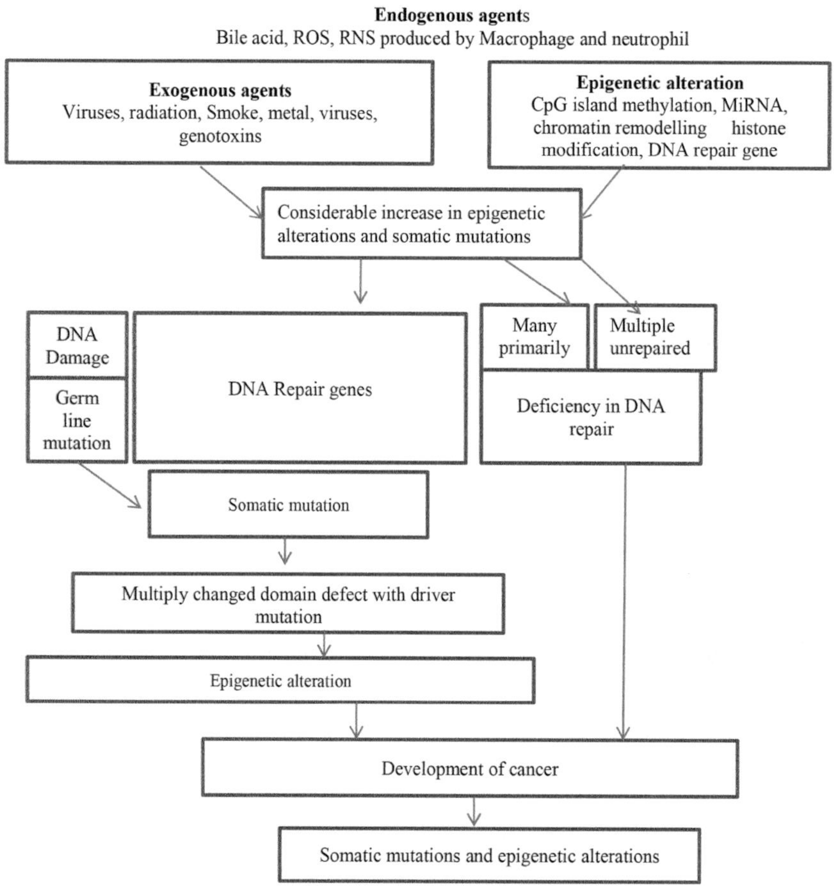

FIGURE 21.1 Effects of exogenous and endogenous agents on DNA damage during carcinogenesis.[18]

21.2.1 TYPES OF TUMORS

- **Benign tumors** are growing masses that press on their primary sites and do not spread to other parts of the body. The tumor cells do not metastasis to other organs, and in most cases, they can be removed but many of them have malignant potential if not removed on time.[26] Examples of benign tumors include papilloma, adenoma, lipoma, osteoma, myoma, angioma, and nevus.
- **Malignant tumors** are usually fast proliferating and spread to other body tissues by entering the bloodstream or lymphatic nodes. The

cancerous cells can spread and damage nearby tissues and organs. This ability of cancer cells to invade other parts of the body is a characteristic feature to proliferate uncontrollably and independently from the primary tissue of malignancy.[10] Metastasis (development of a secondary tumor) is the term used to describe the spread of cancer from the original mass (the primary tumor) to other parts of the body and this is a proof of malignancy. Malignant tumor is generally classified as carcinoma, sarcoma, lymphoma and leukemia, germ cell tumor, and blastoma. The characteristics of malignancy are self-sufficient in growth signaling; it causes insensitivity to antigrowth signals; it can evade apoptosis; it enables limitless in replication potential; it induces and sustains angiogenesis; it leads to activation of metastasis and invasion of tissue.

21.2.2 CAUSES OF CANCER

The causes of particular cancer differ from the other and they are dependent on many factors. It is a very complex process involving both environmental and cellular factors.[10] The causes of most cancers have no known source from environmental, genetic, chemical, or viral organisms while it can also be spontaneous.[18] The possible causes of cancer are listed below:

- Chemicals that are known to cause cancer are called carcinogens. Exposure to certain chemicals, metals, or pesticides predisposes a person to the risk of cancer. Uranium, benzidene, nickel, cadmium, radon, asbestos, benzene, and vinyl chloride are examples common carcinogens. Cigarette smokes a major contributing chemical factor in the cause of lung, mouth, and esophagus cancer.[18]
- Ionizing radiation rays, such as X-rays and ultraviolet rays can cause the DNA of cells to mutate and lead to the incidence of cancerous cells.

Symptoms of cancer

Early symptoms of cancer include:

- Persistent weight loss.
- Frequent bleeding or discharge.
- Continuous indigestion.
- Patches in the mouth or tongue.

21.2.3 CARCINOGENESIS

Carcinogenesis is a complex process that interplays genetics, environmental exposure, and age combined, to create favorable conditions for the development and dissemination of cancer. Several research and experiments on animals have brought us to the knowledge on chemical carcinogenesis which aid in identifying cancer-causing chemicals and how human can reduce the risk of exposure.

Carcinogenesis which is caused by mutation of the genome is also referred to as oncogenesis or tumorigenesis. The uncontrollable growth of cells can cause the development of benign tumor which can become malignant over time if not removed. Many different types of chemical exposure can increase tumors in human before the carcinogenic risk of exposure is manifested. Carcinogenesis is now recognized as more complex. Two distinct classes of genes that are proto-oncogenes and tumor suppressor gene are involved in cancer processes.

21.2.4 STEPS INVOLVED IN CARCINOGENESIS

Cancer development is involved in a multistep processes, as shown in Figure 21.2. The carcinogenesis includes the following steps.

21.2.4.1 TUMOR INITIATION

The inactivation of tumor suppressor gene by DNA methylation of promoter region of gene by a carcinogenic chemical agent leads to gene mutation through the modification of the DNA structure during the synthetic phase, to form adducts between the chemical carcinogen and the DNA. The DNA adducts are known to initiate tumorigenesis by inactivating the tumor suppressor gene or activating proto-oncogene. Initiation also involves damage to the genome, these can be caused by damage to the gene, epigenome (that regulates the use of gene in gene expression), or damage to chromosomes which result from an irreversible genetic alteration. It also needs an initiator, which must be metabolized before becoming active. Initiators are often specified to a particular species or type of tissue.[24] The effect of an initiator is irreversible and susceptible to promotion.[18]

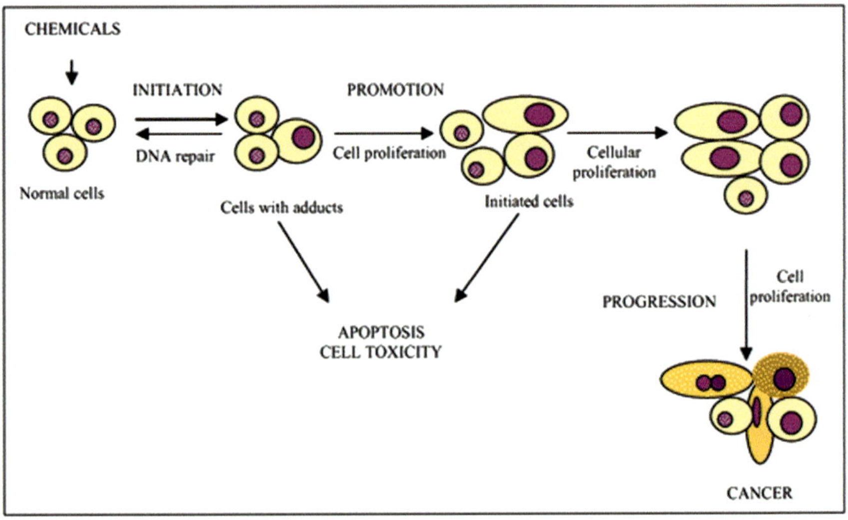

FIGURE 21.2 Steps involved in carcinogenesis.

Source: Reprinted from Ref. [28]. Open access.

21.2.4.2 TUMOR PROMOTION

Tumor promotion is selective clonal expansion of initiated cells. Tumor promoters often mediate their biological effect without metabolic activation but are generally nonmutagenic. Tumor promoters enhance the proliferation of the cells by producing increased number of daughter cells containing the mutated genetic materials.[27] Croton oil bioactive compound, 12-otetradecamoyl phorbol-13 acetate, is used in murine skin carcinogenesis as promoter of tumor growth promoter, by activating protein kinase C (PKC). It is the best-used tumor promoter.

21.2.4.3 TUMOR PROGRESSION

Tumor progression involves the development of malignant cells that possess hostile traits. It also involves tumor growth, clonal growth, tumor angiogenesis, and invasion metastasis.[26] Furthermore, the primary cause of cancer in carcinogenesis is DNA damage and deficiency of DNA repair.

21.2.4.4 MALIGNANT CONVERSION

Malignant conversion (metastasis) occurs spontaneously at a low frequency. It is a process that changes a preneoplastic cell into one that conveys the malignant phenotype.[2] The rate of the cell becoming malignant will be increased proportionally to the rate and quality of dividing cells into benign tumor. The low chance of a tumor to metastasize is further enhanced by exposure of benign cells to carcinogenic agents which activates proto-oncogenes and inactivates tumor suppressor genes.[16]

21.3 BIOACTIVE PHYTOCHEMICALS

Phytochemicals are chemicals produce by plants. They are sometimes called secondary metabolites; these metabolites include carotenoids, saponins, alkaloids tannins, oxalates, phytate, etc.[11] The beneficial medicinal effects of plants, typically result from the secondary products found in plants. Studies on the protective effect of diet rich in vegetables and fruits are giving consideration to the possibility that bioactive plant secondary metabolites possess anticarcinogenic activities.[23] Phytochemicals provide much of the color and flavors of edible plant, having antioxidant, or hormone-like action. Many plant-derived bioactive compounds play an immunomodulatory role.[6] Phytochemicals are beneficial and important to the human system, due to their anti-inflammatory, anticancer, antimicrobial antibacterial activities, etc.

21.3.1 IMPORTANCE OF BIOACTIVE PHYTOCHEMICALS

Researchers have found that phytochemicals have the potential to stimulate the immune system, prevent carcinogenic effects, reduced inflammation, type-2 diabetes through hormonal regulation, and other related diseases which include reducing the rate of cancer, triggers damaged cell to self-destruction (apoptosis) before they can carry out their physiological and biochemical activities. Each phytochemicals act differently in carrying out these activities as we have the alkaloid for antibacterial effect, flavonoids for anti-inflammatory, etc.[7]

21.3.2 *BIOLOGICAL FUNCTIONS OF PHYTOCHEMICALS*[6,7]

- **Antioxidant activity:** The antioxidant activities of most phytochemicals prevent oxidative stress damage in cells, which minimizes the onset of carcinogenesis.
- **Hormonal action:** Isoflavones found in human estrogen help to reduce menopausal symptom and osteoporosis. Stimulation of enzymes in reducing the risk of cancer (breast cancer) by stimulating the enzyme that makes estrogen less effective.
- **Antibacterial** like allicin from garlic aid in disrupting the action of free radicals and reactive oxygen species.
- **Physical action:** Physical action by phytochemicals protects human cells by preventing the binding of pathogenic organisms.
- **Interference with DNA replication** by preventing the multiplication of cancer cells and preventing DNA from carcinogens.

21.4 ROLE OF BIOACTIVE COMPOUNDS IN CANCER CHEMOPREVENTION

Traditional and experimental studies have recommended numerous plants to have anticancer activities. These plants contain several bioactive phytochemicals that exert pro-apoptotic, antiproliferative, antimetastatic, and antiangiogenic effects on tumor cells.[12]

Based on the literature review, the following bioactive compounds have been used in targeting cancer chemoprevention (Figures 21.3 and 21.4):

- **Curcumin** (tumor regression in advanced pancreatic cancer; histological improvement in uterine cervical neoplasm, urinary bladder cancer, and intestinal metaplasia.[15]
- **Camptothecin** (leads to tumor regression in breast, prostate, and melanoma cancer).[19]
- **Resveratrol** (increases apoptosis in malignant colorectal and hepatic metastases, decreased tumor proliferation in colorectal adenocarcinoma).[25,21]
- **Others** include lycopene, panax ginseng, genistein, luteolin epigallocatechin-3-gallate (EGCG),[15] vinblastine, vincristine, sulforaphane, camptothecin,[19] indole-3-carbinol, quercetin, β-sistosterol,[15] and resveratrol.[25,21]

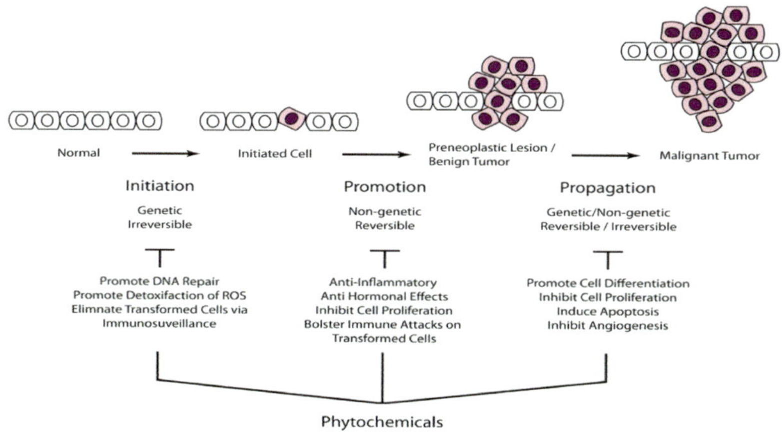

FIGURE 21.3 The exertion of phytochemical chemopreventive effects in the multistep process of carcinogenesis.[15]

Source: Reprinted from Ref. [15]. © 2016 Kotecha et al. https://creativecommons.org/licenses/by/4.0/

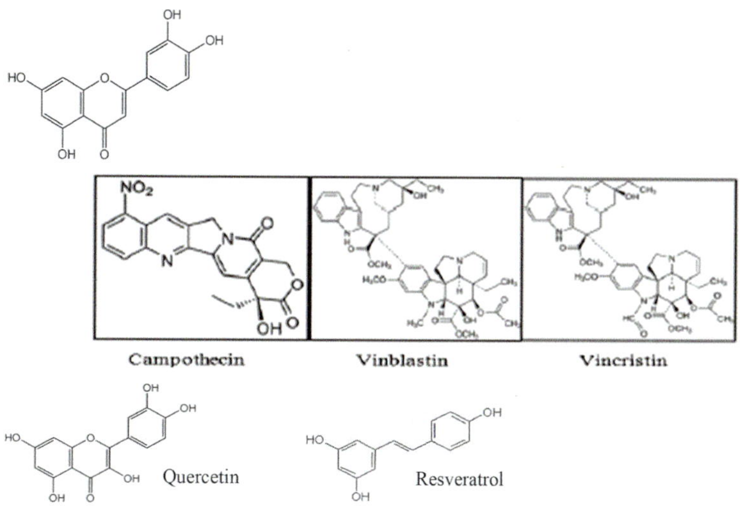

FIGURE 21.4 Chemical structures of various bioactive compounds.

21.4.1 MECHANISMS OF BIOACTIVE COMPOUNDS IN CANCER CHEMOPREVENTION

The phytochemicals anticancer activity involves multitarget mechanisms of action, such as antioxidation, antiproliferation, carcinogen inactivation, cell cycle arrest, apoptotic induction, and angiogenic inhibition.[12]

21.4.1.1 ANTIPROLIFERATIVE MECHANISM BY β-SISTOSTEROL IN CANCER CHEMOPREVENTION

β-sistosterol are found in nuts, grain, and seeds. They are effective against breast and colon tumor, leukemia, prostate cancer, and fibrous sarcoma. β-sitosterol activates caspase 3 and 8, increases mitogen-activated protein kinase (MAPK) activity and Fas levels and modulates the expression of Bcl-2 to prevent tumor cell multiplication and initiate apoptosis. It sensitizes breast cancer cells to induced cytotoxicity by tumor necrosis factor-related apoptosis-inducing ligand (TRAIL).[5]

21.4.1.2 ANTIOXIDATIVE MECHANISM OF SULPHORAPHANE AND ISOTHOCYANATE IN CANCER CHEMPREVENTION

Sulphoraphane and isothiocyanate are found in cruciferous vegetables and green tea. They modulate cytoprotective enzyme activation to prevent the initiation phase of carcinogenesis. This involves regulation of nuclear factor-E2 related factor 2 (Nrf2) that is induced by oxidative stress. The modulation of Nrf2 targets maintains genomic integrity and reduces the effects of oxidative stress on DNA. They inhibit conversion of pro-carcinogens to DNA-damaging species.[15]

21.4.1.3 PREVENTION OF GENOMIC INSTABILITY, ANTIOXIDANT AND ANTI-INFLAMMATORY MECHANISM BY CURCUMIN

Curcumin is harvested from rhizomes of turmeric plants. The molecular multitargets of curcumin are beneficial to all three stages of cancer. It is effective against colon, pancreatic cancer, and myeloma. It also inhibits nuclear factor-kappa B pathway (mediator of inflammation and pro-proliferation of cancer cells), which consequently leads to anti-inflammatory, antioxidant, cellular apoptotic response, and antitumor effects.[5] Curcumin modulates Bcl-2 and up-regulates p53. It suppresses matrix metalloproteinases (MMP) that promote metastatic progression, hence, reducing invasion and subsequent metastasis of cancer cells. It also promotes immune surveillance by modulating T-cells effectors against tumor cells.[14]

21.4.1.4 ANTI-INFLAMMATORY MECHANISM OF LUTEOLIN IN CANCER CHEMOPREVENTION

Luteolin is found in medicinal vegetables and herbs, celery, and pepper. They decreases the incidences of ovarian cancer and breast cancer. Luteolin decreases protein expression of prostaglandin E2 (PGE2) and cyclo-oxygenase-2 (cox-2) in macrophage-like cells to induce anti-inflammatory action. It activates superoxide and hydroxyl radical scavenging which limit reactive oxygen species production that plays a major role in the early stages of tumor development.[14]

21.4.1.5 ANTIMUTAGENIC MECHANISM OF QUERCETIN IN CANCER CHEMOPREVENTION

Quercetin is found in green tea, apple, and onions. It possesses free radical scavenging antioxidant, antimutagenic, and antiproliferative activities. Quercetin inhibits PI3K/Akt survival signaling pathway, Hedgehog signaling, PGE 2, and Cox-2 production. It sensitizes colon, prostate, and lung cancer cells to TRAIL-induced cytotoxicity due to the up-regulation of death receptor 5 and down-regulation of surviving.[25]

21.4.1.6 ANTIMETASTATIC MECHANISM BY EPIGALLOCATECHIN-3-GALLATE, CURCUMINAND RESVERATROL IN CANCER CHEMOPREVENTION

They induce the increased expression of E-cadherin molecule related to epithelial phenotype that inhibits motility and invasiveness of cancer cells by inhibiting epithelial–mesenchymal transition pathway.[25] It inhibits MMP which supports increased dissemination of cancer cells to distant sites.[17]

21.4.1.7 ANTIMETASTATIC MECHANISM BY RESVERATROL IN CANCER CHEMOPREVENTION

Resveratrol is found in grapes, peanuts, berries, and other fruits. It induces both antioxidant and anti-inflammatory effects and inhibits the proliferation of different cancer cells. Antiproliferative effects involve up-regulation of p53, caspases, and Bax, inhibition of transcription factors, down-regulation of survivin, cyclin, and Bcl-2 [5]. Resveratrol is effective in all stages of cancer as illustrated in Figures 21.5 and 21.6.[4,8,25]

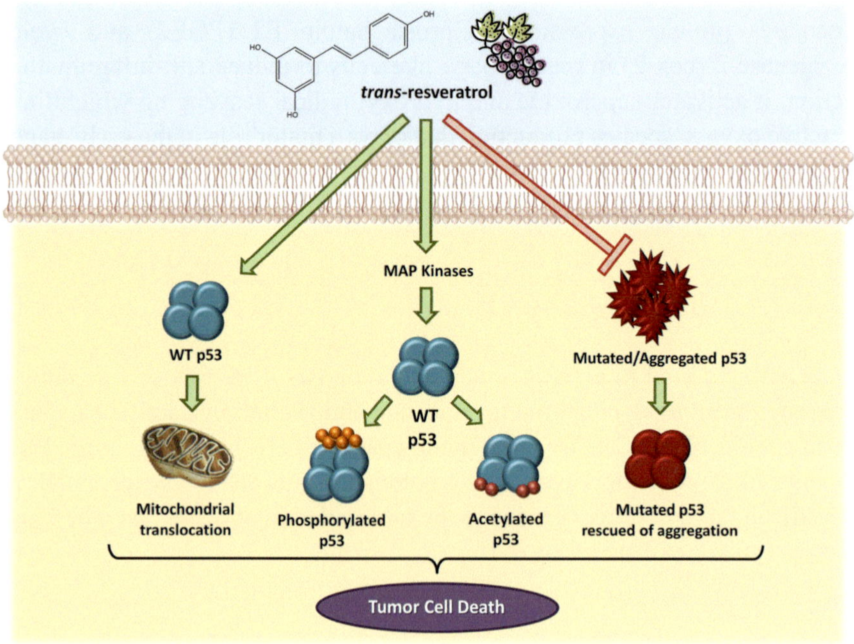

FIGURE 21.5 Activation of p53 pathway in tumor cell death by resveratrol.

Source: Reprinted from Ref. [8]. © 2017 by the authors. Licensee MDPI, Basel, Switzerland. https://creativecommons.org/licenses/by/4.0/

21.4.1.8 IMMUNOSURVEILLANCE MECHANISM OF EPIGALLOCATECHIN-3-GALLATE (EGCG) IN CANCER CHEMOPREVENTION

Immunosurveillance suppression by cancer cells involves the secretion of immunosuppresive cytokines, resistance to apoptosis, reduced expression of immunomodulatory molecules, and MHC class 1 antigens. EGCG antiangiogenic activities inhibit vascular endothelial growth factor (VEGF)/VEGF receptor axis. ECGC increases immune surveillance against tumor, down-regulates the expression of indole-amine 2,3-doxygenase, through antiproliferative and pro-apoptotic effects on T cells in the cancer environment.[17]

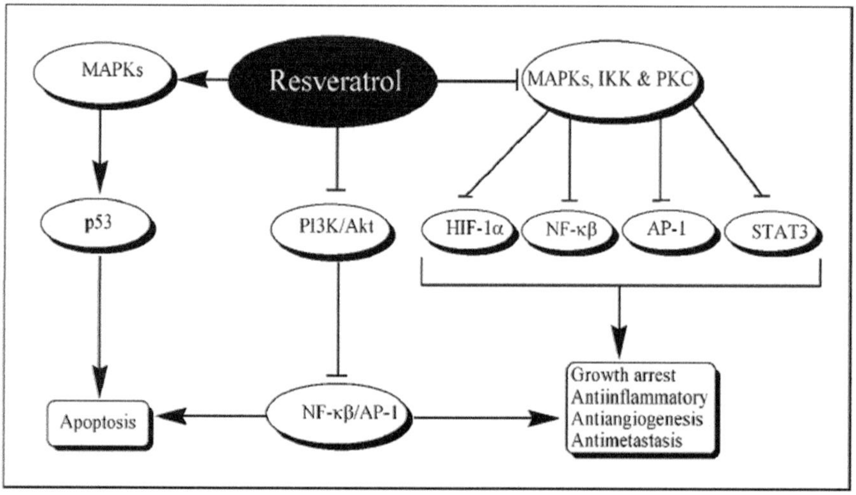

FIGURE 21.6 Inactivation of intracellular signal transduction pathways involved in by resveratrol carcinogenesis.

Source: Reprinted from Ref. [4]. © 2009 American Association for Cancer Research. https://creativecommons.org/licenses/by/4.0/

Much of the anticancer bioactive compounds should be cleared for their activities, drug interaction, standard dosages, prolong safety, and deleterious effects. Characterization of bioactive compounds should be done based on individual genetic makeup. The effect of different polymorphism on the activity of the bioactive compounds should be considered as well. Bioactive compounds have been shown to elicit their anticancer effects by targeting multiple molecular pathways and regulating signaling pathways in several stages of cancer development. Bioactive compounds can act on these molecular pathways with the aim of multitargeted therapies. Hence, bioactive compounds could be used and combined with other therapies for cancer prevention, by limiting the dose and greatly reducing the toxicity effect.

21.5 SUMMARY

The evidence from in vivo and in vitro studies have proven that plant-based diet reduces the risk of chronic diseases due to the presence of biologically active plant compounds or phytochemicals. Cancer is a progressive disease that develops in distinct stages (initiation, promotion, and progression) that are separated by a long period of latency. The bioactive compound's anticancer activity depends on multitarget mechanisms that blocking mutagenic activities of other compounds, reduce macromolecular damage induced by oxidative stress, suppress abnormal cell proliferation, and modulate the immune response, inflammation, cell metabolism, and genome instability. The bioactive compounds used in targeting cancer include curcumin that inhibits nuclear factor-kappa B, induces anti-inflammatory, antioxidant, and inhibits cancer cell growth, quercetin which modulates tumor necrosis factor-related apoptosis induced ligand-induced cytotoxity, epigallocatechin-3-gallate which has antiangiogenic effect. Other bioactive compounds in cancer chemoprevention and treatment include camptothecin, luteolin, lycopene, resveratrol, β-sitosterol, genistein, sulforaphane, indole-3-carbinol, 3,3′-diindolylmethane, vinblastine, vincristine, and piperin. Based on the understanding of the several molecular signaling pathways in cancer chemoprevention by bioactive compounds, a synergistic and additive approach combined with other cancer therapeutics including radiotherapy, immunotherapy, and chemotherapy could limit toxicity in the management and prevention of cancer.

KEYWORDS

- **bioactive compounds**
- **cancer**
- **chemoprevention**
- **mechanisms**
- **molecular targets**
- **phytochemica**

REFERENCES

1. Bernhoft, A. A Brief Review on Bioactive Compounds in Plants. In *Bioactive Compounds in Plants: Benefits and Risks for Man and Animals*; Bernhoft, A., Ed.; The Norwegian Academy of Science and Letters: Oslo, Norway, 2010; pp 11–18.
2. Bignold, L. Mechanism of Clastogen in Chromosomal Aberration. *J. Health Sci.* **2009,** *2*(3), 271–278.
3. Bilecova-Rabajdova, M.; Birkova, A.; Urban, P.; Gregova, K.; Durovcova, E.; Marekova, M. Naturally Occurring Substances and Their Role in Chemopreventive Effects. *Cent. Eur. J. Public Health* **2013,** *21*(4), 213–219.

4. Bishayee, A. Cancer Prevention and Treatment With Resveratrol: From Rodent Studies to Clinical Trials. *Cancer Prev. Res.* **2009,** *2*, 409–418.
5. Catalano, E. Role of Phytochemicals in the Chemoprevention of tumors. [Online] 2016, p 6. https://arxiv.org/ftp/arxiv/papers/1605/1605.04519.pdf (accessed May 30, 2021).
6. Nair, A.; Chattopadhyay, D.; Saha, B. Plant-derived immunomodulators. In New look to Phytomedicine; Academic Press; 2019; pp 435–499
7. Falope, M. O.; Ibrahim, H.; Takeda, Y. Screening of Higher Plants Requested as Pesticides Using the Brine Shrimp Lethality Assay. *Int. J. Pharmacogn.* **1993,** *31*(4), 230–254.
8. Farraz de Costa D. C.; Fialho, E.; Silva, J. L. Cancer Chemoprevention by Resveratrol: The p53 Tumor Suppressor Protein as a Promising Molecular Target. *Molecules* **2017,** *22,* 1–24.
9. Guilford, J. M.; Pezzuto, J. M. In *Cancer Chemoprevention* [Online], 2009; p 7. https://www.eolss.net/sample-chapters/c06/E6-151-10-00.pdf (accessed May 30, 2021).
10. Haung, L.; Snyder, A. R. Radiation-Induced Genomic Instability and its Implications for Radiation Carcinogenesis. *Oncogene* **2003,** *22*(37), 5848–5854.
11. Herouart, D.; Sigaud, S.; Frendo, P.; Puppo, A. Characterization of *Kat* Genes Encoding Catalases in *Sinorhizobium meliloti. Curr. Plant Sci. Biotechnol. Agric.* **1998,** *31,* 497–498.
12. Hosseini, A.; Ghorbani, A. Cancer Therapy With Phytochemicals: Evidence From Clinical Studies. *Avicenna J. Phytomed.* **2015,** *15*(2), 84–97.
13. Huang Y. W.; Baluna R.; Vitetta E. S. Adhesion Molecules as Targets for Cancer Therapy. *Histol. histopathol.* **1997,** *12,* 467–447.
14. Issa, A. Y.; Volate, S. R.; Wargovich, M. J. The Role of Phytochemicals in Inhibition of Cancer and Inflammation: New Directions and Perspectives. *J. Food Compos. Anal.* **2006,** *19*(5), 205–419.
15. Kotecha, R.; Takami, A.; Espinozam L. J. Dietary Phytochemicals and Cancer Chemoprevention: A Review of the Clinical Evidence. *Oncotarget* **2016,** *7*(32), 52518–52529.
16. Lodish, H.; Berk, A.; Matsudaira, P.; Kaiser, C. A.; Krieger, M.; Scott, M. P.; Zipursky, L.; Darnell, J. Eds.; In *Molecular Biology of the Cell*, 5th ed.; W.H. Freeman: New York, NY, 2005; pp 935–943.
17. Milligan, S. A.; Burke, P.; Coleman, D. T.; Bigelow, R. L.; Steffan, J. J.; Carroll, J. L.; Williams, B. J.; Cardelli, J. A. The Green Tea Polyphenol EGCG Potentiates the Antiproliferative Activity of C-Met and Epidermal Growth Factor Receptor Inhibitors in Non-Small Cell Lung Cancer Cells. *Clin. Cancer Res.* **2009,** *15*, 4885–4894.
18. Pitot, H. C.; Goldsworthy, T.; Morgan, S. The Natural History of Carcinogenesis. *J. Supramol. Struct. Cell. Biochem.* **2004,** *17*(2), 133–146.
19. Prakash, O.; Kumar, A.; Kumar, P. A. Anticancer Potential of Plant and Natural Products. *Am. J. Pharmacol. Sci.* **2013,** *1*(6), 104–115.
20. Pshaouha, F. H.; Drusopoulos, K. G.; Doubravska, L. Quercetin Enhances TRAIL-Mediated Apoptosis in Colon Cancer Cells by Inducing the Accumulation of Death Receptors in Lipid Rafts. *Mol. Cancer Ther.* **2007,** *6*, 2591–2599.
21. Sak, K. Chemotherapy and Dietary Phytochemical Agents. *Chemother. Res. Pract.* **2012,** *1*, 1–11.
22. Torpy, J. M.; Lynm, C.; Glass, R. M.; Suyana M. Cancer: The Basics. *J. Am. Med. Assoc.* **2010,** *304*(14), 1628.
23. Tiwari, P.; Kaur, M.; Kaur, H. Phytochemical screening and Extraction: A Review. *Int. Pharm. Sci.* **2011,** *1*, 98–106.

24. Troll, W.; Wiesner, R. The Role of Oxygen Radicals as a Possible Mechanism of Tumor Promotion. *Ann. Rev. Pharmacol. Toxicol.* **1995,** *25*, 509–528.
25. Varoni, E. M.; Lo Faro, A. F.; Sharifi-Rad, J.; Iriti, M. Anticancer Molecular Mechanisms of Resveratrol. *Front. Nutr.* **2016,** *3*, 8–12.
26. Vavder, R. P.; Califoro, J.; Westra, N. Genetic Progression Model for Head and Neck Cancer. *J. Physiol. Res.* **1996,** *6*(11), 2488–2492.
27. Yamagiwa, K.; Ichikawa, K. Experimental Study of the Pathogenesis of Carcinoma. *J. Cancer Res.* **1918,** *3*, 1–29.
28. Oliveira, Paula Alexandra et al. "Chemical carcinogenesis." Anais da Academia Brasileira de Ciencias 79 4 (2007): 593–616.

Index

R

S

T